# Understanding Satellite Communications: The Stochastic Geometry Perspective

Shuai Wang · Jinpeng Song · Xiaqing Miao ·
Gaofeng Pan
Editors

# Understanding Satellite Communications: The Stochastic Geometry Perspective

Springer

*Editors*
Shuai Wang
School of Information and Electronics
Beijing Institute of Technology
Beijing, China

Xiaqing Miao
School of Cyberspace Science
and Technology
Beijing Institute of Technology
Beijing, China

Jinpeng Song
School of Cyberspace Science
and Technology
Beijing Institute of Technology
Beijing, China

Gaofeng Pan
School of Cyberspace Science
and Technology
Beijing Institute of Technology
Beijing, China

ISBN 978-981-97-7101-1 ISBN 978-981-97-7102-8 (eBook)
https://doi.org/10.1007/978-981-97-7102-8



This Springer imprint is published by the registered company Springer Nature Singapore Pte Ltd.
The registered company address is: 152 Beach Road, #21-01/04 Gateway East, Singapore 189721, Singapore

# Preface

Satellite communication technology has become an essential part of the modern world, providing a reliable and efficient means of communication across the globe. With the advent of 5G and 6G technologies, space-aerial-terrestrial integrated networks (SATINs) have emerged as a promising concept and popular research area. On the one hand, the high demand for advanced services and the increase in the number of connected devices have made it necessary for current wireless communication systems to expand to meet the users' needs in terms of quality of service, throughput, latency, connectivity, and security. On the other hand, the coverage of marine and mountainous areas, which constitute a considerable part of the Earth's surface, can be quite challenging due to the unreachability of terrestrial cellular facilities as well as the high cost and instability of deploying aerial-based networks. To this end, employing satellites that are capable of providing wide coverage and supporting more devices and higher data rates is essential for further evolution of communication services. SATINs can cooperate to provide service for remote areas where terrestrial communication networks may be expensive and hard to maintain. They are expected to provide seamless coverage, enabling ubiquitous connectivity over the wireless networks.

However, the availability and performance of SATINs can be greatly affected by channel impairments, such as path loss and small-scale fading. Moreover, due to the openness and broadcast nature of the wireless channels, security issues may also occur. Specifically, transmissions in SATINs can be intercepted or eavesdropped by illegal receivers, which may result in information leakage. Hence, it is significant to analyze the performance of SATINs under imperfect channel conditions and security threats, which can provide insights and guidance for the design of practical SATINs. In this book, we conduct analytical performance analysis and optimization to model a range of scenarios in SATINs and to evaluate the corresponding performance metrics. Particularly, key metrics of communication are analyzed and their accurate/approximate expressions are derived, including outage/coverage probability, secrecy outage probability, energy efficiency, channel capacity/secrecy outage capacity, throughput, and transmission delay. In addition, the relationships and tradeoffs among these metrics are also investigated and clarified.

This book clarifies not only the theoretical framework, but also the practical implications of these principles. It takes a unique approach to study satellite communication systems by employing the perspective of stochastic geometry. By harnessing the principles of this branch of mathematics, it offers a meticulous exploration of the fundamental topics within the field, effectively bridging the divide between theoretical understanding and practical application. This intersectional approach serves as a priceless toolbox for those keen on the development and enhancement of this dynamic and constantly evolving technology.

Whether you're a seasoned professional seeking to broaden your horizons and gain new insights into the intricate relationships within satellite networks, or a budding engineer embarking on a journey of discovery, this comprehensive guide provides profound directions to effectively navigate the world of satellite communication systems. It enables you to visualize and analyze the spatial distributions and patterns within satellite networks, understanding how these characteristics affect performance and reliability.

By incorporating stochastic geometry into its framework, this book not only deepens our understanding of satellite communication systems but also opens up new avenues for innovation and optimization. It arms you with the tools and knowledge to contribute to the further advancement of this crucial technology, driving its progress towards more efficient, reliable, and widespread communication capabilities.

The content overview of each chapter in this book is as follows:

- Chapter 1: In this chapter, the basic concepts, development history and characteristics of satellite communications are introduced.
- Chapter 2: In this chapter, several Satellite-to-terrestrial communication systems are considered, which include HARQ schemes, earth rotation-aware nonstationary satellite communication and secure satellite-terrestrial downlink transmissions.
- Chapter 3: In this chapter, several satellite-aerial communications are considered, which include secure uplink transmissions and s downlink deliveries in presence of spatially random space interference.
- Chapter 4: In this chapter, several satellite-aerial-terrestrial communication systems are considered, which include modelling and optimization and outage analysis in presence of spatially random interference.
- Chapter 5: In this chapter, several end-to-end terrestrial-satellite-terrestrial communication systems are considered, which include latency versus reliability and spatially aerial random interference.

- Chapter 6: In this chapter, all satellite communication systems are considered which include satellite-to-terrestrial communication systems, satellite aerial communications, satellite-aerial-terrestrial communications and end-to end terrestrial-satellite-terrestrial communications.

Beijing, China
January 2024

Shuai Wang
Gaofeng Pan

# Contents

# Chapter 1
# Introduction to Satellite Communications

**Shuai Wang, Jinpeng Song, Xiaqing Miao, and Gaofeng Pan**

This chapter provides a comprehensive examination of satellite communications, addressing key issues essential to optimizing performance and ensuring secure transmission. It begins with a historical overview of the development of satellite communications and progresses to an in-depth analysis of the challenges associated with modeling satellite links, including critical factors such as orbit determination and collision avoidance. The study also examines the characteristics that define satellite communications, including global coverage, low latency, and adaptability in adverse conditions. It also examines the basic concepts and components that makeup satellite systems, highlighting the critical roles played by payloads, propulsion systems, and onboard electronics. Security considerations [1, 2] are thoroughly addressed, with a focus on the robust physical safeguards, access controls, and encryption protocols that are essential to protect against unauthorized access and tampering. This research serves as an invaluable resource for understanding the complexities and potential of satellite communications, providing a holistic view of this indispensable technology.

## 1.1 Overview of Satellite Communications

Communication technologies have become the cornerstone of modern life in an era characterized by increasing demands for connectivity, and satellite communications involve the transmission of signals by satellites to provide connectivity between different points on Earth [3–5]. This chapter discusses the benefits of these technologies, focusing on key areas of satellite communications, including satellite orbits, ground station infrastructure, and airborne communication systems. As the backbone of

S. Wang · J. Song (✉) · X. Miao · G. Pan
Beijing Institute of Technology, Beijing, China
e-mail: jinpeng@bit.edu.cn



S. Wang et al. (eds.), *Understanding Satellite Communications: The Stochastic Geometry Perspective*, https://doi.org/10.1007/978-981-97-7102-8_1

global connectivity, satellite communications systems have revolutionized the way information is shared over long distances. Understanding satellite orbit variations is critical to optimizing coverage, latency, and signal reliability. Ground stations are the most widely used, providing seamless communication between terrestrial networks and satellites [6, 7]. In addition, the integration of airborne communications systems provides a clear paradigm for extending communications reach by utilizing high-altitude platforms. In general, satellites in orbit can be divided into three categories: geostationary, mid-Earth, and low-Earth. Multiple satellites orbiting the Earth relay analog and digital signals, transmitting voice and data to and from one or more locations around the world.

### *1.1.1 Satellite Orbit*

The path a satellite follows is called its orbit, and it is tailored to the capability and purpose of the sensors it carries. Orbiting satellites are classified into different categories based on their functions and altitudes. Some of the satellites in orbit and their characteristics include:

- Geostationary Earth Orbit (GEO): These satellites orbit the Earth at an altitude of about 35,786 km above the equator. They appear to be stationary because they are positioned at a specific point above the Earth's surface. They move from west to east as the Earth rotates, taking 23 h, 56 min, and 4 s by traveling at the same speed as the Earth. Because satellites in this orbit cover a large area, they are most often used for reliable communications and weather monitoring.
- Medium Earth Orbit (MEO): This includes a wide range of orbits between LEO and GEO. It is similar to LEO in that it does not have to take specific paths around the Earth. It is commonly used by navigation systems, especially Global Navigation Satellite Systems (GNSS).
- Low Earth Orbit (LEO): These satellites are relatively close to the Earth's surface. It is usually at an altitude of less than 1000 km but can be as low as 160 km above the Earth, which is low compared to other orbits, but still quite far above the Earth's surface. Unlike satellites in GEO, LEO satellites do not always have to follow the same path around the Earth, so their plane can be moved or tilted. This means that there are more routes available for satellites in LEO, which is one of the reasons why it is a very common orbit. Its proximity to the Earth makes it more useful for satellite imagery, as being close to the surface allows it to take higher-resolution images. They are also commonly used for remote sensing and satellite constellations such as those used for Internet services.

### 1.1.2 Advantages of Satellite Communication

The satellite communications process is a series of carefully orchestrated steps. It begins with the uplink phase, during which ground stations transmit signals to satellites using specific radio frequencies. These signals serve as carriers for a variety of information types, including voice, data, and video. When the uplink signals reach the satellite, they are received and then amplified. This critical step allows the signals to be effectively transmitted back to Earth. Acting as a relay, the satellite ensures that the signals can travel long distances and reach a wider geographic area. In the downlink phase, the amplified signals are then transmitted back to Earth. Here, designated ground stations are strategically positioned to receive and process the signals, completing the communication cycle. This complex process enables seamless and efficient communication over long distances, supporting vital applications such as global Internet coverage, television broadcasting, and various other forms of real-time data exchange.

Satellite communication has many advantages in many fields. Satellite communications offer numerous advantages that span various applications and scenarios. Chief among them is their global network coverage, extending even to the most remote and inaccessible regions of the world. This extensive reach not only ensures connectivity in distant locations but also enables the provision of Internet services to these areas. Furthermore, satellites facilitate versatile connectivity, facilitating connection and data exchange between widely dispersed points, making them ideal for maritime [8, 9], aviation [10], and even remote control applications where accessing and exchanging crucial data is paramount. Additionally, satellite communications play a pivotal role in broadcasting [11], transmitting digital signals worldwide to deliver television and radio broadcasts simultaneously and in real-time to a vast audience. Their mobility support is another key advantage, providing connectivity to platforms in transit such as ships, aircraft, and vehicles, as well as supporting Global Navigation Satellite Systems (GNSS) [12]. Moreover, satellite communications enable broadband capabilities, facilitating videoconferencing, live streaming, and online gaming by transmitting large amounts of data at faster rates without interference. Finally, in times of natural disasters or infrastructure failures [13], satellite communications provide reliable communications as they are less susceptible to ground and weather conditions, ensuring uninterrupted connectivity even in the most challenging environments. These advantages make satellite communications a promising solution for system monitoring, emergency information gathering, and defense systems.

However, despite these advantages, satellite communication also has certain drawbacks. One key issue is latency, as signals transmitted to and from satellites encounter a certain delay due to the vast distances they traverse. This delay can pose problems for real-time applications, such as video conferencing and gaming, where a seamless exchange of information is crucial. Another significant consideration is cost. Establishing the necessary satellite connections, and launching, and maintaining satellites is an expensive endeavor. Some satellites may even require frequent maintenance, given their potentially shorter lifespan, which can further inflate the cost of satellite communication services. This financial barrier can limit the global reach of satellite communications, as areas lacking sufficient funds to support the launch and maintenance of satellites may remain underserved. Finally, there is the complex issue of spectrum allocation. The frequency spectrum available for satellite communications is limited and strictly regulated by international agencies, necessitating efficient and careful management to accommodate the growing demand for wireless communication services. Balancing these challenges with the benefits of satellite communications is crucial for ensuring its widespread adoption and sustainability.

### *1.1.3 Security Issues in Satellite Communication*

Security is also a satellite communications challenge [14–16], which can disrupt the confidentiality, integrity, and availability of information transmitted to and from the satellite, as with any other form of communication, some of the security concerns include the following.

- Jamming and interference is a satellite communications challenge, where satellites are susceptible to interference from both natural sources and intentional interference from malicious actors. Malicious actors may attempt to transmit powerfully on the same frequency to disrupt satellite communications, which can cause jamming. Ensuring secure and reliable communications requires careful and complex anti-jamming measures and robust error correction codes.
- Satellite communications are also vulnerable to physical attacks through unauthorized access and even information tampering if not physically secured. It is also necessary to use measures such as authentication to ensure that only authorized users can access the network.
- Eavesdropping and interception are also a security challenge, where attackers may attempt to intercept the signals transmitted between satellites and the ground station to gather sensitive information. Satellite communications systems are vulnerable to cyber-attacks that can cause data loss or even data corruption through eavesdropping and tampering. Implementing robust encryption protocols and securing communication channels are essential to prevent such cyber-attacks.
- Unauthorized individuals may be able to access satellite systems due to weaknesses in authentication procedures. In addition, poor authorization controls may allow systemic unauthorized actions that could result in changes to transmitted data or even access to highly sensitive data.
- Attackers can flood ground stations or satellite systems with traffic, degrading service or even disrupting connectivity. This can result in a denial of service (DoS) that can slow work productivity or even cause important data to be intercepted and not delivered to the target user.
- Like any computer-based system, satellite systems are vulnerable to hackers. Satellite activities may be subject to malware, ransomware, and other cyber threats.

Therefore, it can be said that satellite communications systems also face cybersecurity challenges that may even lead to data loss or reduce data integrity and confidentiality.

### 1.1.4 Solutions for Secure Satellite Communication

To safeguard satellite communications from various security threats, a comprehensive approach is necessary. Key measures include interference [17] and jamming mitigation, where advanced techniques such as signal filtering and frequency hopping are employed to reduce both natural and intentional sources of interference. Implementing robust error correction codes further ensures data integrity even in the face of disruption. Physical security measures are also crucial, encompassing secure access controls, tamper-evident seals, and physical hardening of satellite components to prevent unauthorized access and tampering. Authentication measures must be strictly enforced to guarantee that only authorized users have access to the satellite network. To prevent eavesdropping and interception, strong encryption techniques are essential in securing communication channels between satellites and ground stations [18], rendering intercepted data unreadable by unauthorized parties. Regular monitoring for suspicious activity and signal behavior adds another layer of defense. Furthermore, intrusion detection and prevention systems (IDS/IPS), coupled with robust firewalls, monitor and filter incoming and outgoing traffic, detecting and blocking suspicious or malicious behavior. This multifaceted approach provides a robust defense against a range of security challenges faced by satellite communications.

In addition, there are also some means used in satellite communication to ensure the security of communication. This involves securely generating, distributing, and rotating encryption keys, ensuring that unauthorized access is prevented and encrypted data remains private. Additionally, regular security audits and penetration tests should be conducted to identify and promptly address any system vulnerabilities. This proactive approach allows for the identification of potential weaknesses before they can be exploited by malicious actors. Moreover, developing a comprehensive incident response plan is crucial for effectively dealing with security events. This plan should outline procedures for promptly identifying, containing, eliminating, recovering from, and documenting any security breaches, ensuring a swift and coordinated response to any potential threats. By applying these additional security measures, satellite communication systems can strengthen their defenses against a wide range of security threats and maintain the integrity, confidentiality, and availability of transmitted data.

### 1.1.5 Space-Aerial-Terrestrial Network Security

Concerning Space-Aerial-Terrestrial Networks (SATNs), the challenges of SATNs can be classified according to the types of nodes, which include orbiting satellites, ground stations, and airborne platforms. These nodes face different difficulties according to their respective types, which are listed below:

First, satellites in orbit face significant obstacles. Here is a brief summary of six common challenges.

- The escalating concerns surrounding space debris and the persistent menace of collisions in orbit have necessitated the diligent implementation of collision avoidance measures and effective debris management strategies. The growing concentration of debris poses a substantial threat to operational satellites, as even minute fragments can inflict significant damage. To counter this potential peril, satellites rely on sophisticated tracking systems and algorithms to forecast potential collision paths.
- Satellite operators meticulously monitor their trajectory and take evasive maneuvers if a potential hazard is detected. The issue of orbital congestion, particularly in LEO, further compounds these challenges, given the remarkable surge in the number of satellites. This congestion not only heightens the risk of collisions but also enhances the likelihood of interference with other satellites or systems. Therefore, coordinating and regulating satellite usage has become imperative to avert disputes and uphold the safe operation of these vital resources. International organizations and regulatory bodies are collaborating to establish norms and best practices for satellite operators, aiming to mitigate orbital congestion and promote global space sustainability.
- Satellites must endure the unforgiving environment of space, enduring extreme temperature fluctuations. Unlike Earth, there is no atmosphere to dissipate heat from the satellite, exposing it to both freezing cold in shaded regions and blistering heat during direct sunlight exposure. Engineers address this by deploying advanced thermal management systems, incorporating specialized insulation, radiators, and heat sinks. These systems assist in regulating temperatures and safeguarding crucial components from thermal stress, ensuring that the satellite remains operational throughout its lifespan.
- The radiation environment in space poses a significant challenge for satellites, as they traverse regions exposed to diverse forms of radiation, including solar radiation and high-energy particles trapped within the Van Allen radiation belts. Over time, this radiation exposure can deteriorate electronic components and material properties, compromising the satellites' performance. To counteract these effects, engineers employ shielding materials and design radiation-resistant components that can withstand harsh conditions. Ensuring the long-term reliability and functionality of satellites in orbit is paramount to the success of space missions.
- Operating and maintaining satellites in space is further complicated by the vast distances involved, necessitating remote monitoring and control. Achieving this

requires robust communication networks and sophisticated automation technologies. Engineers and operators rely on intricate command sequences and protocols to meticulously monitor various aspects of satellite operations, including orbit adjustments and troubleshooting anomalies. Redundant systems and fail-safe mechanisms are incorporated into satellite design to safeguard uninterrupted operation in the event of component failure.
- Satellites encounter a diverse array of obstacles in their orbital paths, ranging from collision risks and space debris management to contending with harsh environmental conditions and radiation exposure. Despite these challenges, engineers and operators demonstrate unwavering dedication to overcoming them and preserving the success and longevity of space missions. This is achieved through innovative engineering solutions, meticulous operational planning, and strict adherence to regulatory standards. Their unceasing efforts ensure the continued reliability and performance of satellites, crucial to the advancement of space exploration and utilization.

Second, there are also major challenges at ground stations. Ground stations, also called earth stations, are critical components of communications systems and are large parabolic antennas shaped like dishes to accurately reflect and capture incoming radio signals. They coordinate communications between satellites in orbit and the terrestrial network. Ground stations, which are an integral part of satellite communications systems, also face the following challenges.

- One of the primary concerns is establishing reliable connectivity, particularly in areas prone to adverse weather conditions. Adverse weather conditions, such as heavy precipitation, snowfall, or fog, have the potential to weaken microwave transmissions, resulting in degraded or lost signal quality. To meet the challenges of inclement weather on ground station operations, engineers must incorporate sophisticated antenna designs, weather-resistant materials, and robust error-correction algorithms. In addition, the strategic placement of ground stations in geographically diverse areas can help maintain redundancy and reliability, especially in the face of extreme weather events.
- Ground stations also place a high value on security. As critical components of satellite communications networks, ground stations are vulnerable to various security risks, such as unauthorized access, information eavesdropping, and cyber-attacks. The use of robust encryption protocols, authentication systems, and firewalls is necessary to mitigate security threats. In addition, it is critical to implement thorough security procedures and adhere to industry standards to ensure the confidentiality and integrity of sensitive data transmitted and received by ground stations.
- Ground stations face a significant challenge in effectively managing the spectrum. The electromagnetic spectrum is a limited resource that is regulated by international authorities. Optimizing spectrum allocation is of paramount importance due to the ever-increasing demand for wireless communication services. To optimize the use of available frequencies, it is imperative that ground stations implement sophisticated spectrum monitoring and management techniques. In addition, the

simultaneous presence of multiple communication systems in close proximity requires careful coordination to mitigate interference and maintain the integrity of communication channels.

- Power control and energy efficiency optimization are of paramount importance for ground stations, especially in remote or off-grid areas. The terminals require the installation of reliable power systems, which typically include a combination of grid power, renewable energy sources, and backup generators. To ensure continuous operation, especially in areas with unreliable or limited access to electrical grids, engineers must effectively deploy power distribution and storage systems that are efficient.
- Ground stations face two equally important challenges: physical security and environmental resilience. These facilities must be designed to withstand a variety of environmental conditions, including but not limited to severe temperature fluctuations, earthquakes, and the occurrence of natural disasters. The physical integrity and resilience of ground stations can be ensured through the use of robust construction materials, the implementation of structural reinforcements, and the development of comprehensive disaster recovery strategies.
- Ground stations play a critical role in satellite communications systems, but they face many unique obstacles. These include ensuring consistent connectivity in adverse weather conditions, implementing strong security measures, efficiently managing spectrum resources, optimizing power consumption, and ensuring the physical resilience of facilities. Ground stations play a critical role in enabling uninterrupted communications between satellites and terrestrial networks by effectively addressing these challenges through the use of innovative engineering solutions and thorough operational planning.

Finally, airborne platforms also face significant challenges. The use of aircraft or airborne vehicles as communications transmission platforms presents unique issues in the area of airborne communications.

- One of the most important factors to consider is the extremely challenging high-altitude conditions in which these platforms are used. In high-altitude regions, environmental conditions can be challenging, with factors such as lower temperatures, reduced atmospheric pressure, and erratic air currents. Developing communications systems that can withstand these conditions while maintaining consistent performance is a remarkable feat of engineering. The use of specialized materials, thermal management systems, and ruggedized components is essential to ensure the durability of communications equipment in this challenging environment.
- Combined with natural obstacles, the ever-changing characteristics of airborne platforms pose a significant problem. Aircraft and other aerial vehicles experience constant motion, vibration, and orientation changes. The fluctuating nature of the environment can affect the stability and alignment of communications equipment. Maintaining precise alignment of antennas and transceivers with satellites

or ground stations is critical to maintaining consistent and reliable communications links. To overcome this difficulty, sophisticated stabilization mechanisms and adaptive tracking systems are used to provide accurate and stable communications even when the platform is moving.
- Power management plays a critical role in aerial platform communications. These platforms are often dependent on limited power resources, such as on-board batteries or generators. The task of balancing the power requirements of communications equipment with other critical systems on the platform, such as avionics and propulsion, requires careful engineering and optimization. Energy-efficient components, power distribution systems, and innovative energy storage technologies are used to ensure continuous operation during a mission.
- The design and performance of antennas are of paramount importance in facilitating communications on aerial platforms. Designing antennas that can effectively maintain precise and consistent links in the face of dynamic platform motion is a significant and challenging task. Maintaining reliable communications links requires achieving accurate antenna pointing and tracking capabilities, even in the presence of platform vibrations and maneuvers. Engineers use advanced antenna technologies, such as phased array antennas and adaptive beamforming, to effectively overcome these difficulties and improve communications efficiency.
- Interference and signal degradation pose additional challenges to airborne platform communications. The platform itself, as well as other on-board electronic systems, can introduce sources of interference. In addition, certain atmospheric phenomena such as precipitation (rain or snow) and atmospheric turbulence have the potential to degrade signal quality. To limit the effects of interference and maintain consistent signal quality, advanced signal processing techniques, robust error correction algorithms, and adaptive modulation schemes are used. In addition, the use of advanced modulation and coding techniques are employed to optimize spectral efficiency and improve data throughput under challenging channel conditions.

## 1.2 System Elements and Basic Concepts

### *1.2.1 System Elements*

As the backbone of global connectivity, satellite communications systems have revolutionized the way information is shared across vast distances. Understanding satellite orbital variations is critical to optimizing coverage, latency, and signal reliability. Ground stations are the most widely used, providing seamless communication between terrestrial networks and satellites. In addition, the integration of airborne communications systems provides a clear paradigm for extending communications reach by utilizing high-altitude platforms. With the exception of satellites, the characteristics of ground stations and airborne platforms are detailed below.

**Fig. 1.1** Ground satellite station

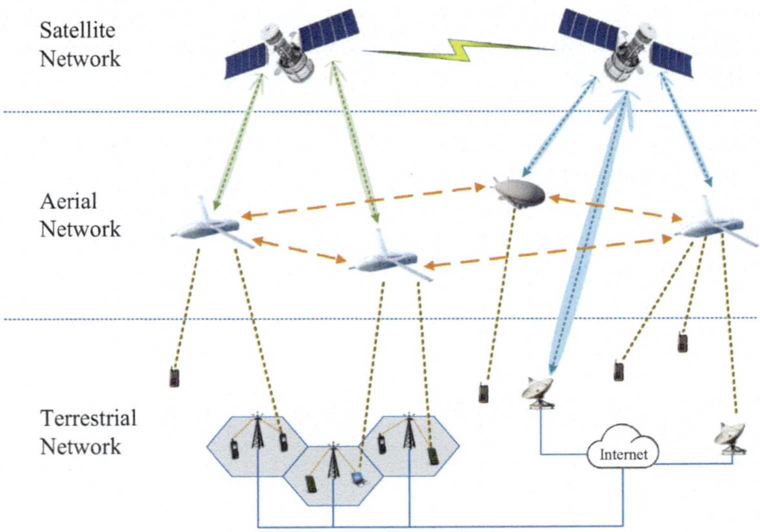


**Fig. 1.2** Space-Aerial-Terrestrial networks

Ground stations and aerial platforms play crucial roles in communication systems. Ground stations, as depicted in Fig. 1.1, are fundamental components that provide high gain and directivity, enabling efficient signal exchange with orbiting satellites. Additionally, as shown in Fig. 1.2, they connect with various satellites to establish an efficient satellite communication network.

On the other hand, aerial platforms leverage high-altitude aircraft or air vehicles as relay stations to enhance communication ranges. Operating at optimal line-of-sight altitudes, these systems employ specialized antennas, transceivers, and adaptive beamforming techniques to transmit and receive signals across multiple frequency bands, including voice, data, and video. Advanced tracking and control systems maintain stable connections, while intelligent power management ensures reliable operation. Security is safeguarded through redundancy, regulatory compliance, and robust security measures. Furthermore, data processing, encryption, and routing capabilities enhance the functionality of these systems.

Essential for various applications such as telecommunications, disaster relief, surveillance, and military communications, aerial platforms must seamlessly integrate with ground infrastructure. They are designed to withstand environmental challenges, making them versatile and crucial components of modern communication networks.

Satellite communication involves multiple basic concepts, including channel, performance, physical layer security, and theory of stochastic geometry [19, 20]. In satellite communication, the channel refers to the channel for information transmission, which affects the communication quality and rate. Performance focuses on the efficiency and reliability of the communication system, such as data transmission rate and bit error rate. Physical layer security emphasizes the encryption and protection of signals during transmission to prevent information leakage and theft. The theory of stochastic geometry is used to study and analyze the spatial layout and coverage of satellite networks, helping to optimize network design and resource allocation. These concepts together constitute the foundation of the satellite communication field, providing important theoretical support for the design, optimization, and security of satellite communication systems.

### *1.2.2 Satellite Communication Channel*

The satellite communication channel is described as the medium through which the radio signals are transmitted between the satellite and the ground station or other communication devices capable of transmitting or receiving radio signals. Due to the distance between satellites and ground stations and the complex propagation environment in the satellite communication system, satellite communication is severely affected by multipath fading and shadowing. These factors seriously affect the quality of communication and the system's spectrum efficiency. Wireless channels can experience several types of impairments, which include the effects of path loss, multipath fading, and Doppler spread. The three common fading models in satellite communication channels include Rician fading, Nakagami $m$ fading, and Rayleigh fading.

- Rician fading occurs when a radio signal follows multiple paths from the transmitter to the receiver. In the context of satellite communications, the direct path is the signal that travels directly from the satellite to the ground station. These paths include a direct line-of-sight path and one or more reflected paths, typically from buildings, terrain, or other objects. The reflected paths can be signals that bounce off buildings, mountains, or other obstacles before reaching the ground station. This results in multiple versions of the same signal being received at slightly different times. The Rician fading model is applicable in areas where there is a strong line-of-sight path and open areas with few obstacles. In the presence of a strong line-of-sight component, the fading process can be represented as the sum of a complex exponential and a narrowband complex Gaussian $g(t)$. When the line-of-sight component arrives at the receiver at a particular angle of arrival (AoA) $\theta$, phase $\phi$, and with the maximum Doppler frequency $f_D$, the fading process in baseband can be described as

$$h(t) = \underbrace{\sqrt{\frac{K\Omega}{K+1}}}_{\text{A:=}} e^{(j2\pi f_D \cos(\theta)t+\phi)} + \underbrace{\sqrt{\frac{\Omega}{K+1}}}_{\text{S:=}} g(t), \tag{1.1}$$

where $A$ and $S$ represent the amplitudes of the line-of-sight component and the scattered components, respectively. $K$ is the Rician factor, which quantifies the strength of the line-of-sight component relative to the scattered components. $\Omega$ is the average received signal power. $e^{(j2\pi f_D \cos(\theta)t+\phi)}$ is a complex exponential term that introduces a phase shift to the line-of-sight component. This term captures the time-varying nature of the channel due to motion or changing angles. $g(t)$ is a complex-valued Gaussian random process that represents the scattered components of the channel. In a Rician fading model, the scattered components are modelled as complex Gaussian random variables due to the central limit theorem.

- Nakagami-$m$ fading is another case of decline. This is a more general statistical model that helps characterize the variations in signal strength experienced in wireless communication channels, especially in environments with significant multipath propagation. It has applications in various communication systems and plays a critical role in the design of robust and reliable wireless networks. It is particularly relevant in environments with significant multipath propagation, where signals reach the receiver via multiple paths due to reflections, diffraction, and scattering. In such scenarios, the received signal strength can experience a wide range of variations. In the Nakagami $m$ fading model, the parameter $m >= 1$ is a positive real number that plays a crucial role in characterizing the severity of fading in a wireless communication channel. It is known as the fading severity parameter. A smaller $m$ value, closer to 1, indicates a more severe fading environment, similar to Rayleigh fading, where the received signal experiences rapid and deep fluctuations. Conversely, as $m$ increases, the fading becomes less severe and approaches more predictable, less volatile behavior. Higher $m$ values indicate a channel with a stronger line-of-sight component and less susceptibility to deep fading. Therefore, $m$ serves as a valuable tool for modeling and predicting signal behavior, allowing engineers to tailor communications systems to specific environmental conditions and optimize performance accordingly.

- Rayleigh fading describes the statistical variation in signal amplitude and phase due to the superposition of multiple scattered signals arriving at the receiver with random phase and amplitude. Due to the random nature of the phases and amplitudes of the individual components, the combined signal experiences fading. Specifically, Rayleigh fading occurs when there is no dominant line-of-sight component, meaning that the received signal is composed entirely of scattered or reflected signals.

Rician fading involves both a strong line-of-sight (LOS) signal component and multiple weaker scattered signals, making it suitable for scenarios with a dominant LOS path and multipath propagation. The Nakagami $m$ model is used to describe variations in signal strength, particularly in environments with multiple signal paths of varying attenuation. The parameter $m$ in Nakagami-$m$ determines the severity of fading, with smaller values resembling Rayleigh fading. Rayleigh fading, on the other hand, applies to scenarios without a dominant LOS component and characterizes statistical variations in signal amplitude and phase due to superimposed scattered signals. Understanding and applying these fading models is critical to designing robust satellite communications systems capable of adapting to varying signal conditions.

### *1.2.3 Satellite Communication Performance*

Satellite communications performance refers to the effectiveness and reliability of data transmission between a satellite and a ground station. It is an important aspect of ensuring that information is transmitted accurately and consistently over the satellite link. Three key indicators are used to evaluate satellite communications performance:

- Outage Probability (OP): Outage probability is a measure of the probability that the received signal will fall below a specified threshold. It quantifies the probability that the communication link will experience a temporary loss of connectivity or degraded performance. In satellite communications, an outage can occur due to factors such as severe atmospheric conditions, interference, or equipment failure. It is primarily defined in terms of a desired data rate or signal-to-noise ratio (SNR) level. It is a measure of the strength of the received signal compared to the background noise. The SNR can be represented as shown in Eq. (1.2):

$$\mathrm{SNR} = \frac{P_s}{P_n} \tag{1.2}$$

In this context, the received signal power $P_s$ is the strength of the signal transmitted by the satellite and received at the ground station or user terminal. The noise power $P_n$ includes all unwanted signals and interference that can affect the clarity and reliability of the communication. A high SNR indicates a strong and clear signal relative to the background noise, resulting in better communication quality and fewer errors. Conversely, a low SNR can result in a weaker signal, making it more susceptible to interference and degradation. Understanding and optimizing SNR is critical when designing satellite communications systems to ensure reliable and efficient data transmission. It helps engineers determine factors such as transmit power, antenna size, modulation scheme, and other parameters to achieve the desired level of performance.

- Ergodic Capacity (EC): Ergodic capacity represents the average achievable data rate over a long period, assuming that channel conditions (e.g., fading, interference) are time-varying and follow certain statistical models. It provides a measure of the long-term performance of the channel in terms of data throughput. Higher ergodic capacity indicates a channel with greater data-carrying capacity.
- Packet Error Rate (PER) or Bit Error Rate (BER): Packet error rate refers to the probability that a transmitted data packet will be received with errors. It is a critical metric, especially for digital communication systems where information is transmitted in discrete packets. Achieving a low PER is essential for reliable data transmission, especially in applications where data integrity is critical.

Although there are certain indicators to evaluate satellite communications performance, there are certain factors that influence satellite communications performance, including:

- Satellite link budget is a critical analysis that includes several key factors that affect the performance of a satellite communications system. These factors include the power at which signals are transmitted, the effectiveness of the antennas at both the transmit and receive ends, the attenuation of the signal as it travels through the medium, and the level of background noise present in the communications environment. By comprehensively evaluating these elements, the link budget serves as a fundamental tool for evaluating and optimizing the strength and quality of the signal in the communications link. This analysis is essential to ensure that the satellite communications system operates at the highest level of efficiency and reliability to meet the requirements of various applications and users.
- Fading and multipath effects are also key factors. Fading, caused by signal reflections and scattering, can cause rapid fluctuations in signal strength. Understanding and mitigating these effects is critical for reliable communications. It is imperative to have a comprehensive understanding of fading effects and implement strategies to mitigate them to ensure reliable and consistent communications. By employing techniques such as diversity reception and adaptive modulation, engineers can effectively counteract the adverse effects of fading and ultimately improve the overall performance and reliability of the communications link.
- Frequency band and modulation are also very important. The choice of frequency band and modulation scheme affects the channel's bandwidth, capacity, and robustness to noise and interference. In addition, the modulation scheme used determines how the information is encoded into the carrier signal. Different modulation schemes have different levels of complexity and efficiency, which affect the channel's ability to transmit data accurately and efficiently. In addition, the selected frequency band and modulation scheme together contribute to the channel's resilience to sources of interference such as noise and jamming, ultimately

affecting the overall quality and reliability of the communications link. Therefore, careful consideration of these factors is essential in the design and optimization of communication systems.
- Antenna characteristics are included in the important factors. The size, type, and orientation of the antennas at both ends of the link play a significant role in the quality of signal reception.
- Atmospheric conditions also have to be considered. Weather conditions such as rain, snow, or atmospheric disturbances can attenuate the signal and degrade performance.

To improve satellite communications performance, engineers use techniques such as adaptive modulation and coding, power control, diversity reception, and error correction coding. In addition, the selection of appropriate frequency bands and orbits based on the specific application can significantly impact performance.

### 1.2.4 Physical Layer Security in Satellite Communication

A key component of physical layer security in satellite communications is to exploit the inherent differences in channel characteristics between legitimate receivers and potential eavesdroppers. This is based on the understanding that the channel quality experienced by an eavesdropper is generally inferior due to factors such as distance, obstructions, and environmental conditions. This distinction forms the basis for establishing secure communications links. In addition, techniques such as artificial noise injection play a critical role. By intentionally introducing controlled noise into the transmitted signal, it becomes much more difficult for eavesdroppers to extract meaningful information. The strategic use of beamforming and directional antennas further enhances security. These techniques allow precise control of the direction in which the signal is transmitted, effectively reducing the likelihood of interception by unauthorized parties. This multi-pronged approach, using channel characteristics, artificial noise, and directional transmission, is the cornerstone of physical layer security in satellite communications. There are two key indicators of physical layer security, which include:

- Secrecy Outage Probability (SOP): SOP is a fundamental physical layer security metric. It quantifies the probability that an eavesdropper can successfully decode the transmitted signal. In essence, it measures the probability that a security breach will occur during a given transmission. This probability is affected by several factors, including the strength of the transmitted signal, the distance between the legitimate receiver and the eavesdropper, and the level of noise and interference in the channel. Achieving a low SOP is critical to ensuring that sensitive information remains confidential.
- Ergodic Secrecy Capacity (ESC): ESC is a more sophisticated measure of physical layer security. It represents the maximum average secrecy rate that can be achieved

over a long period, given that channel conditions are time-varying and subject to fading interference, and other dynamic factors. This metric takes into account the statistical properties of the channel and provides insight into the long-term security performance of the communication link.

Physical layer security techniques are essential for protecting information transmitted over satellite communications channels, especially in environments where potential eavesdroppers may attempt to intercept sensitive data. By exploiting channel characteristics and using advanced techniques such as artificial noise, beamforming, and secure precoding, physical layer security provides an additional layer of protection against security breaches.

### *1.2.5 Theory of Stochastic Geometry*

Stochastic geometry is a mathematical framework for modeling and analyzing spatial patterns and arrangements of objects in a random or uncertain environment. In the context of satellite communications, it is used to study the impact of the deployment of satellites and ground stations on communications performance. Some of the key indicators in stochastic geometry are presented below:

- Point Process Intensity: Point process intensity refers to the average number of points (communication nodes) per unit area. In satellite communications, this parameter helps to describe the spatial density of satellites or ground stations in a given region. It plays a critical role in modeling the deployment and distribution of communication nodes.
- Pair Correlation Function (PCF): The pair correlation function characterizes the spatial clustering or dispersion of points in a point process. It quantifies the probability of finding a point at a given distance from another point relative to a random distribution. In satellite communications, understanding the PCF helps assess the degree of clustering or regularity in the placement of satellites or ground stations.
- K-Function: Related to the pair correlation function, the K-function provides a measure of the expected number of points within a given distance of a randomly selected point. It allows a more detailed analysis of the spatial distribution patterns in the point process. In satellite communications, the K-function can be used to analyze the spatial characteristics of satellite constellations.
- Void Probability: Void Probability represents the probability that a randomly selected point falls within a region free of other points (communication nodes). It provides insight into the presence of "gaps" or unoccupied areas in the spatial distribution. In satellite communications, understanding the void probability helps evaluate coverage gaps or areas with limited satellite connectivity.
- Second-Order Characteristics: Second-order characteristics of a point process refer to statistical properties related to the pairwise interactions between points. These include metrics such as the pair correlation function, K-function, and void probability. Analyzing these characteristics helps to understand how communication

nodes are distributed relative to each other, which is critical for designing reliable and efficient satellite networks.

Stochastic geometry finds extensive applications in analyzing various performance metrics of satellite communication networks. Firstly, it plays a pivotal role in coverage and connectivity analysis. By modeling the distribution of ground terminals and satellites, stochastic geometry enables us to assess the reach and reliability of the network, pinpointing areas with robust communication links as well as potential coverage gaps. Additionally, it aids in interference analysis, crucial in satellite communications where neighboring satellites or ground stations can introduce significant interference, affecting signal quality. Stochastic geometry models the interference distribution, thereby facilitating the evaluation of system performance and the design of interference management strategies.

Moreover, stochastic geometry is indispensable in capacity analysis and resource allocation. It captures the spatial randomness of user terminals, crucial for estimating the maximum data rate supported by the satellite network. This understanding informs network designers in optimizing capacity. Furthermore, it guides the allocation of resources like power, bandwidth, and antenna beamforming, ensuring optimal configurations that maximize capacity and efficiency. Lastly, stochastic geometry informs network planning and deployment decisions. It provides insights into the optimal deployment of satellites and ground stations, guiding the determination of satellite count, orbital positions, and ground terminal placement to meet desired network performance objectives. Additionally, it models the spatial distribution of potential security threats or jammers, enhancing the network's resilience and security.

In summary, stochastic geometry is a powerful tool in satellite communications for modeling and analyzing the spatial aspects of communication networks. By quantifying the spatial distributions, clustering patterns, and coverage characteristics of communication nodes, it aids in the design, optimization, and performance evaluation of satellite systems. It enables engineers and researchers to make informed decisions about satellite deployment, resource allocation, and network planning.

## References

1. H. Guo, J. Li, J. Liu, N. Tian and N. Kato, "A Survey on Space-Air-Ground-Sea Integrated Network Security in 6G," in *IEEE Commun. Surv. Tutorials*, vol. 24, no. 1, pp. 53–87, Firstquarter 2022.
2. I. Ahmad, J. Suomalainen, P. Porambage, A. Gurtov, J. Huusko and M. Höyhtyä, "Security of Satellite-Terrestrial Communications: Challenges and Potential Solutions," in *IEEE Access*, vol. 10, pp. 96038–96052, 2022.
3. L. You, K. -X. Li, J. Wang, X. Gao, X. -G. Xia and B. Ottersten, "Massive MIMO Transmission for LEO Satellite Communications," in *IEEE J. Sel. Areas Commun.*, vol. 38, no. 8, pp. 1851–1865, Aug. 2020.
4. Q. Yan, J. Jiao, Y. Wang, L. An, R. Lu and Q. Zhang, "Age of Information Minimization for Short-Packet Communications RSMA in Satellite-based IoT," in *Proceedings of the 2023 IEEE 98th Vehicular Technology Conference (VTC2023-Fall)*, Hong Kong, Hong Kong, 2023, pp. 1–5.

5. L. You, A. Liu, W. Wang and X. Gao, "Outage Constrained Robust Multigroup Multicast Beamforming for Multi-Beam Satellite Communication Systems," in *IEEE Wireless Commun. Lett.*, vol. 8, no. 2, pp. 352–355, April 2019.
6. C. N. Efrem and A. D. Panagopoulos, "Globally Optimal Selection of Ground Stations in Satellite Systems With Site Diversity," in *IEEE Wireless Commun. Lett.*, vol. 9, no. 7, pp. 1101–1104, July 2020.
7. Y. Liu, Y. Chen, Y. Jiao, H. Ma and T. Wu, "A Shared Satellite Ground Station Using User-Oriented Virtualization Technology," in *IEEE Access*, vol. 8, pp. 63923–63934, 2020.
8. T. Wei, W. Feng, Y. Chen, C. -X. Wang, N. Ge and J. Lu, "Hybrid Satellite-Terrestrial Communication Networks for the Maritime Internet of Things: Key Technologies, Opportunities, and Challenges," in *IEEE Internet Things J.*, vol. 8, no. 11, pp. 8910–8934, 1 June1, 2021.
9. C. Chen, S. Yan and J. Mei, "A Maritime Target Location Method Using Geometric Semantic Constraints Based on Spaceborne GNSS-R," in *IEEE Geosci. Remote Sens. Lett.*, vol. 21, pp. 1–5, 2024, Art no. 1500705.
10. J. Wang, C. Jiang, L. Kuang and R. Han, "Satellite Multi-Beam Collaborative Scheduling in Satellite Aviation Communications," in *IEEE Trans. Wireless Commun.*, vol. 23, no. 3, pp. 2097–2111, March 2024.
11. Z. Li, S. Wang, S. Han and C. Li, "Non-Orthogonal Broadcast and Unicast Joint Transmission for Multibeam Satellite System," in *IEEE Trans. Broadcast.*, vol. 69, no. 3, pp. 647–660, Sept. 2023.
12. M. Sever, T. Y. Erkec and C. Hajiyev, "Comparison of EKF & UKF for GNSS Based Micro Satellite Orbital State Estimation," in *Proceedings of the 2023 10th International Conference on Recent Advances in Air and Space Technologies (RAST)*, Istanbul, Turkiye, 2023, pp. 01–07.
13. S. Scrocciolani, V. Messina, R. M. G. Alarcia, J. Sindermann and A. Golkar, "Advancing Satellite Network Performance: Network Analysis for Federated Satellite Systems," in *IEEE Access*, vol. 12, pp. 45616–45630, 2024.
14. G. Jang, B. You and H. Jung, "A Survey on Physical Layer Security Schemes in Satellite Networks," in *Proceedings of the 2022 13th International Conference on Information and Communication Technology Convergence (ICTC)*, Jeju Island, Korea, Republic of, 2022, pp. 1213–1215.
15. D. Song, Z. Yang, G. Pan, S. Wang and J. An, "RIS-Assisted Covert Transmission in Satellite-Terrestrial Communication Systems," in *IEEE Internet Things J.*, vol. 10, no. 22, pp. 19415–19426, 15 Nov.15, 2023.
16. M. Wang, W. Yang, X. Lu, C. Hu, B. Liu and X. Lv, "Channel Inversion Power Control Aided Covert Communications in Uplink NOMA Systems," in *IEEE Wireless Commun. Lett.*, vol. 11, no. 4, pp. 871–875, April 2022.
17. W. Abdurrokhman, S. Utama, P. R. Hakim, N. Najati, N. S. Y. Hasanah and D. A. Rahayu, "Investigating LAPAN-A3 Satellite X-Band Data Transmission Interference in Orbital Approaches," in *Proceedings of the 2023 IEEE International Conference on Aerospace Electronics and Remote Sensing Technology (ICARES)*, Bali, Indonesia, 2023, pp. 1–4.
18. S. Jeon, J. Kwak and J. P. Choi, "Cross-Layer Encryption of CFB-AES-TURBO for Advanced Satellite Data Transmission Security," in *IEEE Trans. Aerosp. Electron. Syst.*, vol. 58, no. 3, pp. 2192–2205, June 2022.
19. D. -H. Jung, J. -G. Ryu, W. -J. Byun and J. Choi, "Performance Analysis of Satellite Communication System Under the Shadowed-Rician Fading: A Stochastic Geometry Approach," in *IEEE Trans. Commun.*, vol. 70, no. 4, pp. 2707–2721, April 2022.
20. R. Wang, M. A. Kishk and M. -S. Alouini, "Evaluating the Accuracy of Stochastic Geometry Based Models for LEO Satellite Networks Analysis," in *IEEE Commun. Lett.*, vol. 26, no. 10, pp. 2440–2444, Oct. 2022.

# Chapter 2
# Satellite-to-Terrestrial Communications

**Gaofeng Pan, Jia Ye, Yongqiang Zhang, and Mohamed-Slim Alouini**

## 2.1 On HARQ Schemes in Satellite-Terrestrial Communications

### *2.1.1 System Model*

In this work, an satellite-terrestrial transmission (STT) system, which consists of a satellite transmitter (S), and a group of terrestrial receivers ($D_i$, $1 \leq i \leq N$), is considered, as presented in Fig. 2.1. It is assumed that S and all terrestrial receivers are equipped with a single transmit and a single receiving antenna.

#### 2.1.1.1 The Distribution Model of Terrestrial Receivers

For tractability purposes, in this work we also treat the coverage area of the beam of S as a circle, $\mathcal{S}$, with radius $L$, where the projection of S is located at the centre of the circle. The location of terrestrial receivers is modeled as a homogeneous Poisson Point process (PPP) with density $\lambda$. The number of terrestrial receivers, which denoted by $N$ ($N \geq 1$), is Poisson distributed i.e., $P\{N = k\} = \left(\mu_S^k/k!\right) \exp(-\mu_S)$, where $\mu_S = \pi L^2 \lambda$ is the mean measure. Then, the $N$ ($N \geq 1$) terrestrial receivers can be modeled as a set of independent and identically distributed points in the circle, $\mathcal{S}$, denoted by $\mathcal{W}$. Therefore, the distance between terrestrial receivers and

G. Pan
Beijing Institute of Technology, Beijing, China

J. Ye (✉)
Chongqing University, Chongqing, China
e-mail: yejiaft@163.com

Y. Zhang · M.-S. Alouini
King Abdullah University of Science and Technology (KAUST), Thuwal, Makkah Province, Saudi Arabia



S. Wang et al. (eds.), *Understanding Satellite Communications: The Stochastic Geometry Perspective*, https://doi.org/10.1007/978-981-97-7102-8_2

**Fig. 2.1** STT system

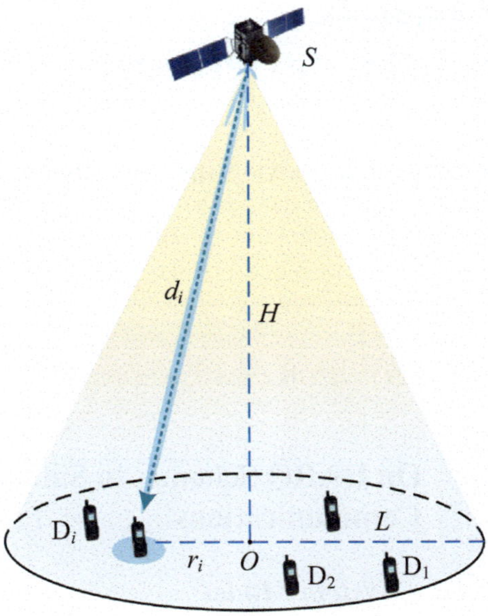


the projection of S ($O$) can be calculated from $\mathcal{W}$, the probability density function (PDF) of which can be given by [1–4]

$$f_W(w) = \frac{\lambda}{\mu(\mathcal{S})} = \frac{1}{\pi L^2}. \tag{2.1}$$

So, the cumulative distribution function (CDF) of $r_i$ $(1 \le i \le N)$ can be derived as

$$F_{r_i}(x) = \int_S f_W(w)dw = \int_0^{2\pi}\int_0^x \frac{1}{\pi L^2} r dr d\theta = \frac{x^2}{L^2}. \tag{2.2}$$

Therefore, the PDF $r_i$ $(1 \le i \le N)$ can be obtained as

$$f_{r_i}(x) = \frac{2x}{L^2}. \tag{2.3}$$

Further, we can derive the PDF of $d_i = \sqrt{{H_1}^2 + r_i^2}$ $(1 \le i \le N)$ as

$$f_{d_i}(x) = \frac{2x}{L^2}, \tag{2.4}$$

where $H_1 \le x \le \sqrt{{H_1}^2 + L^2}$.

#### 2.1.1.2 Hybrid Automatic Repeat Request (HARQ) Schemes

In this work, S is designed to deliver information messages to terrestrial receivers ($\mathrm{D}_i$, $1 \le i \le N$) by using a multicast transmission. The time axis is divided into slots with duration $T$ seconds, and all terrestrial receivers are assumed to be synchronized so they are with a common time reference.

Under the automatic repeat request (ARQ) protocol, the information messages at S, the length of which is $b$ bits, are encoded into $K$ subblocks, which are of the length of $L$. In other words, S encodes its information message of $b$ bits each into a channel code $\boldsymbol{c}$ of length $LK$ over the complex numbers chosen from the codebook, $C \in \mathbb{C}^{LK}$. This codeword can be divided into $K$ subblocks of length $L$. For each subblock $\boldsymbol{c}_j \in \{\boldsymbol{c}_1, \boldsymbol{c}_2, \ldots, \boldsymbol{c}_K\}$, all information for successfully decoding the original information message is included. Consequently, at each receiver, the original message can be rebuilt by using each subblock or any combination of these subblocks, while the receiving signal-to-noise ratio (SNR) is large enough.

At the start of the $j$th $(1 \le j \le K)$ time slot, S encapsulates the subblock $\boldsymbol{c}_j$ into one packet, which takes slightly less than $T$ seconds to allow S to process the feedback information from the terrestrial receivers, and transmits the packet to all terrestrial receivers for the $j$th $(1 \le j \le K)$ round of multicasting the $b$ bits of information message. In each round of delivering $b$ bits of information message, if a terrestrial user is unable to decode the message, it sends a negative acknowledgment (NACK) message through an error-free and zero-delay feedback channel to S. If the decoding is successful, the terrestrial receiver remains silent and does not feedback any messages. This NACK message is received by S at the end of each time slot. Then, when at least one NACK message is obtained at S, S will deliver another packet containing $\boldsymbol{c}_{j+1}$ to all terrestrial receivers. If S does not receive a NACK message via the feedback channel, it is regarded that the transmission of the bits information message is complete, and then S starts to deliver another $b$ bits information message again following the aforementioned process. Furthermore, if at least one NLCK message is overheard by S at the end of the $K$ time slot, the transmitted message will be discarded by S and an outage event occurs for the whole system.

In the $j$th $(1 \le j \le K)$ time slot, upon receiving the packet containing $\boldsymbol{c}_j$, the terrestrial receiver will decode the transmitted message by using the received signals from the subblocks $\left\{\boldsymbol{c}_{(j-M+1)^+}, \ldots, \boldsymbol{c}_j\right\}$, where $M$ is the memory size at the terrestrial receiver, and $(j - M + 1)^+ = \max\{1, j - M + 1\}$. Namely, the $M$ most recently received packets are adopted to decode the original transmitted message at each terrestrial receiver. Moreover, generally, if a terrestrial receiver successfully decodes the transmitted message, it will stay silent in the next time slot.

#### 2.1.1.3 Signal Model

In this work, it is assumed that all channels experience independent and identically distributed (i.i.d.) block-fading. That is to say, the channel gain for each channel remains constant over a block/slot but varies independently from one block/slot to another [5]. The received signal at the $i$th ($1 \le i \le N$) terrestrial receiver, $D_i$, in the $j$th ($1 \le j \le K$) time slot is given by

$$y_{D_i}[j] = \sqrt{P_S d_i^{-n}} h_{SD_i}[j] x[j] + n_{D_i}, \tag{2.5}$$

where $P_S$ is the transmit power at S, $d_i$ is the distance between S and $D_i$, $n$ is the path-loss factor over S-$D_i$ link, $h_{SD_i}[j]$ is the channel gain for S-$D_i$ link in the $j$th ($1 \le j \le K$) time slot, $x[j]$ is the transmitted subblock in the $j$th ($1 \le j \le K$) time slot, and $n_{D_i}$ denotes the complex additive white Gaussian noise (AWGN) at $D_i$.

Furthermore, in this work Shadowed-Rician (SR) model is adopted to describe the statistical distribution of the STT link from S to $D_i$, which has been proved to be an accurate, practical and applicable tool to evaluate performance for satellite propagation environments in various frequency bands, e.g., the UHF-band, L-band, S-band, Ku-band, and Ka-band.

Without loss of generality, the PDF of the power gain, $|h_{SD_i}[j]|^2$, for S-$D_i$ link in SR channel is given as [6]

$$f_{|h_{SD_i}[j]|^2}(x) = \alpha \exp(-\beta x)\, {}_1F_1(m; 1; \delta x), \quad x \ge 0, \tag{2.6}$$

where $\alpha = \left(\frac{2bm}{2bm+\Omega}\right)^m/(2b)$, $\beta = \frac{1}{2b}$ and $\delta = \frac{\Omega}{2b(2bm+\Omega)}$, $\Omega$ and $2b$ are the average power of the line-of-sight (LOS) and multipath components, respectively. $m$ is the fading severity parameter (When $m = 0$, it denotes the case without LOS when $0 < m < \infty$, it stands for the case of both small-scale fading and LOS when $m = \infty$, it represents the LOS case), and ${}_1F_1(\cdot;\cdot;\cdot)$ is the confluent hypergeometric function of first kind [7].

Then, it is easy to obtain the received SNR at $D_i$ in the $j$th ($1 \le j \le K$) time slot as

$$\gamma_{ij} = \frac{P_S}{\sigma^2} \frac{|h_{SD_i}[j]|^2}{d_i^n} = \frac{\lambda_j}{d_i^n}, \tag{2.7}$$

where $\lambda_j = P_S |h_{SD_i}[j]|^2/\sigma^2 = \bar{\lambda}|h_{SD_i}[j]|^2$, $\bar{\lambda} = P_S/\sigma^2$ is the transmit SNR, and $\sigma^2$ is the average power of the AWGN at $D_i$.

By variable substitution and applying Kummer's transform, the PDF and CDF of $\lambda_j$ can be presented as

$$f_{\lambda_j}(x) = \alpha \sum_{k=0}^{m-1} \frac{\varsigma(k)}{\bar{\lambda}^{k+1}} x^k \exp\left(-\frac{\beta-\delta}{\bar{\lambda}}x\right) \tag{2.8}$$

and

$$F_{\lambda_j}(x) = 1 - \alpha \sum_{k=0}^{m-1} \frac{\varsigma(k)}{\bar{\lambda}^{k+1}} \sum_{p=0}^{k} \frac{k!}{p!} \left(\frac{\beta-\delta}{\bar{\lambda}}\right)^{-(k+1-p)} x^p \exp\left(-\frac{\beta-\delta}{\bar{\lambda}}x\right), \tag{2.9}$$

respectively, where $\varsigma(k) = \frac{(-1)^k(1-m)_k\delta^k}{(k!)^2}$ and $(t)_k = t(t+1)\ldots(t+k-1)$ is the Pochhammer symbol [7].

Then, according to the ARQ protocol presented in the previous subsection and adopting maximum ratio combining scheme at each receiver to process the multiple received signals, the accumulated mutual information at $\mathrm{D}_i$ used for decoding in the $j$th $(1 \le j \le K)$ time slot is then given by

$$I_{ij} = \sum_{s=(j-M+1)^+}^{j} \log_2(1+\gamma_{is}) = \sum_{s=(j-M+1)^+}^{j} \log_2\left(1+\frac{\lambda_s}{d_i^n}\right). \tag{2.10}$$

#### 2.1.1.4 Performance Metric

In this work, we mainly concern the impacts of the adopted HARQ schemes on the performance of the considered STT system. Specifically, the outage performance and diversity gain will be studied in the following parts of this work, which are respectively defined as

(1) Outage probability (OP): The probability that an originally transmitted message cannot be successfully decoded by at least one terrestrial receiver after $K$ rounds of deliveries.

(2) Diversity gain: Multiple copies of the transmitted signal can be exploited during the decoding process.

To facilitate the following analysis, some assumptions are made as follows:

(1) Infinite-size buffer is available at S, which means that there always exist information messages at S to be delivered to terrestrial receivers;

(2) The processing delay at S and terrestrial receivers and transmission delays are omitted.

### 2.1.2 Outage Analysis

In this subsection, outage analysis will be carried out for three hybrid-ARQ schemes: generalized slotted ALOHA (GSA), repetition time diversity (RTD), and general incremental redundancy (IR) schemes, respectively.

In each round of the transmission of the information message, the outage event occurs at $D_i$ when the accumulated mutual information is less than the transmission rate, $R$, resulting in the failure of the decoding. So, the OP at $D_i$ in the $j$th $(1 \leq j \leq K)$ time slot can be represented as

$$P_{\text{out},ij} = \Pr\{\text{decoding error}\} = \Pr\left\{I_{ij} < R\right\}. \tag{2.11}$$

**Theorem 2.1** *As there are $N$ $(N \geq 1)$ terrestrial receivers within the coverage area of S in each system realization, and all links between S and the terrestrial receivers experience i.i.d. fading, the OP for the whole system can be presented as*

$$P_{\text{out}} = 1 - \left(1 - \rho^K\right)^N, \tag{2.12}$$

*where $\rho = P_{\text{out},ij}$.*

***Proof*** For terrestrial receiver $D_i$, the probability that it can successfully decode the transmitted information message in the $j$th $(1 \leq j \leq K)$ time slot is written as

$$P_{\text{su},ij} = {P_{\text{out},ij}}^{j-1}\left(1 - P_{\text{out},ij}\right). \tag{2.13}$$

So, the probability that $D_i$ can successfully decode the transmitted information message during the $K$ rounds of the transmissions can be given as

$$P_{\text{su},i} = \sum_{j=1}^{K} P_{\text{su},ij} = \sum_{j=1}^{K} {P_{\text{out},ij}}^{j-1}\left(1 - P_{\text{out},ij}\right). \tag{2.14}$$

Considering that all links between S and the terrestrial receivers experience i.i.d. fading, we have $P_{\text{out},ij} = P_{\text{out},pq}$ for $\{(i, j, p, q)\,|\{i, j, p, q\} \in \{1, 2, \ldots, N\}\}$. Then, the OP for the whole system can be derived as

$$\begin{aligned} P_{\text{out}} &= 1 - \prod_{i=1}^{N} P_{\text{su},i} = 1 - \prod_{i=1}^{N}\sum_{j=1}^{K} {P_{\text{out},ij}}^{j-1}\left(1 - P_{\text{out},ij}\right) \\ &= 1 - \prod_{i=1}^{N}\left(1 - \rho^K\right) = 1 - \left(1 - \rho^K\right)^N, \end{aligned} \tag{2.15}$$

where in this work we let $\rho = P_{\text{out},ij}$ for simplification.

Then, the proof is completed. □

#### 2.1.2.1 GSA Scheme

Under the GSA scheme, the memory size at the terrestrial receivers $M = 1$. In other words, all terrestrial receivers only adopt the latest received signal to decode the transmitted message, i.e., in the $j$th $(1 \leq j \leq K)$ time slot, the received signal of the packet is only used for decoding.

Therefore, the accumulated mutual information at $\mathrm{D}_i$ in the $j$th $(1 \leq j \leq K)$ time slot is

$$I_{ij} = \log_2\left(1 + \gamma_{ij}\right) = \log_2\left(1 + \frac{\lambda_j}{d_i^n}\right). \tag{2.16}$$

Accordingly, the OP under GSA scheme in the $j$th $(1 \leq j \leq K)$ time slot can be written as

$$P_{\mathrm{out},ij} = \Pr\left\{\log_2\left(1 + \frac{\lambda_j}{d_i^n}\right) < R\right\} = \Pr\left\{\frac{\lambda_j}{d_i^n} < \varepsilon\right\}, \tag{2.17}$$

where $\varepsilon = 2^R - 1$.

**Theorem 2.2** *Considering the randomness of the channel gain between S and $\mathrm{D}_i$, and the randomness of the position of $\mathrm{D}_i$, the OP under GSA scheme in the $j$th $(1 \leq j \leq K)$ time slot can be derived as*

$$\begin{aligned} P_{\mathrm{out},ij} = 1 - &\frac{2\alpha}{nL^2} \sum_{k=0}^{m-1} \frac{\varsigma(k)}{\bar{\lambda}^{k+1}} \sum_{p=0}^{k} \frac{k!}{p!} \left(\frac{\beta - \delta}{\bar{\lambda}}\right)^{-(k+\frac{2}{n}+1)} \\ &\times \varepsilon^{-\frac{2}{n}} \left[\gamma\left(p + \frac{2}{n}, y_{\max}\right) - \gamma\left(p + \frac{2}{n}, y_{\min}\right)\right], \end{aligned} \tag{2.18}$$

*where $y_{\min} = \frac{\varepsilon(\beta-\delta)}{\bar{\lambda}} H_1^n$, $y_{\max} = \frac{\varepsilon(\beta-\delta)}{\bar{\lambda}} \left({H_1}^2 + L^2\right)^{\frac{n}{2}}$, and $\gamma(a, x) = \int_0^x t^{a-1} \exp(-t)\, dt$ is the lower incomplete gamma function.*

***Proof*** Please refer to Appendix I. □

Therefore, the OP for the whole system under the GSA scheme can be obtained by substituting (2.18) into (2.12).

#### 2.1.2.2 RTD Scheme

Under the RTD scheme, the information messages are encoded by using repetition coding, leading to the fact that all the packets transmitted in different time slots of the $K$ rounds are the same. So, under RTD scheme, the accumulated mutual information at $\mathrm{D}_i$ used for decoding in the $j$th $(1 \leq j \leq K)$ time slot is given by

$$
\begin{aligned}
I_{ij} &= \log_2\left(1+\sum_{s=(j-M+1)^+}^{j}\gamma_{is}\right)\\
&= \log_2\left(1+\sum_{s=(j-M+1)^+}^{j}\frac{\lambda_s}{d_i^n}\right)\\
&= \log_2\left(1+\frac{1}{d_i^n}\sum_{s=(j-M+1)^+}^{j}\lambda_s\right).
\end{aligned}
\tag{2.19}
$$

Thus, the OP under RTD scheme in the $j$th $(1 \le j \le K)$ time slot can be written as

$$
\begin{aligned}
P_{\text{out},ij} &= \Pr\left\{\log_2\left(1+\frac{1}{d_i^n}\sum_{s=(j-M+1)^+}^{j}\lambda_s\right) < R\right\}\\
&= \Pr\left\{\frac{1}{d_i^n}\sum_{s=(j-M+1)^+}^{j}\lambda_s < \varepsilon\right\}.
\end{aligned}
\tag{2.20}
$$

**Theorem 2.3** *Considering the randomness of the channel gain between S and* $\mathrm{D}_i$*, and the randomness of the position of* $\mathrm{D}_i$*, the OP under RTD scheme in the* $j$*th* $(1 \le j \le K)$ *time slot can be derived as*

$$
P_{\text{out},ij} = \alpha^{\mathcal{M}}\sum_{l=0}^{c}\binom{c}{l}\beta^{c-l}\left[H(l,d)+\nu\delta H(l,d+1)\right],
\tag{2.21}
$$

*where* $\mathcal{M} = j-[j-M]^+$, $c=\left[d-\mathcal{M}\right]^+$, $\nu = m\mathcal{M}-d$, $d=\max\left\{\mathcal{M}, \lfloor m\mathcal{M}\rfloor\right\}$, $\lfloor x\rfloor$ *denotes the largest integer not larger than x,* $[x]^+$ *indicates that if* $x \le 0$*, then use* $x=0$, $t_{\max}=\left({H_1}^2+L^2\right)^{\frac{n}{2}}$, $t_{\min}={H_1}^n$, $J=\frac{\varepsilon(\beta-\delta)}{\bar{\lambda}}$,

$$
\begin{aligned}
H(l,d) =& \frac{2(\beta-\delta)^{\frac{l-d-1}{2}}}{nL^2\Gamma(d)}\left(\frac{\varepsilon}{\bar{\lambda}}\right)^{\frac{d-l-1}{2}}\left\{t_{\max}^{\frac{d-l-1}{2}+\frac{2}{n}}\right.\\
&\times G_{2,3}^{1,2}\left[Jt_{\max}\left|\begin{matrix}\frac{3-d-l}{2},\frac{3-d+l}{2}-\frac{2}{n}\\ \frac{d-l+1}{2},\frac{1-d+l}{2}-\frac{2}{n},\frac{1-d+l}{2}\end{matrix}\right.\right]-t_{\min}^{\frac{d-l-1}{2}+\frac{2}{n}}\\
&\left.\times G_{2,3}^{1,2}\left[Jt_{\min}\left|\begin{matrix}\frac{3-d-l}{2},\frac{3-d+l}{2}-\frac{2}{n}\\ \frac{d-l+1}{2},\frac{1-d+l}{2}-\frac{2}{n},\frac{1-d+l}{2}\end{matrix}\right.\right]\right\},
\end{aligned}
\tag{2.22}
$$

*and* $G_{p,q}^{m,n}\left[x\left|\begin{matrix}a_1,\ldots,a_p\\ b_1,\ldots,b_q\end{matrix}\right.\right]$ *is the Meijer's G-function, as defined by [7, Eq. (9.301)].*

***Proof*** Please refer to Appendix II. □

Then, the OP for the whole system under the RTD scheme can be obtained by substituting (2.21) into (2.12).

#### 2.1.2.3 HARQ-IR Scheme

Under the HARQ-IR scheme, a general IR code at S is adopted, which indicates that the transmitted packets in different timeslots of the $K$ rounds of the transmission will be different from each other. At the receiver, in the $j$th $(1 \leq j \leq K)$ time slot, the signal of the previous $j - (j - M + 1)^+$ received packets are adopted to decode the transmitted information message. So, under this case, the accumulated mutual information at $\mathrm{D}_i$ used for decoding in the $j$th $(1 \leq j \leq K)$ time slot can be given by (2.10).

Accordingly, the OP under HARQ-IR scheme is then given by

$$\begin{aligned} P_{\mathrm{out},ij} &= \Pr\left\{ \sum_{s=(j-M+1)^+}^{j} \log_2(1+\gamma_{is}) < R \right\} \\ &= \Pr\left\{ \log_2 \prod_{s=(j-M+1)^+}^{j} (1+\gamma_{is}) < R \right\} \\ &= \Pr\left\{ \prod_{s=(j-M+1)^+}^{j} (1+\gamma_{is}) < 2^R \right\}. \end{aligned} \tag{2.23}$$

Observing the above equation, one can see that it is impossible to derive the exact closed-form analytical expression for the CDF of $\prod_{s=(j-M+1)^+}^{j}\left(1+\frac{\lambda_s}{d_i^n}\right)$, while considering the complexity of the statistical characteristics of $\lambda_s$ shown by (2.8) and (2.9), as well as the complexity of $\prod_{s=(j-M+1)^+}^{j}\left(1+\frac{\lambda_s}{d_i^n}\right)$. In the following, we will investigate the OP in high SNR region for the considered system. Then, using $1+\gamma_{is} \approx \gamma_{is}$, (2.23) can be further rewritten as

$$\begin{aligned} P_{\mathrm{out},ij} &= \Pr\left\{ \prod_{s=(j-M+1)^+}^{j} \left(1+\bar{\lambda}\frac{|h_{\mathrm{SD}_i}[s]|^2}{d_i^n}\right) < 2^R \right\} \\ &\approx \Pr\left\{ \prod_{s=(j-M+1)^+}^{j} \bar{\lambda}\frac{|h_{\mathrm{SD}_i}[s]|^2}{d_i^n} < 2^R \right\} \end{aligned}$$

$$= \Pr\left\{\frac{\bar{\lambda}^{\mathcal{M}}}{d_i^{n\mathcal{M}}} \prod_{s=(j-M+1)^+}^{j} \left|h_{\mathrm{SD}_i}[s]\right|^2 < 2^R\right\}$$
$$= \Pr\left\{\frac{Y}{Z} < 2^R\right\}, \tag{2.24}$$

where $Y = \prod_{s=(j-M+1)^+}^{j} \left|h_{\mathrm{SD}_i}[s]\right|^2$, $Z = \frac{d_i^{n\mathcal{M}}}{\bar{\lambda}^{\mathcal{M}}}$, and $z_{\min} = \frac{H_1^{n\mathcal{M}}}{\bar{\lambda}^{\mathcal{M}}} \le Z \le z_{\max} = \frac{(H_1{}^2+L^2)^{\frac{n\mathcal{M}}{2}}}{\bar{\lambda}^{\mathcal{M}}}$.

**Theorem 2.4** *Considering that all links between S and the terrestrial receivers suffer i.i.d. SR fading, the CDF of $W = Y/Z$ (or equivalently the OP under HARQ-IR scheme) can be derived as* (2.25)*, where $\jmath = \sqrt{-1}$ is the imaginary unit,* $\mathrm{Im}\{\cdot\}$ *represents the imaginary part of a complex number, and ${}_2F_1(\cdot,\cdot;\cdot;\cdot)$ denotes the hypergeometric function [7].*

$$P_{\mathrm{out},ij} = F_W\left(2^R\right)$$
$$= \frac{1}{2} - \int_0^{\infty} \frac{\mathrm{Im}\left\{2^{-\jmath Rt}\,\jmath\,\frac{2\bar{\lambda}^{-\frac{2}{n}}}{L^2}\,\frac{z_{\max}^{\frac{2}{n\mathcal{M}}-\jmath t} - z_{\min}^{\frac{2}{n\mathcal{M}}-\jmath t}}{2\jmath + n\mathcal{M}t}\left[\alpha\beta^{-(\jmath t+1)}\Gamma(\jmath t+1)\cdot {}_2F_1\left(m, \jmath t+1; 1; \frac{\delta}{\beta}\right)\right]^{\mathcal{M}}\right\}}{\pi t}\,dt \tag{2.25}$$

***Proof*** Please refer to Appendix III. □

Because of the complexity of the statistical characteristics of the product of multiple SR fading variables, there exists an integral operation in (2.25). However, it can be easily solved by the built-in function in numerical software, such as *int* in MATLAB. So, the obtained analytical result can readily serve as a useful tool to study the outage performance of the HARQ-IR scheme in STT scenarios.

Therefore, the OP for the whole system under the GSA scheme can be obtained by substituting (2.25) into (2.12).

### *2.1.3 Diversity Gain*

Clearly, HARQ schemes exploit either coding gain or time diversity to improve the reliability of the transmissions between the communication pairs. In this work, the diversity order of the considered system is given by

$$\Delta = -\lim_{\bar{\lambda}\to\infty} \frac{\log P_{\mathrm{out},ij}}{\log \bar{\lambda}}, \tag{2.26}$$

where $\bar{\lambda}$ is defined after (2.7).

#### 2.1.3.1 GSA Scheme

**Lemma 2.1** *The diversity order of the considered system under GSA scheme is unitary.*

***Proof*** The OP under GSA scheme can be written as

$$\Pr\left\{\lambda_j \leq \varepsilon d_i^n\right\} = \mathbb{E}_{d_i}\left\{F_{\lambda_j}\left(\varepsilon d_i^n\right)\right\}. \tag{2.27}$$

From [8, Eq. (8)], the asymptotic CDF of $\lambda_j$ is given by

$$F_{\lambda_j}(x) \overset{\bar{\lambda}\to\infty}{\simeq} \alpha x \bar{\lambda}^{-1}. \tag{2.28}$$

Substituting (2.28) into (2.27) yields

$$P_{\text{out},ij} = \Pr\left\{\lambda_j \leq \varepsilon d_i^n\right\} \overset{\bar{\lambda}\to\infty}{\simeq} \bar{\lambda}^{-1}\mathbb{E}_{d_i}\left\{\alpha\left(\varepsilon d_i^n\right)\right\}. \tag{2.29}$$

From the definition of diversity order in (2.26), the corresponding diversity order of the OP under GSA scheme is easily obtained as

$$\Delta_{\text{GSA}} = -\lim_{\bar{\lambda}\to\infty} \frac{\log\left(\bar{\lambda}^{-1}\right) + \log \mathbb{E}_{d_i}\left\{\alpha\left(\varepsilon d_i^n\right)\right\}}{\log \bar{\lambda}} = 1, \tag{2.30}$$

which completes the proof. □

#### 2.1.3.2 RTD Scheme

**Lemma 2.2** *The diversity order of the considered system under RTD scheme is $\mathcal{M}$.*

***Proof*** We first investigate the diversity order of the CDF of $\max\{\lambda_1, \lambda_2, \ldots, \lambda_{\mathcal{M}}\}d_i^{-n}$. It is easy to obtain the CDF of $\max\{\lambda_1, \lambda_2, \ldots, \lambda_{\mathcal{M}}\}d_i^{-n}$, given by

$$\begin{aligned}\Pr\left\{\max\left\{\lambda_1, \lambda_2, \ldots, \lambda_{\mathcal{M}}\right\} \leq x d_i^n\right\} &= \mathbb{E}_{d_i}\left\{\left[F_{\lambda_1}(x)\right]^{\mathcal{M}}\right\} \\ &\overset{\bar{\lambda}\to\infty}{\simeq} \bar{\lambda}^{-\mathcal{M}}\mathbb{E}_{d_i}\left\{\left(\alpha x d_i^n\right)^{\mathcal{M}}\right\}.\end{aligned} \tag{2.31}$$

The corresponding diversity order is

$$\Delta = -\lim_{\bar{\lambda}\to\infty} \frac{\log\left(\bar{\lambda}^{-\mathcal{M}}\right) + \log \mathbb{E}_{d_i}\left\{\left(\alpha x d_i^n\right)^{\mathcal{M}}\right\}}{\log \bar{\lambda}} = \mathcal{M}. \tag{2.32}$$

Then, we can easily see that $\mathcal{M}\max\{\lambda_1, \lambda_2, \ldots, \lambda_{\mathcal{M}}\}d_i^{-n}$ has the same order as $\max\{\lambda_1, \lambda_2, \ldots, \lambda_{\mathcal{M}}\}d_i^{-n}$. Finally, in view of the following relationship,

$$\frac{\max\{\lambda_1, \lambda_2, \ldots, \lambda_{\mathcal{M}}\}}{d_i^n} \leq \frac{\lambda_1 + \lambda_2 + \cdots + \lambda_{\mathcal{M}}}{d_i^n} \leq \frac{\mathcal{M}\max\{\lambda_1, \lambda_2, \ldots, \lambda_{\mathcal{M}}\}}{d_i^n}, \tag{2.33}$$

we come to the conclusion that the diversity order of the OP under RTD scheme is $\mathcal{M}$. □

#### 2.1.3.3 HRAQ-IR Scheme

**Lemma 2.3** *The diversity order of the considered system under HRAQ-IR scheme is* $\mathcal{M}$.

***Proof*** Let $X_i = \bar{\lambda}|h_{\text{SD}i}[s]|^2$. It is obvious that the OP under HRAQ-IR scheme is the CDF of $(X_1 X_2 \ldots X_{\mathcal{M}})d_i^{-n\mathcal{M}}$, which means that we can alternatively calculate the diversity order of the CDF of $(X_1 X_2 \ldots X_{\mathcal{M}})d_i^{-n\mathcal{M}}$.

First, the CDF of $\frac{[\min\{X_1 X_2 \ldots X_{\mathcal{M}}\}]^{\mathcal{M}}}{d_i^{n\mathcal{M}}}$ can be derived as

$$\begin{aligned}\Pr\left\{\frac{[\min\{X_1 X_2 \ldots X_{\mathcal{M}}\}]^{\mathcal{M}}}{d_i^{n\mathcal{M}}} \leq x\right\} &= 1 - \Pr\left\{\min\{X_1 X_2 \ldots X_{\mathcal{M}}\} > d_i^n x^{\frac{1}{\mathcal{M}}}\right\} \\ &= 1 - \mathbb{E}_{d_i}\left\{\left[1 - F_{X_1}\left(d_i^n x^{\frac{1}{\mathcal{M}}}\right)\right]^{\mathcal{M}}\right\}.\end{aligned} \tag{2.34}$$

By using $1 - (1-x)^{\mathcal{M}} \overset{x\to 0}{\approx} x^{\mathcal{M}}$ and the asymptotic CDF of $X_1$ ($F_{X_1}(x) \approx \alpha\bar{\lambda}^{-1}x$ for $\bar{\lambda} \to \infty$ [8]), the probability can be further written as

$$\begin{aligned}\Pr\left\{\frac{[\min\{X_1 X_2 \ldots X_{\mathcal{M}}\}]^{\mathcal{M}}}{d_i^{n\mathcal{M}}} \leq x\right\} &\approx \mathbb{E}_{d_i}\left\{\left[F_{X_1}\left(d_i^n x^{\frac{1}{\mathcal{M}}}\right)\right]^{\mathcal{M}}\right\} \\ &= \mathbb{E}_{d_i}\left\{\left[\alpha\bar{\lambda}^{-1} d_i^n x^{\frac{1}{\mathcal{M}}}\right]^{\mathcal{M}}\right\} \\ &= \bar{\lambda}^{-\mathcal{M}}\mathbb{E}_{d_i}\left\{\left[\alpha d_i^n x^{\frac{1}{\mathcal{M}}}\right]^{\mathcal{M}}\right\}.\end{aligned} \tag{2.35}$$

The corresponding diversity order can be easily derived as

$$\begin{aligned}\Delta_{\min} &= -\lim_{\overline{\lambda}\to\infty}\frac{\log\Pr\left\{\frac{[\min\{X_1X_2\ldots X_{\mathcal{M}}\}]^{\mathcal{M}}}{d_i^{n\mathcal{M}}}\le x\right\}}{\log\overline{\lambda}}\\ &= -\lim_{\overline{\lambda}\to\infty}\frac{\log\left(\overline{\lambda}^{-\mathcal{M}}\right)+\log\mathbb{E}_{d_i}\left\{\left[\alpha d_i^n x^{\frac{1}{\mathcal{M}}}\right]^{\mathcal{M}}\right\}}{\log\overline{\lambda}}\\ &= \mathcal{M}. \end{aligned} \tag{2.36}$$

Then, the CDF of $\frac{[\max\{X_1X_2\ldots X_{\mathcal{M}}\}]^{\mathcal{M}}}{d_i^{n\mathcal{M}}}$ is given by

$$\begin{aligned}\Pr\left\{\frac{[\max\{X_1X_2\ldots X_{\mathcal{M}}\}]^{\mathcal{M}}}{d_i^{n\mathcal{M}}}\le x\right\} &= \mathbb{E}_{d_i}\left\{\left[F_{X_1}\left(d_i^n x^{\frac{1}{\mathcal{M}}}\right)\right]^{\mathcal{M}}\right\}\\ &\overset{\overline{\lambda}\to\infty}{\approx} \overline{\lambda}^{-\mathcal{M}}\mathbb{E}_{d_i}\left\{\left[\alpha d_i^n x^{\frac{1}{\mathcal{M}}}\right]^{\mathcal{M}}\right\},\end{aligned} \tag{2.37}$$

which implies that the corresponding diversity order is $\Delta_{\max} = \mathcal{M}$.

Finally, according to the following relationship,

$$\frac{[\min\{X_1X_2\ldots X_{\mathcal{M}}\}]^{\mathcal{M}}}{d_i^{n\mathcal{M}}} \le \frac{X_1X_2\ldots X_{\mathcal{M}}}{d_i^{n\mathcal{M}}} \le \frac{[\max\{X_1X_2\ldots X_{\mathcal{M}}\}]^{\mathcal{M}}}{d_i^{n\mathcal{M}}}, \tag{2.38}$$

one can conclude that the diversity order of the OP under HRAQ-IR scheme is $\mathcal{M}$. □

### 2.1.4 Numerical Results

In this subsection, numerical results will be provided to study the outage performance of the considered HARQ schemes, as well as to verify the proposed analytical models. In the following, we use D instead of $D_i$ for simplification, and the path-loss factor for the S-D link is set as 2. Furthermore, we run $1 \times 10^5$ times of the realizations of the considered system and $1 \times 10^6$ trials of Monte-Carlo simulations, to model the randomness of the positions of the considered terminals and channel gains over each link. Unless otherwise explicitly specified, the main parameters adopted in this subsection are set as $n = 2$, $R = 0.01$ bps, $K = 10$, $\mathcal{M} = 10$, $H = 300$ km, $L = 50$ km, $m = 3$, $b = 2.56$, $\Omega = 0.123$.

#### 2.1.4.1 Outage Performance Under GSA and RTD Schemes

In this subsubsection, the outage performance of GSA and RTD schemes in STT scenarios will be studied, while $L = 20$ km.

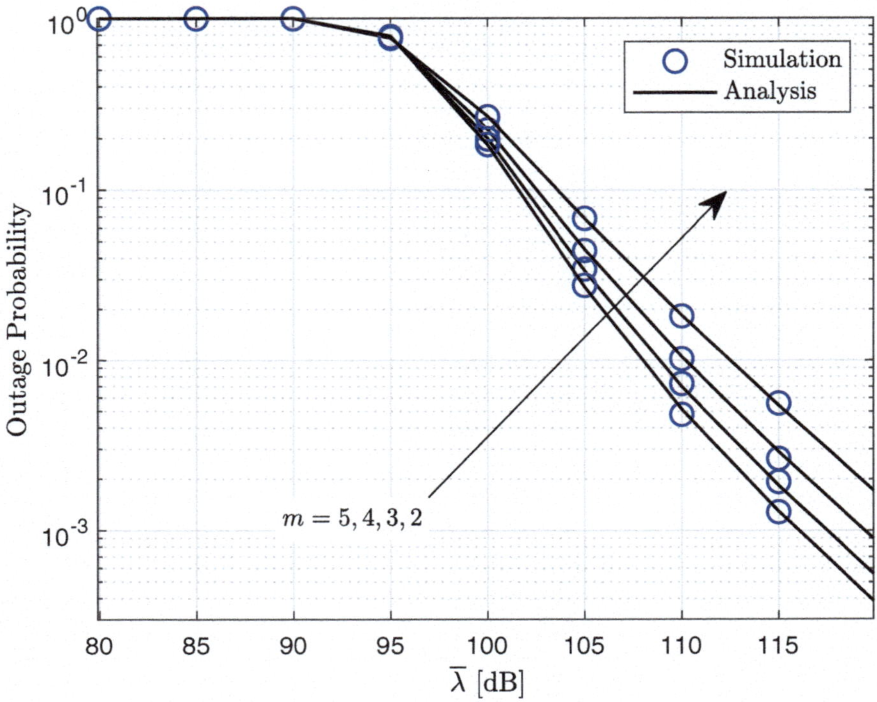


**Fig. 2.2** Outage probability versus transmit SNR $\overline{\lambda}$ for various values of the fading severity parameter $m$ under GSA scheme

In Figs. 2.2 and 2.3, the OP under GSA scheme is presented for various $m$ and $H$, respectively, while the one under RTD scheme is given in Figs. 2.4 and 2.5.

It is easy to see from Figs. 2.2 and 2.4 that the OP under GSA and RTD schemes with a large $m$ outperforms the one with a low $m$. This observation can be explained by the fact that a large $m$ represents fewer obstructions of the LOS components between the satellite and the terrestrial receivers. In other words, a large $m$ means a better channel condition for the transmissions over STT links.

In Figs. 2.3 and 2.5, the OP under GSA and RTD schemes is respectively depicted for various values of $H$ from 300 to 400 km. Clearly, the OP under the GSA scheme degrades with the increment of $H$. That is to say, the OP of the GSA scheme gets worse while the height of the orbit enlarges, which leads to increased path-loss. Recalling the truth that, compared with small-scale fading, path-loss mainly domains the fading effects suffered by the transmitted signals when the transmission distance is large enough, the finding here is easy to understand from the fact that in STT scenarios the transmission distance ranges from hundreds to thousands of kilometers.

Furthermore, one can also see that the outage performance under both schemes can be improved, while the transmit SNR, $\overline{\lambda}$, increases, which indicates that the received SNR at the terrestrial receiver can be improved.

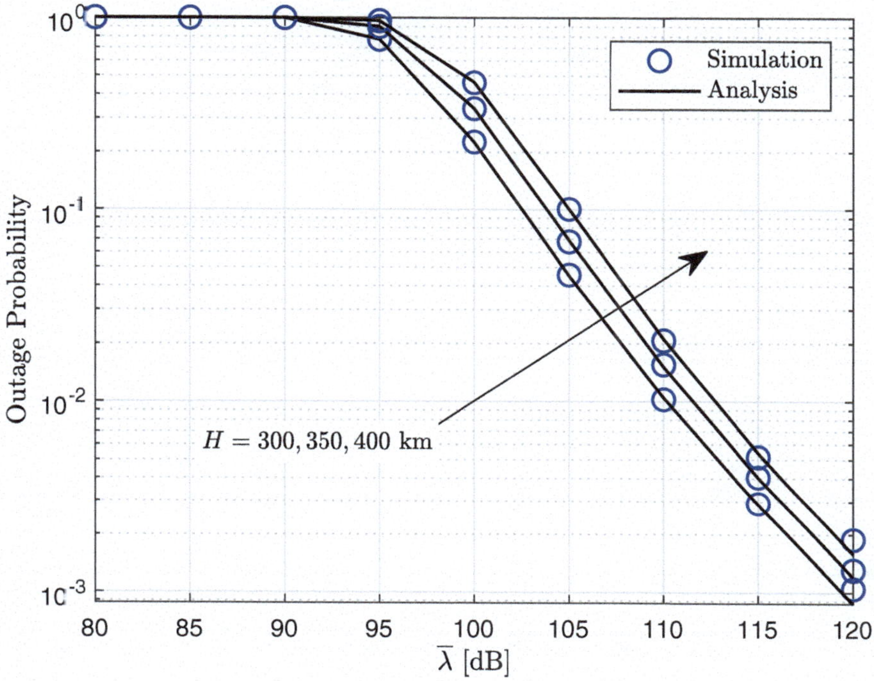


**Fig. 2.3** Outage probability versus transmit SNR $\overline{\lambda}$ for various values of the satellite altitude $H$ under GSA scheme

#### 2.1.4.2 Outage Performance Under HARQ-IR Scheme

The outage performance of the HARQ-IR scheme will be investigated in this subsubsection while considering the effects of the randomness of small-scale fading and the positions of the terrestrial receivers.

Observing the principles achieved in the previous subsection, we can readily reach some similar conclusions with the HARQ-IR scheme, namely, the ones addressing the effects of $m$, $H$, and $\overline{\lambda}$ on the outage performance, as the same transmission scenario shown in Fig. 2.1 is considered for these three schemes. Therefore, to more completely understand how the system parameters affect the outage performance of the target system and save page space, the effects of some other system parameters, i.e., $\mathcal{M}$ and $R$, will be studied in this subsubsection, instead of considering the cases of different $m$ and $H$.

Obviously, the OP under HARQ-IR scheme with a low $\mathcal{M}$ outperforms the one with a large $\mathcal{M}$ when $\overline{\lambda}$ is less than 123 dB. On the contrary, an opposite observation is found while $\overline{\lambda}$ is larger than 123 dB. What is presented here means that, when $\overline{\lambda}$ is not large enough, the outage performance under the HARQ-IR scheme cannot be improved via exploiting more copies of the transmitted signal during the decoding process.

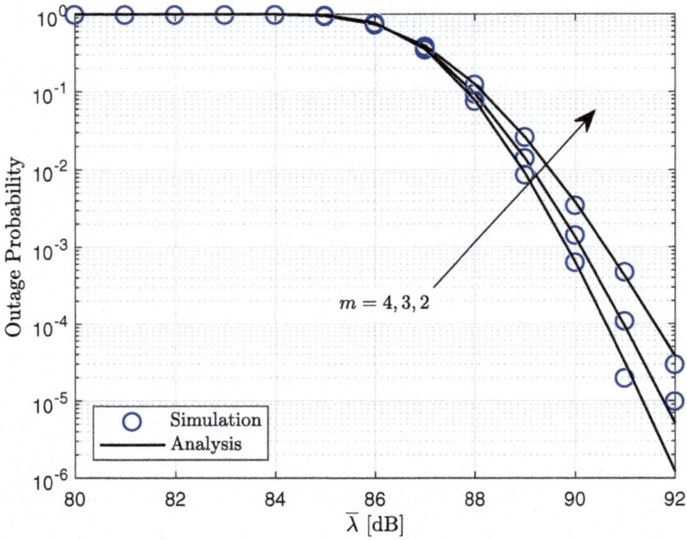


**Fig. 2.4** Outage probability versus transmit SNR $\overline{\lambda}$ for various values of the fading severity parameter $m$ under RTD scheme

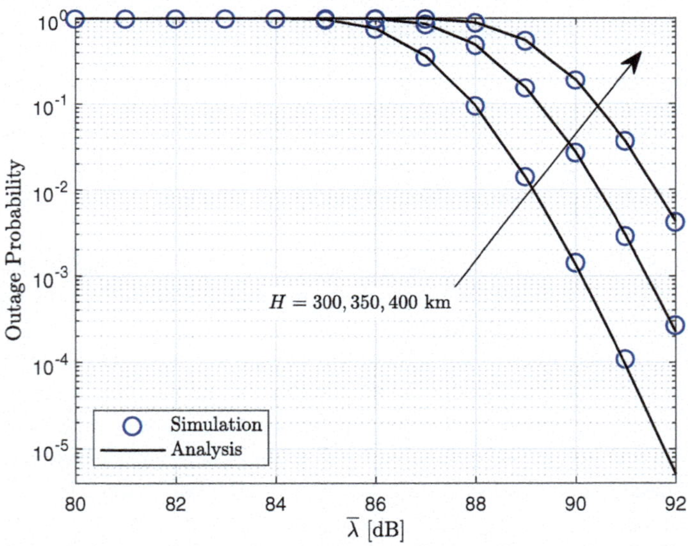


**Fig. 2.5** Outage probability versus transmit SNR $\overline{\lambda}$ for various values of the satellite altitude $H$ under RTD scheme

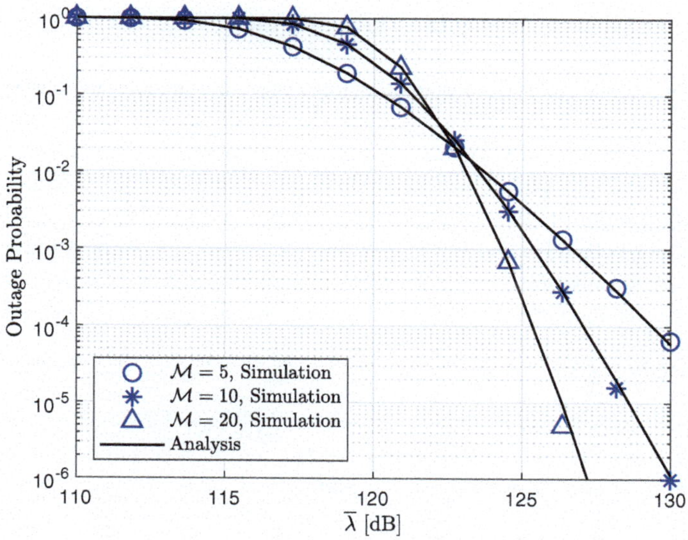


**Fig. 2.6** Outage probability versus transmit SNR $\overline{\lambda}$ for various values of $\mathcal{M}$ under HARQ-IR scheme for $2^R = 0.01, m = 2.56, b = 0.0158, \Omega = 8.97 \times 10^{-4}$ (i.e., infrequent light shadowing)

**Fig. 2.7** Outage probability versus transmit SNR $\overline{\lambda}$ for various values of $R$ under HARQ-IR scheme for $m = 19.4, b = 0.158, \Omega = 1.29$ (i.e., infrequent light shadowing)

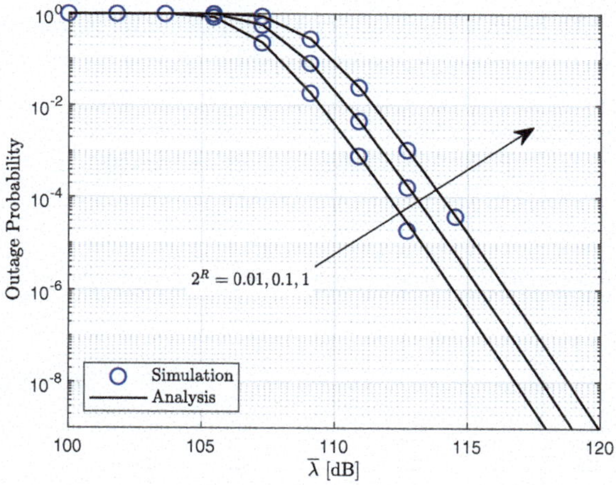


As presented in Fig. 2.7, the influence of the outage threshold, $R$, is investigated. As expected, enlarging $R$ shows a negative effect on the OP of the HARQ-IR scheme, since a large outage threshold represents a large probability of outage events. Furthermore, we can also conclude that the outage threshold exhibits a similar effect on the outage performance of the two other schemes, GSA and RTD schemes.

Further, one can easily see from Figs. 2.2, 2.3, 2.4, 2.5, 2.6 and 2.7 that simulation and numerical results can match with each other well in various cases, which verify the correctness of our proposed analytical models.

#### 2.1.4.3 Comparisons Among the Three Schemes

In this subsubsection, we will compare the outage performance of these three schemes in various cases.

As depicted in Figs. 2.8 and 2.9, RTD, and HARQ-IR schemes with a same $\mathcal{M}$ show the same outage performance, while outperforming GSA scheme. Because the diversity gain in the time domain, which is shown in (2.19) and (2.23), can be achieved under both RTD and HARQ-IR schemes, leading to improved outage performance, while only one copy of the received signal is adopted for decoding under GSA scheme, as suggested by (2.16). Moreover, recalling (2.7), (2.19), and (2.23), one can have $\log_2\left(1+\sum_{s=(j-\mathcal{M}+1)^+}^{j}\gamma_{is}\right)\approx\sum_{s=(j-\mathcal{M}+1)^+}^{j}\log_2\left(1+\gamma_{is}\right)$ while $\overline{\lambda}$ is sufficiently large, which can be used to explain the finding that RTD and HARQ-IR schemes with a same $\mathcal{M}$ exhibit similar outage performance in Figs. 2.8 and 2.9. This finding can also be verified by the theoretical analysis on the diversity gain presented in Sect. IV, namely, the diversity gains of RTD and HARQ-IR schemes are $\mathcal{M}$, as suggested by Lemmas 2.2 and 2.3.

As the three different kinds of HARQ schemes exhibit different outage performance and require different levels of resource consumption, they can be chosen and applied in practical applications by considering the hardware resources at terrestrial receivers, e.g., power budgets and computational capability, and the application requirements, e.g., the working lifetime of the terrestrial receivers.

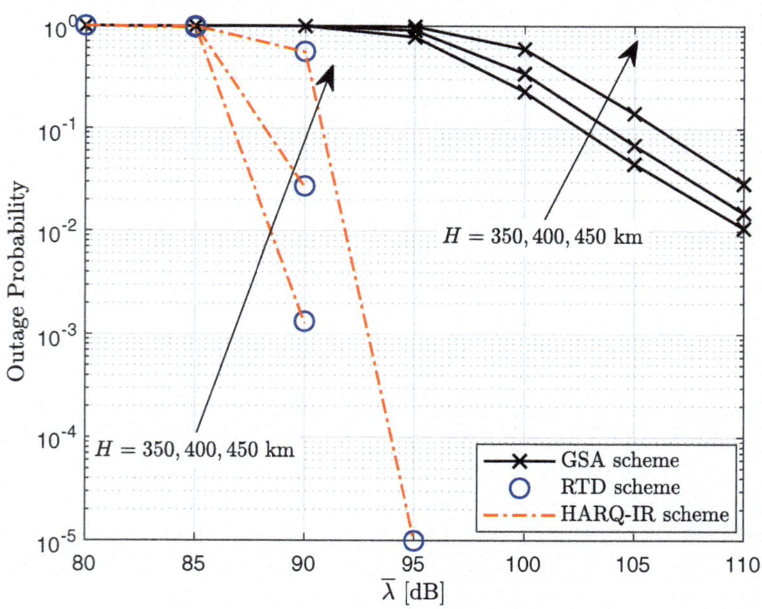


**Fig. 2.8** Outage probability for various values of the satellite altitude $H$ under the three schemes

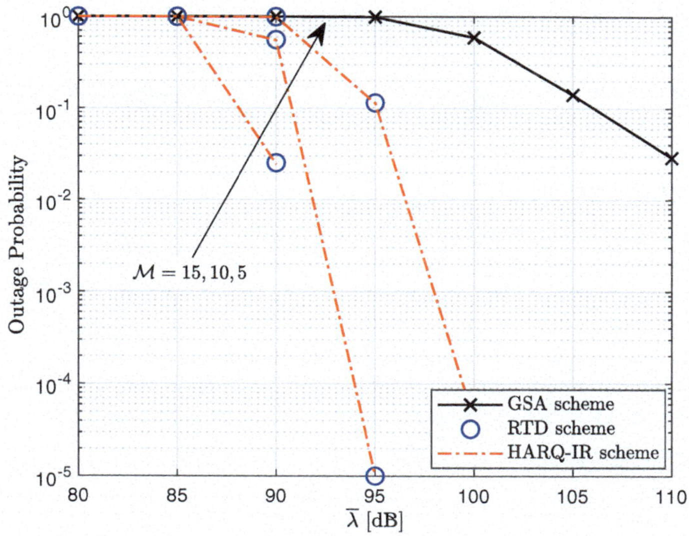


**Fig. 2.9** Outage probability for various values of $\mathcal{M}$ under the three schemes with $H = 450$ km

### *2.1.5 Conclusion*

In this work, we mainly study the outage performance and diversity gain under three different HARQ schemes, GSA, RTD, and HARQ-IR schemes, in STT scenarios, while considering the randomness of the power gain of the wireless channels and the positions of the terrestrial receivers. Analytical modes are presented for the OP under these three schemes, via which some useful conclusions are obtained as follows:

1. Path-loss plays a key role in affecting the outage performance of these HARQ schemes, compared with small-scale fading over STT channels;

2. The transmission distance shows a negative effect on the outage performance, while the shape parameter, $m$, of SR channel, exhibits a positive influence on the outage performance;

3. When the transmit SNR is sufficiently large, RTD and HARQ-IR schemes are with similar outage performance;

4. The outage performance of both RTD and HARQ-IR schemes outperforms the one of GSA, profiting from the adopted diversity gain in the time domain;

5. The diversity orders of GSA, RTD, and HARQ-IR schemes are 1, $\mathcal{M}$, and $\mathcal{M}$, respectively.

In this work, we only focus on the impacts of channel fading and the distribution of terrestrial receivers on the outage performance of HARQ schemes. In the future, the continuity of the transmission behaviors on the timeline will be considered to further understand the performance of these HARQ schemes.

## *Appendix I: Proof of Theorem 2.2*

Considering the randomness of $\lambda_j$ and using (2.9), the OP under GSA scheme in the $j$th $(1 \le j \le K)$ time slot can be rewritten as

$$P_{\mathrm{out},ij} = \Pr\left\{\lambda_j < \varepsilon d_i^n\right\} \\ = 1 - \alpha \sum_{k=0}^{m-1} \frac{\varsigma(k)}{\bar{\lambda}^{k+1}} \sum_{p=0}^{k} \frac{k!}{p!} \left(\frac{\beta-\delta}{\bar{\lambda}}\right)^{-(k+1-p)} \varepsilon^p d_i^{np} \exp\left(-\frac{\varepsilon(\beta-\delta)}{\bar{\lambda}} d_i^n\right). \tag{2.39}$$

Then, taking the randomness of the position of $\mathrm{D}_i$ into account, the OP under GSA scheme in the $j$th $(1 \le j \le K)$ time slot can be finally calculated as (2.40), where $(a)$ denotes $y = \frac{\varepsilon(\beta-\delta)}{\bar{\lambda}} x^n \Rightarrow x = y^{\frac{1}{n}} \left(\frac{\varepsilon(\beta-\delta)}{\bar{\lambda}}\right)^{-\frac{1}{n}} \Rightarrow dx = \frac{1}{n}\left(\frac{\varepsilon(\beta-\delta)}{\bar{\lambda}}\right)^{-\frac{1}{n}} y^{\frac{1}{n}-1} dy$, $y_{\min} = \frac{\varepsilon(\beta-\delta)}{\bar{\lambda}} H_1^n$, $y_{\max} = \frac{\varepsilon(\beta-\delta)}{\bar{\lambda}} \left({H_1}^2 + L^2\right)^{\frac{n}{2}}$, and $\gamma(a, x) = \int_0^x t^{a-1} \exp(-t)\, dt$ is the lower incomplete gamma function.

$$\begin{aligned} P_{\mathrm{out},ij} &= 1 - \frac{2\alpha}{L^2} \sum_{k=0}^{m-1} \frac{\varsigma(k)}{\bar{\lambda}^{k+1}} \sum_{p=0}^{k} \frac{k!}{p!} \left(\frac{\beta-\delta}{\bar{\lambda}}\right)^{-(k+1-p)} \\ &\quad \times \varepsilon^p \int_{H_1}^{\sqrt{{H_1}^2+L^2}} x^{np+1} \exp\left(-\frac{\varepsilon(\beta-\delta)}{\bar{\lambda}} x^n\right) dx \\ &\overset{(a)}{=} 1 - \frac{2\alpha}{nL^2} \sum_{k=0}^{m-1} \frac{\varsigma(k)}{\bar{\lambda}^{k+1}} \sum_{p=0}^{k} \frac{k!}{p!} \left(\frac{\beta-\delta}{\bar{\lambda}}\right)^{-(k+\frac{2}{n}+1)} \\ &\quad \times \varepsilon^{-\frac{2}{n}} \int_{y_{\min}}^{y_{\max}} y^{p+\frac{2}{n}-1} \exp(-y)\, dy \\ &= 1 - \frac{2\alpha}{nL^2} \sum_{k=0}^{m-1} \frac{\varsigma(k)}{\bar{\lambda}^{k+1}} \sum_{p=0}^{k} \frac{k!}{p!} \left(\frac{\beta-\delta}{\bar{\lambda}}\right)^{-(k+\frac{2}{n}+1)} \\ &\quad \times \varepsilon^{-\frac{2}{n}} \left[\gamma\left(p+\frac{2}{n}, y_{\max}\right) - \gamma\left(p+\frac{2}{n}, y_{\min}\right)\right] \end{aligned} \tag{2.40}$$

With this result, the proof of Theorem 2.2 is completed.

## Appendix II: Proof of Theorem 2.3

As suggested in [9], the CDF of $\Lambda = \sum_{s=(j-M+1)^+}^{j} \lambda_s$ can be given as

$$F_\Lambda(x) = \alpha^{\mathcal{M}} \sum_{l=0}^{c} \binom{c}{l} \beta^{c-l} \left[ G(x, l, d) + \nu\delta G(x, l, d+1) \right], \tag{2.41}$$

where $\mathcal{M} = j - [j - M]^+$, $c = [d - \mathcal{M}]^+$, $\nu = m\mathcal{M} - d$, $d = \max\{\mathcal{M}, \lfloor m\mathcal{M} \rfloor\}$, $\lfloor x \rfloor$ denotes the largest integer not larger than $x$, $[x]^+$ indicates that if $x \leq 0$, then use $x = 0$, and

$$G(x, l, d) = \frac{(\beta - \delta)^{\frac{l-d-1}{2}}}{\bar{\lambda}^{\frac{d-l-1}{2}} \Gamma(d - l - 1)} x^{\frac{d-l-1}{2}} \exp\left( -\frac{\beta - \delta}{2\bar{\lambda}} x \right) M_{\frac{d-l-1}{2}, \frac{d-l}{2}} \left( \frac{\beta - \delta}{\bar{\lambda}} x \right), \tag{2.42}$$

where $M_{\lambda,\mu}(z)$ is Whittaker functions, as defined as [7, Eq. (9.220.2)].

Using [10, Eq. (8.4.43.1)], $G(x, l, d)$ can be rewritten as

$$G(x, l, d) = \frac{(\beta - \delta)^{\frac{l-d-1}{2}}}{\bar{\lambda}^{\frac{d-l-1}{2}} \Gamma(d)} x^{\frac{d-l-1}{2}} G_{1,2}^{1,1} \left[ \frac{\beta - \delta}{\bar{\lambda}} x \,\middle|\, \begin{matrix} \frac{3-d-l}{2} \\ \frac{d-l+1}{2}, \frac{1-d+l}{2} \end{matrix} \right], \tag{2.43}$$

where $G_{p,q}^{m,n}\left[ x \,\middle|\, \begin{matrix} a_1, \ldots, a_p \\ b_1, \ldots, b_q \end{matrix} \right]$ is the Meijer's $G$-function, as defined by [7, Eq. (9.301)].

Considering the randomness of $\Lambda = \sum_{s=(j-M+1)^+}^{j} \lambda_s$, the OP under RTD scheme in the $j$th $(1 \leq j \leq K)$ time slot can be further written as

$$\begin{aligned} P_{\text{out},ij} &= \Pr\left\{ \sum_{s=(j-M+1)^+}^{j} \lambda_s < \varepsilon d_i^n \right\} \\ &= \alpha^{\mathcal{M}} \sum_{l=0}^{c} \binom{c}{l} \beta^{c-l} \left[ G\left(\varepsilon d_i^n, l, d\right) + \nu\delta G\left(\varepsilon d_i^n, l, d+1\right) \right]. \end{aligned} \tag{2.44}$$

Next, taking the randomness of the position of $\text{D}_i$ into account, the OP under RTD scheme in the $j$th $(1 \leq j \leq K)$ time slot can be calculated as (2.45).

$$
\begin{aligned}
P_{\mathrm{out},ij} &= \int_{H_1}^{\sqrt{{H_1}^2+L^2}} \alpha^{\mathcal{M}} \sum_{l=0}^{c} \binom{c}{l} \beta^{c-l} \left[ G\left(\varepsilon d_i^n, l, d\right) + \nu\delta G\left(\varepsilon d_i^n, l, d+1\right) \right] f_{d_i}(x)\, dx \\
&= \alpha^{\mathcal{M}} \sum_{l=0}^{c} \binom{c}{l} \beta^{c-l} \\
&\times \left[ \underbrace{\int_{H_1}^{\sqrt{{H_1}^2+L^2}} G\left(\varepsilon d_i^n, l, d\right) f_{d_i}(x)\, dx}_{I_1} + \nu\delta \underbrace{\int_{H_1}^{\sqrt{{H_1}^2+L^2}} G\left(\varepsilon d_i^n, l, d+1\right) f_{d_i}(x)\, dx}_{I_2} \right]
\end{aligned}
\tag{2.45}
$$

Using (2.4) and [11, Eq. (26)], $I_1$ can be derived as (2.46), where $(b)$ denotes $t = x^n \Rightarrow x = t^{\frac{1}{n}} \Rightarrow dx = \frac{1}{n} t^{\frac{1}{n}-1} dt$, $t_{\max} = \left({H_1}^2 + L^2\right)^{\frac{n}{2}}$, and $t_{\min} = {H_1}^n$.

$$
\begin{aligned}
I_1 &= \int_{H_1}^{\sqrt{{H_1}^2+L^2}} G\left(\varepsilon d_i^n, l, d\right) f_{d_i}(x)\, dx \\
&= \frac{2(\beta-\delta)^{\frac{l-d-1}{2}}}{L^2 \Gamma(d)} \left(\frac{\varepsilon}{\bar{\lambda}}\right)^{\frac{d-l-1}{2}} \int_{H_1}^{\sqrt{{H_1}^2+L^2}} x^{\frac{n(d-l-1)}{2}+1} G_{1,2}^{1,1}\left[ \frac{\varepsilon(\beta-\delta)}{\bar{\lambda}} x^n \,\middle|\, \begin{matrix} \frac{3-d-l}{2} \\ \frac{d-l+1}{2}, \frac{1-d+l}{2} \end{matrix} \right] dx \\
&\overset{(b)}{=} \frac{2(\beta-\delta)^{\frac{l-d-1}{2}}}{n L^2 \Gamma(d)} \left(\frac{\varepsilon}{\bar{\lambda}}\right)^{\frac{d-l-1}{2}} \int_{t_{\min}}^{t_{\max}} t^{\frac{d-l-1}{2}+\frac{2}{n}-1} G_{1,2}^{1,1}\left[ \frac{\varepsilon(\beta-\delta)}{\bar{\lambda}} t \,\middle|\, \begin{matrix} \frac{3-d-l}{2} \\ \frac{d-l+1}{2}, \frac{1-d+l}{2} \end{matrix} \right] dt \\
&= \frac{2(\beta-\delta)^{\frac{l-d-1}{2}}}{n L^2 \Gamma(d)} \left(\frac{\varepsilon}{\bar{\lambda}}\right)^{\frac{d-l-1}{2}} \left\{ t_{\max}^{\frac{d-l-1}{2}+\frac{2}{n}} G_{2,3}^{1,2}\left[ \frac{\varepsilon(\beta-\delta)}{\bar{\lambda}} t_{\max} \,\middle|\, \begin{matrix} \frac{3-d-l}{2}, \frac{3-d+l}{2} - \frac{2}{n} \\ \frac{d-l+1}{2}, \frac{1-d+l}{2} - \frac{2}{n}, \frac{1-d+l}{2} \end{matrix} \right] \right. \\
&\left. - t_{\min}^{\frac{d-l-1}{2}+\frac{2}{n}} G_{2,3}^{1,2}\left[ \frac{\varepsilon(\beta-\delta)}{\bar{\lambda}} t_{\min} \,\middle|\, \begin{matrix} \frac{3-d-l}{2}, \frac{3-d+l}{2} - \frac{2}{n} \\ \frac{d-l+1}{2}, \frac{1-d+l}{2} - \frac{2}{n}, \frac{1-d+l}{2} \end{matrix} \right] \right\}
\end{aligned}
\tag{2.46}
$$

Similarly, $I_2$ in (2.45) can also be calculated as (2.47).

$$
\begin{aligned}
I_2 &= \int_{H_1}^{\sqrt{{H_1}^2+L^2}} G\left(\varepsilon d_i^n, l, d+1\right) f_{d_i}(x)\, dx \\
&= \frac{2(\beta-\delta)^{\frac{l-d-2}{2}}}{n L^2 \Gamma(d+1)} \left(\frac{\varepsilon}{\bar{\lambda}}\right)^{\frac{d-l}{2}} \left\{ t_{\max}^{\frac{d-l}{2}+\frac{2}{n}} G_{2,3}^{1,2}\left[ \frac{\varepsilon(\beta-\delta)}{\bar{\lambda}} t_{\max} \,\middle|\, \begin{matrix} \frac{2-d}{2}, \frac{2+l-d}{2} - \frac{2}{n} \\ \frac{d-l+2}{2}, \frac{l-d}{2} - \frac{2}{n}, \frac{l-d}{2} \end{matrix} \right] \right. \\
&\left. - t_{\min}^{\frac{d-l}{2}+\frac{2}{n}} G_{2,3}^{1,2}\left[ \frac{\varepsilon(\beta-\delta)}{\bar{\lambda}} t_{\min} \,\middle|\, \begin{matrix} \frac{3-d}{2}, \frac{2+l-d}{2} - \frac{2}{n} \\ \frac{d-l+2}{2}, \frac{l-d}{2} - \frac{2}{n}, \frac{l-d}{2} \end{matrix} \right] \right\}
\end{aligned}
\tag{2.47}
$$

Finally, the OP under RTD scheme in the $j$th $(1 \le j \le K)$ time slot can be achieved as (2.21) by substituting (2.46) and (2.47) into (2.45).

## *Appendix III: Proof of Theorem 2.4*

First, we calculate the complex moment of $\left|h_{\mathrm{SD}_i}[s]\right|^2$. By using the PDF of $\left|h_{\mathrm{SD}_i}[s]\right|^2$, we can have $\Theta=\mathbb{E}\left\{\left(\left|h_{\mathrm{SD}_i}[s]\right|^2\right)^{\jmath t}\right\}$ as

$$\begin{aligned}\Theta&=\alpha\int_0^{\infty}x^{\jmath t}\exp\left(-\beta x\right)\cdot {}_1F_1\left(m;1;\delta x\right)dx\\&=\alpha\beta^{-(\jmath t+1)}\Gamma\left(\jmath t+1\right)\cdot {}_2F_1\left(m,\jmath t+1;1;\frac{\delta}{\beta}\right),\quad t\geq 0.\end{aligned}\tag{2.48}$$

Then, by using (2.4), the PDF of $Z=\frac{d_i^{nM}}{\bar{\lambda}^M}$ $(z_{\min}\leq z\leq z_{\max})$ can be derived as

$$f_Z\left(z\right)=\frac{\bar{\lambda}^M}{nM}\frac{2}{L^2}\left(\bar{\lambda}^M z\right)^{\frac{2}{nM}-1}=\frac{2\bar{\lambda}^{\frac{2}{n}}}{nL^2M}z^{\frac{2}{nM}-1}.\tag{2.49}$$

The complex $(-\jmath t)$th moment of $Z$ is

$$\begin{aligned}\mathbb{E}\left\{Z^{-\jmath t}\right\}&=\frac{2\bar{\lambda}^{\frac{2}{n}}}{nML^2}\int_{z_{\min}}^{z_{\max}}z^{\frac{2}{nM}-\jmath t-1}dz\\&=\jmath\frac{2\bar{\lambda}^{\frac{2}{n}}}{L^2}\frac{z_{\max}^{\frac{2}{nM}-\jmath t}-z_{\min}^{\frac{2}{nM}-\jmath t}}{2\jmath+nMt}.\end{aligned}\tag{2.50}$$

Further, by considering the virtues of logarithm function, let $W'=\log\left(\frac{Y}{Z}\right)$. The characteristic function (CF) of $W'$ can be derived as

$$\begin{aligned}\mathrm{CF}_{W'}\left(t\right)&=\mathbb{E}\left\{\exp\left(\jmath t\log\left(\frac{\prod_{s=(j-M+1)^+}^{j}\left|h_{\mathrm{SD}_i}[s]\right|^2}{Z}\right)\right)\right\}\\&=\left[\mathbb{E}\left\{\left(\left|h_{\mathrm{SD}_i}[s]\right|^2\right)^{\jmath t}\right\}\right]^M\mathbb{E}\left\{Z^{-\jmath t}\right\}.\end{aligned}\tag{2.51}$$

According to Gil-Pelaez Theorem, the CDF of $W'$ can be deduced from the CF of $W'$, given by

$$F_{W'}\left(w'\right)=\frac{1}{2}-\frac{1}{\pi}\int_0^{\infty}\frac{\mathrm{Im}\left\{\exp\left(-\jmath tw'\right)\mathrm{CF}_{W'}\left(t\right)\right\}}{t}dt.\tag{2.52}$$

It is easy to see that the relationship between the CDFs of $W$ and $W'$ is

$$F_W(w) = \Pr\{\log W \le \log w\} = F_{W'}(\log w) = \frac{1}{2} - \frac{1}{\pi}\int_0^{\infty} \frac{\operatorname{Im}\left\{w^{-jt}\mathrm{CF}_{W'}(t)\right\}}{t}dt. \tag{2.53}$$

Finally, combining (2.48), (2.50), (2.51) and (2.53) yields (2.25).

## 2.2 Earth Rotation-Aware Non-stationary Satellite Communications

### *2.2.1 ECEF Satellite Coordinate Introduction*

To demonstrate the effects of the rotations of the Earth and the satellite, we first introduce the Earth-centered inertial (ECI) frame and the Earth-centered Earth-fixed (ECEF) frame [12, 13]. Cartesian coordinates are commonly used to show the location of an object on the Earth by using the ECI. In ECI frame, the $x-y$ plane coincides with the equatorial plane of the Earth; the $x$-axis is permanently fixed in a direction relative to the celestial sphere, which does not rotate as the Earth does; and the $z$-axis lies at 90° to the equatorial plane and extends through the North Pole. Due to the forces that the sun and moon exert, the Earth's equatorial plane moves with respect to the celestial sphere. The Earth rotates, while ECI coordinate system does not. In contrast to the ECI frame, the ECEF frame is another coordinate that remains fixed concerning the Earth's surface in its rotation. This means that the ECEF frame rotates with the Earth, and therefore the coordinates of a point fixed on the surface of the Earth do not change.

As shown in Fig. 2.10, if we take the plane of the equator as the $x-y$ coordinate plane and the axis of the Earth's rotation tilted at an angle of 23.5° relative to the orbital plane as the $z$-axis, we could set up a spherical coordinate based on the Earth center, which is the ECI coordinate. Assuming the Earth rotates an angle of $\theta$ after a while, the new ECEF coordinate will be composed of $x_T$-axis, $y_T$-axis, and $z$-axis.

Assuming all satellite orbits take the Earth's core as the coordinate origin, all orbits could be obtained from a rotation of the satellite orbits set in the same plane as the equator. For example, one of the orbit coordinates could be obtained by rotating $\Omega$ degree about the $z$-axis and $i$ degree about the $x$-axis in three dimensions by using the right-hand rule, as seen in Fig. 2.10. The obtained coordinate is composed of $P$-axis, $Q$-axis, and $W$-axis. However, to compute the distance between the satellites and users, we should put them in the same coordinates. To achieve this goal, we first re-express the position of the satellite in ECI coordinate through coordinate transformation. Assuming there is a satellite with coordinate $\mathbf{s}_{0,S} = (H_S \cos(v), H_S \sin(v), 0)$, where $v$ is the angle between the $P$-axis and the line $ES$, $H_S$ is the height from the satellite S to the center of the Earth. The new coordinate of S on the ECI coordinate

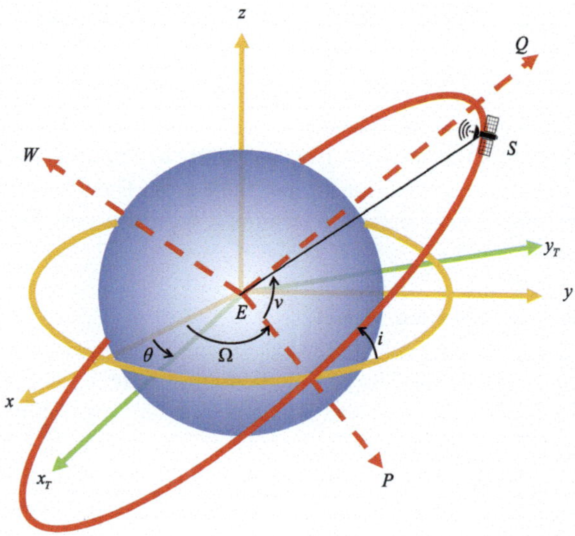


**Fig. 2.10** Spherical coordinates

can be obtained by multiplying the unitary rotation matrices $\mathbf{R}_x(-i)$ and $\mathbf{R}_z(-\Omega)$, which can be expressed as

$$\mathbf{s}_{0,ECI} = \mathbf{R}_z(-\Omega)\mathbf{R}_x(-i)\mathbf{s}_{0,S}, \tag{2.54}$$

where $\mathbf{R}_x(\cdot)$ and $\mathbf{R}_z(\cdot)$ are the rotation matrices defined as

$$\mathbf{R}_x(\alpha) = \begin{bmatrix} 1 & 0 & 0 \\ 0 & \cos(\alpha) & -\sin(\alpha) \\ 0 & \sin(\alpha) & \cos(\alpha) \end{bmatrix} \tag{2.55}$$

and

$$\mathbf{R}_z(\alpha) = \begin{bmatrix} \cos(\alpha) & -\sin(\alpha) & 0 \\ \sin(\alpha) & \cos(\alpha) & 0 \\ 0 & 0 & 1 \end{bmatrix}. \tag{2.56}$$

Taking the rotation of the Earth into consideration, we should transform the coordinate in ECI coordinate further to ECEF coordinate. It means the coordinate of S in ECEF-coordinate with $\theta$ Earth's rotation is

$$\mathbf{s}_{ECEF} = \mathbf{R}_z(\theta)\mathbf{s}_{0,ECI}. \tag{2.57}$$

Assuming the coordinate of the ground user (U) is denoted as $\mathbf{u}_{ECEF}$, the distance between S and U can be calculated as

$$d_E = \|\mathbf{s}_{ECEF} - \mathbf{u}_{ECEF}\|_2 = \left\|\mathbf{R}_z(\theta - \Omega)\mathbf{R}_x(-i)\mathbf{s}_{0,S} - \mathbf{u}_{ECEF}\right\|_2. \tag{2.58}$$

## 2.2.2 System Model and Channel Model

In this subsection, we will introduce the considered non-stationary satellite communication system model as well as the channel model. Although we focus on the downlink transmission from a satellite to a ground user, the analysis can be extended to uplink transmissions in the same way. The ground user considered here can also be other kinds of ground terminals, like base stations.

### 2.2.2.1 System Model

In a realistic satellite communication system, a satellite can only serve the terminals located in its coverage. A simplified coverage area of a satellite is presented in Fig. 2.11, where S only can serve U in the circle with maximal achievable angle $\beta_{\max}$ composed by the direct line from S to U, and the line from S to the center of the Earth. $\beta_{\max}$ depends on many aspects, such as hardware implementation, propagation environment, system design, and so on. The distance between S to the Earth surface $H_{SE}$ and $\beta_{\max}$ decides the maximal communication distance, which can be calculated as

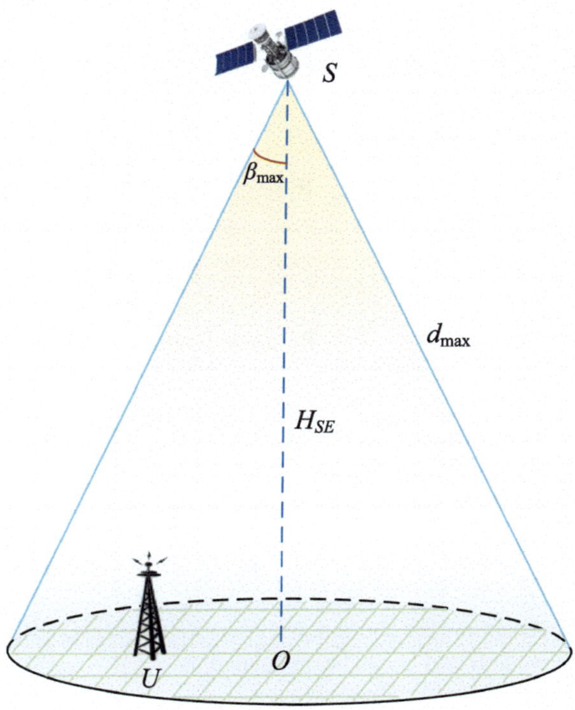


**Fig. 2.11** Coverage area

$$d_{\max} = \frac{H_{SE}}{\cos(\beta_{\max})}, \tag{2.59}$$

where $H_{SE} = H_S - H_E$, $H_E$ is a fixed value that denotes the distance from the center of the Earth to the Earth's surface. It also denotes the distance between S and the edge of the coverage area. Assuming U locates in the coverage area of S, whose position could be denoted by

$$\begin{aligned} \mathbf{u}_{ECEF} &= [u_1, u_2, u_3]^T \\ &= [H_E \sin(\theta_u)\cos(\varphi_u), H_E \sin(\theta_u)\sin(\varphi_u), H_E \cos(\theta_u)]^T, \end{aligned} \tag{2.60}$$

where $\theta_u$ is the angle between the zenith direction and the line from the Earth's center to U, and $\varphi_u$ is the signed angle measured from the azimuth reference direction to the orthogonal projection of the line segment from the Earth's center to U on the reference plane. It satisfies $\|\mathbf{u}_{ECEF}\|_2 = H_E$ in the ECEF coordinate. Since the Earth rotates 360 degrees every 24 h, it rotates about $\omega_e = 0.0042$ degrees every second. Assuming that S rotates about $\omega_s$ degrees every second on its orbit, it is located at $\mathbf{s}_{0,S}$ when U falls into the coverage of S. Since the communication only starts when U falls into S's coverage area, the initial distance between S and U can be denoted as

$$d_0 = \left\|\mathbf{s}_{0,ECEF} - \mathbf{u}_{ECEF}\right\|_2 = d_{\max}. \tag{2.61}$$

Considering the rotations of S and the Earth within rotation time $t$, the coordinate of U does not change in ECEF coordinate, while the position of S after $t$ seconds in $P$-$Q$-$W$-axis is $\mathbf{s}_{t,S} = \mathbf{R}_z(\omega_s t)\mathbf{s}_{0,S}$ . From $P$-$Q$-$W$-axis to ECEF coordinate, the position can be denoted as

$$\mathbf{s}_{t,ECEF} = \mathbf{R}_z(\omega_e t - \Omega)\mathbf{R}_x(-i)\mathbf{R}_z(\omega_s t)\mathbf{s}_{0,S}, \tag{2.62}$$

where $\mathbf{s}_{t,ECEF} = \left[s_1\ s_2\ s_3\right]^T$ can be separately calculated as

$$s_1 = \alpha_1 \cos(\omega_a t + \beta_1) + \alpha_2 \cos(\omega_m, t + \beta_2), \tag{2.63}$$

$$s_2 = \alpha_1 \sin(\omega_a t + \beta_1) + \alpha_2 \sin(\omega_m t + \beta_2), \tag{2.64}$$

and

$$s_3 = H_S \sin(-i)\sin(\omega_s t + v), \tag{2.65}$$

where $\alpha_1 = \frac{H_S}{2}[1 + \cos(i)]$, $\omega_a = \omega_e + \omega_s$ , $\beta_1 = v - \Omega$, $\alpha_2 = \frac{H_S}{2}[1 - \cos(i)]$ , $\omega_m = \omega_e - \omega_s$, $\beta_2 = -v - \Omega$ .

Thus, the distance between U and S can be written as

$$\begin{aligned} d(t)^2 &= \left\| \mathbf{s}_{t,ECEF} - \mathbf{u}_{ECEF} \right\|_2^2 \\ &= \left\| \mathbf{R}_z(\omega_e t - \Omega)\mathbf{R}_x(-i)\mathbf{R}_z(\omega_s t)\mathbf{s}_{0,S} - \mathbf{u}_{ECEF} \right\|_2^2 \\ &= \left\| \mathbf{R}_z(\omega_e t - \Omega)\mathbf{R}_x(-i)\mathbf{R}_z(\omega_s t)\mathbf{s}_{0,S} \right\|_2^2 + \|\mathbf{u}_{ECEF}\|_2^2 \\ &\quad - 2\mathbf{u}_{ECEF}^T \mathbf{R}_z(\omega_e t - \Omega)\mathbf{R}_x(-i)\mathbf{R}_z(\omega_s t)\mathbf{s}_{0,S} \\ &= H_S^2 + H_E^2 - 2\left(u_1 s_1 + u_2 s_2 + u_3 s_3\right) \\ &= H + \mu_1 \cos\left(\omega_a t + \eta_1\right) + \mu_2 \cos\left(\omega_m t + \eta_2\right) + \mu_3 \sin\left(\omega_s t + v\right), \end{aligned} \quad (2.66)$$

where $H = H_S^2 + H_E^2$, $\mu_1 = -2H_E \sin(\theta_u)\alpha_1$, $\mu_2 = -2H_E \sin(\theta_u)\alpha_2$, $\mu_3 = -2u_3 H_S \sin(-i)$, $\eta_1 = \beta_1 - \varphi_u$ and $\eta_2 = \beta_2 - \varphi_u$.

The communication time between S and U starts once U falls into the coverage area and ends when U is out of the coverage area of S, which is denoted as $t_s$ and $t_e$, respectively. In the following, we denote the satellite visibility window as the period that starts when the ground terminal falls into the coverage area of the satellite and ends when the ground terminal is out of the coverage area of the satellite.

#### 2.2.2.2 Channel Model

Similar to traditional terrestrial communications, the terrestrial propagation between the satellite and ground users can be divided into LoS case and NLoS case, which depends on the location and environment of the ground user [14]. The LoS link is established when the connected ground terminal tends to be located in an open area without many objects surrounding it, while the NLoS link is established when the ground terminals are surrounded by objects, for example, the indoor terminals and terminals deployed in dense urban areas. Since satellite communications are expected to ensure service availability anywhere and to enable 5G network scalability by providing efficient multicast/broadcast resources for data delivery towards the network edges and user terminals, the system performance under both LoS and NLoS cases should be considered and investigated. In the following, we introduce the channel model, which is able to describe both NLoS and LoS propagation by setting appropriate parameters.

In mobile satellite communications, the channel is always modeled as a combination of one major non-blocked LoS path and multiple NLoS paths [15], given by

$$h(t,\tau) = \alpha_0 \exp\left[j\left(2\pi f_{D,0} t + \theta_0\right)\right]\delta(\tau) + \sum_{l=1}^{L_c-1} \alpha_l \exp\left[j\left(2\pi f_{D,l} t + \theta_l\right)\right]\delta\left(\tau - \tau_l\right), \quad (2.67)$$

where $\alpha_0$ and $\alpha_l$ are the complex value path gains of LoS link and $l$th NLoS link, $f_{D,l}, l \in \{0, \ldots, L_c\}$ denotes the Doppler shift, $\theta_0$ represents an initial phase, while $\theta_l$ is uniformly distributed within $[-\pi, \pi]$, $\tau_l$ is the delay of the $l$th path. Actually, the

Doppler shifts caused by the motion of the satellite can be assumed to be identical for different propagation paths due to the relatively high altitude of the satellite [16], that is, $f_{D,l} = f_D$. It should also be noted that there are long propagation delays in satellite communication systems compared to terrestrial wireless channels, which leads to a much smaller delay spread of the satellite channels than that of the terrestrial wireless channels as observed in measurement results. By taking advantage of the Doppler and delay properties of the satellite propagation channels and performing appropriate time and frequency synchronization at the satellite and the ground user [16], the Doppler shift and time delay can be well compensated [17–19].

It should be noted that the LoS component's amplitude can be assumed to be a gamma variable following Nakagami distribution due to the complete or partial blockage of the LoS by buildings, trees, hills, mountains, etc. Therefore, taking the almost perfect Doppler and time delay compensation into consideration, the channel can be simplified to a combination of an NLoS Rayleigh distributed component and an LoS Nakagami distributed component, which is also named SR model presented in [6]. The SR channel model has already been proved as an accurate, practical, and applicable tool to evaluate the performance of the satellite propagation environments in various frequency bands, e.g., the UHF-band, L-band, S-band, Ku-band, and Ka-band. Without loss of generality, the PDF of the power gain, $|h(t)|^2$ following SR fading is given as [6],

$$f_{|h(t)|^2}(x) = \alpha(t) \exp(-\beta(t)x)\, {}_1F_1(m(t); 1; \delta(t)x),\ x \geq 0, \tag{2.68}$$

where $\alpha(t) = \left(\frac{2b(t)m(t)}{2b(t)m(t)+\Omega_h(t)}\right)^{m(t)} / (2b(t))$, $\beta(t) = \frac{1}{2b(t)}$, and $\delta(t) = \frac{\Omega_h(t)}{2b(t)(2b(t)m(t)+\Omega_h(t))}$, $\Omega_h(t)$ and $2b(t)$ are the average power of the LoS and multi-path components at time $t$, respectively, $m(t)$ is the fading severity parameter and ${}_1F_1(\cdot;\cdot;\cdot)$ is the confluent hypergeometric function of first kind [7, Eq. (9.14.1)].

The special cases considering infrequent light shadowing, as well as frequent heavy shadowing, can also be reflected in the SR channel model. For example, by setting $b = 0.158$, $m = 19.4$, and $\Omega_h = 1.29$, the infrequent shadowing can be represented, while $b = 0.063$, $m = 0.739$, and $\Omega_h = 8.87 \times 10^{-4}$ for heavy shadowing. The channels corresponding to suburban, rural, urban, and open areas can be represented appropriately by setting suitable values for $m$, where $m = 0$ corresponds to urban areas with complete obstruction of the LoS, while $m = \infty$ corresponds to open areas with no obstruction of the LoS. It should be noted that the atmosphere, such as rainfall, cloud, and various gases, has a strong impact on the space-to-ground transmissions in satellite communication systems, especially for high-frequency bands, like Ka-band. Among them, rain attenuation is [20] the dominant factor, whose power gain commonly follows a log-normal distribution. Specifically, the distribution of the power gain resulted by rain attenuation in dB $\xi_{\mathrm{dB}}$ can be written as $\ln(\xi_{\mathrm{dB}}) \sim \mathcal{CN}(\mu_\xi, \sigma_\xi^2)$, where $\mu_\xi$ and $\sigma_\xi$ are both in dB and represent the log-normal location and scale parameter, respectively. Since this work focuses on investigating the impacts of the satellite's and the Earth's rotations, communication distance, and communication time on the system performance, we will treat the rain attenuation as a constant during the communication process for simplification. The research on the influence of the attenuation factor will be left for our future work.

It is easy to obtain the received SNR at U as

$$\gamma(t) = \frac{P_S \xi^2 |h(t)|^2}{\sigma^2 C d(t)^2}, \tag{2.69}$$

where $C = \left(\frac{4\pi}{\lambda}\right)^2$ with signal wavelength $\lambda$, $P_S$ and $\sigma^2$ are the transmit power at S and the average power of the AWGN at U, respectively.

### 2.2.3 *System Performance*

In this subsection, we will evaluate the system performance, while considering the position variation on instantaneous OP, instantaneous channel capacity, and throughput.

#### 2.2.3.1 Instantaneous OP

By variable substitution, the PDF and CDF of the instantaneous SNR can be obtained as

$$f_\gamma(x,t) = \alpha(t) \sum_{k=0}^{m(t)-1} \frac{\varsigma(k)}{\bar{\lambda}^{k+1}} d(t)^{2(k+1)} x^k \exp\left(-\frac{\beta(t)-\delta(t)}{\bar{\lambda}} d(t)^2 x\right), \tag{2.70}$$

and

$$F_\gamma(x,t) = 1 - \alpha(t) \sum_{k=0}^{m(t)-1} \varsigma(k) \sum_{p=0}^{k} \frac{k!}{p!} \frac{d(t)^{2p}}{\bar{\lambda}^p} x^p \times (\beta(t)-\delta(t))^{-(k+1-p)} \exp\left(-\frac{\beta(t)-\delta(t)}{\bar{\lambda}} d(t)^2 x\right), \tag{2.71}$$

respectively, where $\bar{\lambda} = \frac{P_S \xi}{\sigma^2 C}$, $\mathrm{s}(k) = \frac{(-1)^k (1-m(t))_k \delta^k}{(k!)^2}$ and $(n)_k = n(n+1)\ldots(n+k-1)$ is the Pochhammer symbol [7].

Thus, the system OP, $\Pr\{\gamma(t) \leq \gamma_{th}\}$, can be presented as

$$F_{\gamma}(\gamma_{th}, t) = 1 - \alpha(t) \sum_{k=0}^{m(t)-1} \varsigma(k) \sum_{p=0}^{k} \frac{k!}{p!} \frac{d(t)^{2p}}{\bar{\lambda}^{p}} \gamma_{th}^{p} \times (\beta(t) - \delta(t))^{-(k+1-p)} \exp\left(-\frac{\beta(t) - \delta(t)}{\bar{\lambda}} d(t)^2 \gamma_{th}\right). \quad (2.72)$$

For the special case $m(t) = \infty$, the SR channel model will reduced to LoS propagation case. Therefore, OP will only depend on the communication distance that varies with the time, which can be given as

$$\Pr\{\gamma(t) \leq \gamma_{th}\} = \Pr\left\{d(t)^2 \geq \frac{P_S |h(t)|^2}{\sigma^2 C \gamma_{th}}\right\}, \quad (2.73)$$

which will be 0 or 1 at each time $t$.

### 2.2.4 *Instantaneous Channel Capacity*

We assume that throughout the time horizon $[t_s, t_e]$, S keeps sending information to U with a constant transmit power $P_S$. As a result, the instantaneous information bits can be expressed as

$$T(t, |h(t)|^2) = B \log_2\left(1 + \frac{\bar{\lambda} |h(t)|^2}{d(t)^2}\right). \quad (2.74)$$

By re-presenting ${}_1F_1(\cdot; \cdot; \cdot)$ and taking expectation of the channel, the average instantaneous information bit is

$$\begin{aligned} T(t) &= \mathbb{E}_{h(t)}\{R(t, h(t))\} \\ &= \int_0^{\infty} R(t, h(t)) f_{|h(t)|^2}(x) dx \\ &= B \int_0^{\infty} \log_2\left(1 + \frac{\bar{\lambda} x}{d(t)^2}\right) \alpha(t) \exp(-\beta(t) x) {}_1F_1(m; 1; \delta(t) x) dx \\ &= \frac{B\alpha(t)}{\ln 2} \sum_{k=0}^{m(t)-1} s(k) \int_0^{\infty} \ln\left(1 + \frac{\bar{\lambda} x}{d(t)^2}\right) x^k \exp(-(\beta(t) - \delta(t)) x) dx. \end{aligned} \quad (2.75)$$

To facilitate the derivations, we use the Meijer-$G$ function [11] to represent the logarithmic function and exponential function by $\ln(1+x) = G_{22}^{12}\left(x \middle| \begin{matrix} 1, 1 \\ 1, 0 \end{matrix}\right)$ and $\exp(-x) = G_{01}^{10}\left(x \middle| \begin{matrix} - \\ 0 \end{matrix}\right)$. Using [21], (2.75) can be calculated as

$$T(t) = \frac{B\alpha(t)}{\ln 2}\sum_{k=0}^{m(t)-1} s(k)\int_0^\infty x^k G_{22}^{12}\left(\frac{\bar{\lambda}x}{d(t)^2}\middle|\begin{matrix}1,\ 1\\ 1,\ 0\end{matrix}\right) G_{01}^{10}\left((\beta(t)-\delta(t))\,x\middle|\begin{matrix}-\\ 0\end{matrix}\right)dx$$
$$= \frac{B\alpha(t)}{\ln 2}\sum_{k=0}^{m(t)-1} s(k)\left(\frac{\bar{\lambda}}{d(t)^2}\right)^{-(k+1)}$$
$$\times G_{23}^{31}\left(\frac{(\beta(t)-\delta(t))\,d(t)^2}{\bar{\lambda}}\middle|\begin{matrix}-(k+1),\ -k\\ 0,\ -(k+1),\ -(k+1)\end{matrix}\right). \quad (2.76)$$

Similar to the OP, considering the special case $m = \infty$, the instantaneous channel capacity can be rewritten as

$$T(t) = B\log_2\left(1 + \frac{\bar{\lambda}}{d(t)^2}\right), \quad (2.77)$$

which will varies with the communication distance as the time passes.

#### 2.2.4.1 Throughput

Considering the total communication time, the total amount of the information bits transferred from S to U can be expressed as

$$T = \int_{t_s}^{t_e} T(t)dt$$
$$= \int_{t_s}^{t_e} \frac{B\alpha(t)}{\ln 2}\sum_{k=0}^{m(t)=m-1} s(k)\left(\frac{\bar{\lambda}}{d(t)^2}\right)^{-(k+1)}$$
$$\times G_{23}^{31}\left(\frac{(\beta(t)-\delta(t))\,d(t)^2}{\bar{\lambda}}\middle|\begin{matrix}-(k+1),\ -k\\ 0,\ -(k+1),\ -(k+1)\end{matrix}\right)dt. \quad (2.78)$$

The integration is not solvable considering time-varying SR channels. Thus, we only derive the results for the special case assuming the SR channel is stationary within the communication, that is, $m(t) = m$ $\Omega_h(t) = \Omega_h$ and $2b(t) = 2b$. Combining the expression of $d(t)^2$ into (2.78) and using the Chebyshev–Gauss quadrature in the first case, which is presented as $\int_{-1}^{1}\frac{f(x)}{\sqrt{1-x^2}}dx \approx \sum_{i=1}^{W} w_i f(x_i)$ with $x_i = \cos\left(\frac{2i-1}{2W}\pi\right)$ and the weight $w_i = \frac{\pi}{W}$, the system throughput within the coverage area of S can be rewritten as

$$T = \int_{t_s}^{t_e} \frac{B\alpha}{\ln 2}\sum_{k=0}^{m-1} s(k)\left(\frac{\bar{\lambda}}{d(t)^2}\right)^{-(k+1)}$$

$$\times G_{23}^{31}\left(\frac{(\beta-\delta)\,d\,(t)^2}{\bar{\lambda}}\middle|\begin{matrix}-(k+1)\,,\ -k\\ 0,\ -(k+1)\,,\ -(k+1)\end{matrix}\right)dt$$

$$=c_1\int_{-1}^{1}\frac{B\alpha}{\ln 2}\sum_{k=0}^{m-1}\mathrm{s}(k)\left(\frac{\bar{\lambda}}{d\,(c_1x+c_2)^2}\right)^{-(k+1)}$$

$$\times G_{23}^{31}\left(\frac{(\beta-\delta)\,d\,(c_1x+c_2)^2}{\bar{\lambda}}\middle|\begin{matrix}-(k+1)\,,\ -k\\ 0,\ -(k+1)\,,\ -(k+1)\end{matrix}\right)dx$$

$$=c_1\sum_{i=1}^{W_I}w_i\sqrt{1-\varepsilon_i^2}\frac{B\alpha}{\ln 2}\sum_{k=0}^{m-1}\mathrm{s}(k)\left(\frac{\bar{\lambda}}{d\,(c_1\varepsilon_i+c_2)^2}\right)^{-(k+1)}$$

$$\times G_{23}^{31}\left(\frac{(\beta-\delta)\,d\,(c_1\varepsilon_i+c_2)^2}{\bar{\lambda}}\middle|\begin{matrix}-(k+1)\,,\ -k\\ 0,\ -(k+1)\,,\ -(k+1)\end{matrix}\right), \tag{2.79}$$

where $w_i=\frac{\pi}{W_I}$, $\varepsilon_i=\cos\left[(2i-1)\pi/W_I\right]$, $c_1=\frac{t_e-t_s}{2}$ and $c_2=\frac{t_e+t_s}{2}$.

When only considering LoS propagation (namely, $m=\infty$), the throughput can be re-calculated as

$$T=\int_{t_s}^{t_e}B\log_2\left(1+\frac{\bar{\lambda}}{d\,(t)^2}\right)dt$$

$$=Bc_1\sum_{i=1}^{W_L}w_l\sqrt{1-\epsilon_i^2}\log_2\left(1+\frac{\bar{\lambda}}{d\,(c_1\epsilon_i+c_2)^2}\right), \tag{2.80}$$

where $w_l=\frac{\pi}{W_L}$, and $\epsilon_i=\cos\left[(2i-1)\pi/W_L\right]$.

#### 2.2.4.2 Simplified Case for Fixed U

Since a low Earth orbit typically takes 90 min to revolve around the Earth, which is much faster than the Earth's rotation, we could ignore the position variation during the communication time. In this case, we only need ECI coordinates to demonstrate the positions of S and U, that is, $\omega_e=0$. Thus, the coordinate of the satellite with ECI coordinate is

$$\mathbf{s}_{t,ECI}=\mathbf{R}_z(-\Omega)\mathbf{R}_x(-i)\mathbf{s}_{t,S}=\mathbf{R}_z(-\Omega)\mathbf{R}_x(-i)\mathbf{R}_z(\omega_s t)\mathbf{s}_{0,S}, \tag{2.81}$$

where each element of the vector $\mathbf{s}_{t,ECI}=\left[\,s_1'\ s_2'\ s_3\,\right]^T$ can be calculated as

$$s_1'=H_S\cos(-\Omega)\cos(\omega_s t+v)-H_S\cos(i)\sin(-\Omega)\sin(\omega_s t+v)\,, \tag{2.82}$$

and

$$s_2'=H_S\sin(-\Omega)\cos(\omega_s t+v)+H_S\cos(-\Omega)\cos(i)\sin(\omega_s t+v)\,. \tag{2.83}$$

Thus, the distance at $t$ is given as

$$d'(t)^2 = H - 2\left(u_1 s_1' + u_2 s_2' + u_3 s_3\right) = H + \Upsilon \sin(\omega_s t + \beta_s), \tag{2.84}$$

where $b_1 = -2u_1 H_S \cos(-\Omega) - 2u_2 H_S \sin(-\Omega)$, $b_2 = 2u_1 H_S \cos(i)\sin(-\Omega) - 2u_2 H_S \cos(-\Omega)\cos(i) - 2u_3 H_S \sin(-i)$, $\Upsilon = -\sqrt{b_1^2 + b_2^2}$ and $\beta_s = \arctan\left(\frac{b_1}{b_2}\right) + v$.

*(1) General Case:* Therefore, the instantaneous OP, instantaneous channel capacity and throughput ignoring the mobility of U could be expressed by replacing $d(t)^2$ by $d'(t)^2$ in (2.71), (2.76), (2.79), respectively.

*(2) Special Case for $\gamma(t) < 1$:* Moreover, we derive the system throughput for a special case when $\gamma(t) < 1$. By re-expressing $\ln(1+x) = \sum\limits_{n=0}^{\infty}(-1)^{n+1}\frac{x^n}{n}$, the throughput can be re-calculated as

$$\begin{aligned}
T &= \int_{t_s}^{t_e} \frac{B\alpha}{\ln 2}\sum_{k=0}^{m-1} \mathrm{s}(k)\int_0^{\infty} \ln\left(1+\frac{\bar{\lambda}x}{d(t)^2}\right) x^k \exp\left(-(\beta-\delta)x\right) dx dt \\
&= \int_{t_s}^{t_e} \frac{B\alpha}{\ln 2}\sum_{k=0}^{m-1} \mathrm{s}(k)\int_0^{\infty} \sum_{n=0}^{\infty}(-1)^{n+1}\frac{\left(\frac{\bar{\lambda}}{d(t)^2}\right)^n}{n} x^{n+k}\exp\left(-(\beta-\delta)x\right) dx dt \\
&= \frac{B\alpha}{\ln 2}\sum_{k=0}^{m-1} \mathrm{s}(k)\sum_{n=0}^{\infty}(-1)^{n+1}\frac{\bar{\lambda}^n}{n}(n+k)!\,(\beta-\delta)^{-(n+k+1)} \\
&\quad\times \underbrace{\int_{t_s}^{t_e}\frac{1}{\left[H+\Upsilon\sin(\omega_s t+\beta)\right]^n} dt}_{\Theta}.
\end{aligned} \tag{2.85}$$

By adopting [7, Eq. 2.552.3], $\Theta$ can be calculated as follows:

$$\begin{aligned}
\Theta &= \int_{t_s}^{t_e}\frac{1}{\left[H+\Upsilon\sin(\omega_s t+\beta_s)\right]^n} dt \\
&= \int_{\omega_s t_s+\beta_s}^{\omega_s t_e+\beta_s}\frac{1}{\omega_s\left[H+\Upsilon\sin(y)\right]^n} dy \\
&= \frac{1}{\omega_s} F_1\left(\omega_s t_e+\beta_s, \omega_s t_s+\beta_s, n\right),
\end{aligned} \tag{2.86}$$

where $F_1(x_1, x_2, n)$ and $F_2(x_1, x_2, n, A_n, B_n)$ can be expressed as

$$F_1(x_1, x_2, n) = \frac{1}{(n-1)\left(H^2-\Upsilon^2\right)}\left[\frac{\Upsilon\cos x_1}{(H+\Upsilon\sin x_1)^{n-1}} - \frac{\Upsilon\cos x_2}{(H+\Upsilon\sin x_2)^{n-1}}\right.$$

$$+F_2\left(x_1, x_2, n-1, (n-1)H, -(n-2)\Upsilon\right)\right], \tag{2.87}$$

$$F_2\left(x_1, x_2, n, A_n, B_n\right) = \frac{1}{(n-1)\left(H^2-\Upsilon^2\right)}\left[\frac{\left(A_n\Upsilon - HA_n\right)\cos x_1}{\left(H+\Upsilon\sin x_1\right)^{n-1}} - \frac{\left(A_n\Upsilon - HB_n\right)\cos x_2}{\left(H+\Upsilon\sin x_2\right)^{n-1}} + F_2\left(x_1, x_2, n-1, A_{n-1}, B_{n-1}\right)\right], \tag{2.88}$$

where $A_{n-1} = (A_n H - B_n \Upsilon)(n-1)$, $B_{n-1} = (HB_n - \Upsilon A_n)(n-2)$ and $F_2\left(x_1, x_2, 1, A_1, B_1\right)$ is shown as

$$F_2\left(x_2, x_2, 1, A_1, B_1\right) = \frac{B_1}{\Upsilon}\left(x_1 - x_2\right) + O\begin{cases} \frac{2}{\sqrt{H^2-\Upsilon^2}}\left(\arctan\frac{H\tan\frac{x_1}{2}+\Upsilon}{\sqrt{H^2-\Upsilon^2}} - \arctan\frac{H\tan\frac{x_2}{2}+\Upsilon}{\sqrt{H^2-\Upsilon^2}}\right), & H^2 > \Upsilon^2 \\ \frac{1}{\sqrt{\Upsilon^2-H^2}}\left(\ln\frac{H\tan\frac{x_1}{2}+\Upsilon-\sqrt{\Upsilon^2-H^2}}{H\tan\frac{x_1}{2}+\Upsilon+\sqrt{\Upsilon^2-H^2}} - \ln\frac{H\tan\frac{x_2}{2}+\Upsilon-\sqrt{\Upsilon^2-H^2}}{H\tan\frac{x_2}{2}+\Upsilon+\sqrt{\Upsilon^2-H^2}}\right), & H^2 < \Upsilon^2 \end{cases}, \tag{2.89}$$

where $O = \frac{A_1\Upsilon - HB_1}{\Upsilon}$.

The throughput under LoS propagation case ($m = \infty$) can also be recalculated as

$$T = \frac{B}{\ln 2}\int_{t_s}^{t_e} (-1)^{n+1}\frac{\left(\frac{\bar{\lambda}}{d(t)^2}\right)^{n+1}}{n}dt = \frac{B}{\ln 2}\sum_{n=0}^{\infty}(-1)^{n+1}\frac{\bar{\lambda}^n}{n}\Theta. \tag{2.90}$$

### 2.2.5 *Asymptotic Results*

Since the proposed exact expressions for OP, capacity, and throughput in practical system designs and optimizations are a little bit complicated, we will derive some asymptotic closed-form expressions for the instantaneous OP, the instantaneous capacity, and the throughput in simpler forms to gain more insights. The asymptotic closed-form expressions are shown to involve much simpler functions and thus can readily provide direct solutions to some resource optimization problems.

#### 2.2.5.1 Asymptotic Instantaneous OP

As shown in [22], $F_1(m; 1; \frac{\delta(t)x}{\bar{\lambda}}) \to 1$ for very large values of $\bar{\lambda}$. Thus, the asymptotic PDF and CDF of $\bar{\gamma} = \bar{\lambda}\left|h(t)\right|^2$ becomes

$$f_{\bar{\gamma}}^{A}(x,t) = \frac{\alpha(t)}{\bar{\lambda}}\exp\left(-\frac{\beta(t)x}{\bar{\lambda}}\right) \tag{2.91}$$

and

$$F_{\bar{\gamma}}^{A}(x,t)=\frac{\alpha(t)}{\beta(t)}\left[1-\exp\left(-\frac{\beta(t)\,x}{\bar{\lambda}}\right)\right], \tag{2.92}$$

respectively.

Thus, the asymptotic instantaneous OP can be expressed as

$$F_{\gamma}^{A}(\gamma_{th},t)=\frac{\alpha(t)}{\beta(t)}\left[1-\exp\left(-\frac{\beta(t)\,d(t)^{2}\,\gamma_{th}}{\bar{\lambda}}\right)\right]. \tag{2.93}$$

#### 2.2.5.2 Asymptotic Instantaneous Channel Capacity

For instantaneous channel capacity, $\ln(1+x)$ could be approximated by $\ln(x)$ when $x$ goes to infinity. Using [7, Eq. 4.352.2], the asymptotic instantaneous channel capacity is

$$\begin{aligned}
T^{A}(t)&=\frac{B\alpha(t)}{\ln 2}\sum_{k=0}^{m(t)-1}\mathrm{s}(k)\int_{0}^{\infty}\ln\left(\frac{\bar{\lambda}x}{d(t)^{2}}\right)x^{k}\exp\left(-\left(\beta(t)-\delta(t)\right)x\right)dx\\
&=\frac{B\alpha(t)}{\ln 2}\sum_{k=0}^{m(t)-1}\mathrm{s}(k)\frac{d(t)^{2(k+1)}}{\bar{\lambda}^{k+1}}\\
&\quad\times\int_{0}^{\infty}\ln(y)\,y^{k}\exp\left(-\frac{d(t)^{2}\left(\beta(t)-\delta(t)\right)}{\bar{\lambda}}y\right)dy\\
&=\frac{B\alpha(t)}{\ln 2}\sum_{k=0}^{m(t)-1}\frac{\mathrm{s}(k)k!}{\left(\beta(t)-\delta(t)\right)^{k+1}}\left[C_{k}-\ln\left(\frac{d(t)^{2}\left(\beta(t)-\delta(t)\right)}{\bar{\lambda}}\right)\right],
\end{aligned} \tag{2.94}$$

where $C_k=1+\frac{1}{2}+\frac{1}{3}+\cdots+\frac{1}{k}-C_M$, $C_M=0.5772156649015328606065 12\ldots$ is Euler's constant.

From the expression of the asymptotic instantaneous OP and channel capacity, we can see that it is much simpler than the exact instantaneous OP and channel capacity, which just involves the exponential function and logarithm function separately. It will be much easier to apply to optimization problems, like transmit power minimization, which will be discussed in Sect. 2.2.6.

#### 2.2.5.3 Asymptotic Throughput

Accordingly, the asymptotic throughput could be calculated as

$$\begin{aligned}T^A &= \frac{B\alpha}{\ln 2}\sum_{k=0}^{m-1}\frac{s(k)k!}{(\beta-\delta)^{k+1}}\int_{t_s}^{t_e}\left[C_k - \ln\left(\Psi d\,(t)^2\right)\right]\mathrm{d}t\\ &= \frac{B\alpha}{\ln 2}\sum_{k=0}^{m-1}\frac{s(k)k!}{(\beta-\delta)^{k+1}}\left[C_k\Delta t - \int_{t_s}^{t_e}\ln\left(\Psi d\,(t)^2\right)\mathrm{d}t\right],\end{aligned} \tag{2.95}$$

where $\Psi = \frac{(\beta-\delta)}{\lambda}$, $\Delta t = t_e - t_s$.

*(1) Simplified Case:* Since the expression of $d\,(t)^2$ in the general case is too complicate to calculate, we first derive an asymptotic throughput for the simplified case as

$$\begin{aligned}T^A &= \frac{B\alpha}{\ln 2}\sum_{k=0}^{m-1}\frac{s(k)k!}{(\beta-\delta)^{k+1}}\left[C_k\Delta t - \int_{t_s}^{t_e}\ln\left(\Psi H + \Psi\Upsilon\sin\left(\omega_s t + \beta_s\right)\right)\mathrm{d}t\right]\\ &= \frac{B\alpha}{\ln 2}\sum_{k=0}^{m-1}\frac{s(k)k!}{(\beta-\delta)^{k+1}}\left[C_k\Delta t - \ln\left(\Psi H\right)\Delta t - \frac{1}{\omega_s}\int_{\omega_s t_s+\beta_s}^{\omega_s t_e+\beta_s}\ln\left(1+\frac{\Upsilon}{H}\sin\left(y\right)\right)\mathrm{d}y\right].\end{aligned} \tag{2.96}$$

Since $d'\,(t)^2 = H + \Upsilon\sin\left(\omega_s t + \beta_s\right) > 0$, we can obtain $H + \Upsilon > 0$. As $-\sqrt{b_1^2 + b_2^2} < 0$, we have $\left|\frac{\Upsilon}{H}\right| < 1$. In this condition, we could take advantage $\ln\left(1+x\right) = \sum\limits_{n=0}^{\infty}\left(-1\right)^{n+1}\frac{x^n}{n}$ to simplify the calculation. Thus, the asymptotic throughput could be further calculated as

$$\begin{aligned}T^A &= \frac{B\alpha}{\ln 2}\sum_{k=0}^{m-1}\frac{s(k)k!}{(\beta-\delta)^{k+1}}\left[C_k\Delta t - \ln\left(\Psi H\right)\Delta t\right.\\ &\qquad\left. - \underbrace{\frac{1}{\omega_s}\sum_{n=1}^{\infty}(-1)^{n+1}\frac{\Upsilon^n}{H^n n}\int_{\omega_s t_s+\beta_s}^{\omega_s t_e+\beta_s}\sin\left(y\right)^n\mathrm{d}y}_{I_1}\right]\\ &= \frac{B\alpha}{\ln 2}\sum_{k=0}^{m-1}\frac{s(k)k!}{(\beta-\delta)^{k+1}}\left[\left(C_k - \ln\Psi H\right)\Delta t\right.\\ &\qquad\left. - \sum_{n=1}^{\infty}\frac{(-1)^{n+1}}{\omega_s}\frac{\Upsilon^n}{H^n n}\left(F_3\left(\omega_s t_e + \beta_s, n\right) - F_3\left(\omega_s t_s + \beta_s, n\right)\right)\right],\end{aligned} \tag{2.97}$$

where $F_3(x)$ is shown as

$$F_3(x,l) = \begin{cases} -\frac{\cos x}{2l}\left\{\sin^{2l-1}x + \sum_{k=1}^{l-1}\frac{(2l-1)(2l-3)...(2l-2k+1)}{2^k(l-1)(l-2)...(l-k)}\sin^{2l-2k-1}x\right\} + \frac{(2l-1)!!}{2l!}x, k \text{ is even} \\ -\frac{\cos x}{2l+1}\left\{\sin^{2l}x + \sum_{k=0}^{l-1}\frac{2^{k+1}l(l-1)...(l-k)}{(2l-1)(2l-3)...(2l-2k-1)}\sin^{2l-2k-2}x\right\}, k \text{ is odd.} \end{cases} \tag{2.98}$$

*(2) General Case:* Similarly, for the general case, we could derive the asymptotic throughput as

$$T^A = \frac{B\alpha}{\ln 2}\sum_{k=0}^{m-1} s(k)k!\Bigg[(C_k - \ln(\Psi H))\,\Delta t$$
$$-\frac{1}{\omega_s}\underbrace{\int_{t_s}^{t_e}\ln\left(1+\frac{\mu_1}{H}cos\,(\omega_a t+\eta_1)+\frac{\mu_2}{H}\cos\,(\omega_m t+\eta_2)+\frac{\mu_3}{H}\sin\,(\omega_s t+v)\right)\mathrm{d}t}_{I_2}\Bigg]. \tag{2.99}$$

To simplify the computation, we only take the first two order of the expansion of the logarithm function as $\lim_{x\to 0}\ln(1+x) = x - \frac{x^2}{2}$, which could be treated as a lower bound for the throughput. Thus, $I$ could be further calculated as

$$I_2 = \underbrace{\int_{t_s}^{t_e}\left[\frac{\mu_1}{H}\cos\,(\omega_a t+\eta_1)+\frac{\mu_2}{H}\cos\,(\omega_m t+\eta_2)+\frac{\mu_3}{H}\sin\,(\omega_s t+v)\right]\mathrm{d}t}_{I_0}$$
$$-\underbrace{\int_{t_s}^{t_e}\frac{\mu_1^2}{2H^2}cos^2\,(\omega_a t+\eta_1)\,\mathrm{d}t}_{I_1}-\underbrace{\int_{t_s}^{t_e}\frac{\mu_2^2}{2H^2}\cos^2\,(\omega_m t+\eta_2)\,\mathrm{d}t}_{I_2}$$
$$-\underbrace{\int_{t_s}^{t_e}\frac{\mu_3^2}{2H^2}\sin^2\,(\omega_s t+v)\,\mathrm{d}t}_{I_3}-\underbrace{\int_{t_s}^{t_e}\frac{\mu_1\mu_2}{H^2}cos\,(\omega_a t+\eta_1)\cos\,(\omega_m t+\eta_2)\,\mathrm{d}t}_{I_4}$$
$$-\underbrace{\int_{t_s}^{t_e}\frac{\mu_1\mu_3}{H^2}cos\,(\omega_a t+\eta_1)\sin\,(\omega_s t+v)\,\mathrm{d}t}_{I_5}$$
$$-\underbrace{\int_{t_s}^{t_e}\frac{\mu_2\mu_3}{H^2}\cos\,(\omega_m t+\eta_2)\sin\,(\omega_s t+v)\,\mathrm{d}t}_{I_6}, \tag{2.100}$$

where $I_0$ can be easily obtained as

$$\begin{aligned} I_0 = &-\frac{\mu_1}{H\omega_a}\left[\sin\left(\omega_a t_e + \eta_1\right] - \sin\left[\omega_a t_s + \eta_1\right)\right] \\ &-\frac{\mu_2}{H\omega_m}\left[\sin\left(\omega_m t_e + \eta_2\right] - \sin\left[\omega_m t_s + \eta_2\right)\right] \\ &-\frac{\mu_3}{H\omega_s}\left[\cos\left(\omega_s t_e + v\right) - \sin\left(\omega_s t_s + v\right)\right]. \end{aligned} \tag{2.101}$$

By taking advantage of [7, Eqs. 2.513.5, Eq. 2.513.11], $I_1$–$I_3$ can be calculated as

$$\begin{aligned} I_1 &= \int_{\omega_a t_s + \eta_1}^{\omega_a t_e + \eta_1} \frac{\mu_1^2}{2H^2\omega_a} cos^2\left(y\right) \mathrm{d}y \\ &= \frac{\mu_1^2}{8H^2\omega_a}\left[\sin(2v_{\max}^1) - \sin(2v_{\min}^1) + 2v_{\max}^1 - 2v_{\min}^1\right], \end{aligned} \tag{2.102}$$

$$\begin{aligned} I_2 &= \int_{\omega_m t_s + \eta_2}^{\omega_m t_e + \eta_2} \frac{\mu_2^2}{2H^2} \cos^2\left(y\right) \mathrm{d}y \\ &= \frac{\mu_2^2}{8H^2\omega_m}\left[\sin(2v_{\max}^2) - \sin(2v_{\min}^2) + 2v_{\max}^2 - 2v_{\min}^2\right] \end{aligned} \tag{2.103}$$

and

$$\begin{aligned} I_3 &= \int_{\omega_s t_s + v}^{\omega_s t_e + v} \frac{\mu_3^2}{2H^2} \sin^2\left(y\right) \mathrm{d}y \\ &= \frac{\mu_3^2}{8H^2\omega_s}\left[\sin(2v_{\min}^3) - \sin(2v_{\max}^3) + 2v_{\max}^3 - 2v_{\min}^3\right], \end{aligned} \tag{2.104}$$

where $v_{\min}^1 = \omega_a t_s + \eta_1$, $v_{\max}^2 = \omega_m t_e + \eta_2$, $v_{\min}^2 = \omega_m t_s + \eta_2$, $v_{\max}^3 = \omega_s t_e + v$ and $v_{\min}^3 = \omega_s t_s + v$.

Also, $I_4$ could be calculated as

$$\begin{aligned} I_4 &= \int_{t_s}^{t_e} \frac{\mu_1\mu_2}{H^2} cos\left(\omega_a t + \eta_1\right) \cos\left(\omega_m t + \eta_2\right) \mathrm{d}t \\ &= \frac{\mu_1\mu_2}{H^2}\left[F_4(\omega_a, \eta_1, \omega_m, \eta_2, t_e) - F_4(\omega_a, \eta_1, \omega_m, \eta_2, t_s)\right], \end{aligned} \tag{2.105}$$

where $F_4\left(a, b, c, d, x\right)$ is given by [7, 2.532.3] as

$$F_4\left(a, b, c, d, x\right) = \frac{\sin[(a-c)x + b - d]}{2(a-c)} - \frac{\sin[(a+c)x + b + d]}{2(a+c)}. \tag{2.106}$$

Furthermore, in the same way $I_5$–$I_6$ could be calculated as

$$I_5 = \frac{\mu_1\mu_3}{H^2}\left[F_5(\omega_s, v, \omega_a, \eta_1, t_e) - F_5(\omega_s, v, \omega_a, \eta_1, t_s)\right] \tag{2.107}$$

and

$$I_6 = \frac{\mu_2\mu_3}{H^2}\left[F_5(\omega_s, v, \omega_m, \eta_2, t_e) - F_5(\omega_s, v, \omega_m, \eta_2, t_s)\right], \tag{2.108}$$

where $F_5\left(a, b, c, d, x\right)$ is given by [7, 2.532.2] as

$$F_5\left(a, b, c, d, x\right) = -\frac{\cos[(a-c)x + b - d]}{2(a-c)} - \frac{\cos[(a+c)x + b + d]}{2(a+c)}. \tag{2.109}$$

Substituting (2.101) (2.102), (2.103), (2.104), (2.105), (2.107) and (2.108) into (2.100), we could obtain the asymptotic throughput for the general case.

From the asymptotic results shown in (2.97) and (2.99), we can see that the minimum transmit power that satisfies a certain throughput threshold can be derived easily. The closed-form solution will be given in Sect. 2.2.6.

## 2.2.6 *Applications and Future Directions*

In this subsection, we give some applications for our derived results in satellite communication, which is also the future directions.

### 2.2.6.1 Resource Allocation Problems

The resource allocation problems are mainly solved based on two kinds of assumptions about CSI, that is, statistical CSI [23], and perfect CSI [24]. Due to the communication distance variation in our considered satellite communication system, we focus on the discussion about the resource allocation under the assumption that perfect CSI is available for better demonstration purpose. The system model and derived results can provide an easier way to allocate resources. For example, the transmit power can be adjusted to satisfy the QoS assuming available CSI by solving the following optimization problems:

$$\begin{aligned} (\mathbf{P1}): \quad & \text{Given}: \ \text{CSI}, \gamma_{th}, \mathbf{s}_{0,S}, \mathbf{u} \\ & \text{Minimize}: P_S\left(t\right) \\ & \text{Subject to}: F\left(\gamma_{th}, t\right) \le P_o, \\ & \qquad 0 < P_S\left(t\right) \le P_{\max}, \end{aligned} \tag{2.110}$$

where $P_o$ is the OP threshold and $P_{\max}$ denotes the maximum transmit power.

$$\begin{aligned} (\mathbf{P2}):\quad & \text{Given}: \text{CSI}, T_{th}, \mathbf{s}_{0,S}, \mathbf{u} \\ & \text{Minimize}: P_S(t) \\ & \text{Subject to}: T(t) \geq T_{th}, \\ & \qquad 0 < P_S(t) \leq P_{\max}, \end{aligned} \tag{2.111}$$

where $T_{th}$ denotes the capacity threshold.

It should be noted that the minimal transmit power for the above two optimization problems occurs when $F(\gamma_{th}, t) = P_o$ and $T(t) = T_{th}$. Based on the derived instantaneous OP and capacity shown in (2.71) and (2.76), the proposed optimization problem can be transformed to the problems of solving equations $F(\gamma_{th}, t) = P_o$ and $T(t) = T_{th}$. Although we cannot derive the closed-form for the optimal solution, its numerical results can be easily solved by many software, like MATLAB.

By taking advantage of the asymptotic results, the sub-optimal solution in closed-form expression can be obtained. Let the asymptotic OP shown in (2.93) equals to $P_o$, one can derive the minimum transmit power satisfying the outage requirement as

$$P_S(t) = \frac{\sigma^2 C \beta(t) d(t)^2 \gamma_{th}}{-\ln\{1 - \frac{P_o \beta(t)}{\alpha(t)}\}}. \tag{2.112}$$

Similarly, let the asymptotic capacity shown in (2.94) equals to $T_{th}$, we can derive the minimum transmit power that satisfies the required capacity as

$$P_S = \frac{\sigma^2 C d(t)^2 (\beta(t) - \delta(t))}{\exp\left\{\frac{1}{m}\sum_{k=0}^{m-1} \frac{s(k)k!}{m(\beta(t)-\delta(t))^{k+1}} C_k - \frac{\ln 2}{B\alpha(t)} T_{th}\right\}}. \tag{2.113}$$

Moreover, the minimum transmit power can also be calculated to achieve the required throughput within the communication coverage by solving the following optimization problem:

$$\begin{aligned} (\mathbf{P3}):\quad & \text{Given}: \text{CSI}, T_{th}, \mathbf{s}_{0,S}, \mathbf{u} \\ & \text{Minimize}: P_S \\ & \text{Subject to}: T \geq T_{th}, \\ & \qquad 0 < P_S \leq P_{\max}, \end{aligned} \tag{2.114}$$

Similarly, the proposed optimization problems could be solved easily by software based on the satellite coordinate and the derived closed-form expressions shown in (2.79). It should be noted that the derived transmit power is not instantaneous but for a satellite visibility window.

By observing the asymptotic throughput for the simplified case and general case shown in (2.97) and (2.99), the minimal transmit power can also be obtained in closed-form as

$$P_S = \frac{\sigma^2 CH\left(\beta-\delta\right)}{\exp\left\{\frac{1}{m\Delta t}\sum_{k=0}^{m-1}\frac{s(k)k!}{(\beta-\delta)^{k+1}}\left[C_k\Delta t - I_1\right] - \frac{\ln 2}{mB\alpha\Delta t}T_{th}\right\}}, \tag{2.115}$$

and

$$P_S = \frac{\sigma^2 CH\left(\beta-\delta\right)}{\exp\left\{\frac{1}{m\Delta t}\sum_{k=0}^{m-1} s(k)k!\left[C_k\Delta t - \frac{1}{\omega_s}I_2\right] - \frac{\ln 2T_{th}}{mB\alpha\Delta t}\right\}}. \tag{2.116}$$

It should be noted that the closed-form solutions approach the optimal solution as the overall system performance improves.

#### 2.2.6.2 Multiple Satellite Communication System

Recently, mega-constellations systems have been proposed and being developed as a promising way to construct global connectivity [25]. Such satellite communication systems are designed to be composed of thousands of mass-produced small satellites working in combination with ground transceivers. Therefore, multi-satellite communication systems are an interesting and promising research direction. Although we only analyze the system performance while considering one satellite and one ground user, the analysis can be readily expanded to multi-satellite scenarios. In such a case, the ground device can be served by several satellites simultaneously in different orbits or the same orbits (e.g., low Earth orbit (LEO) satellites). In other words, under that case, the ground device is located in the intersection area of multiple satellites with different altitudes and rotation speeds or the same altitude/rotation speed. The device on the ground can adopt combining schemes to improve its communication quality when several satellites are active and send the same information bits. For example, the ground user is located in a coverage area, where it can be served by $N\left(t\right)$ satellites simultaneously. The number of serving satellites $N\left(t\right)$ varies over time because of the mobility of the satellite and the Earth rotation. The received SNR at the ground user for the $i$th satellite can be expressed as

$$\gamma_i\left(t\right) = \frac{P_{S,i}\left(t\right)\left|h_i\left(t\right)\right|^2}{\sigma^2 d_i\left(t\right)^2}. \tag{2.117}$$

To improve the communication quality, the ground user can adopt the maximal ratio combining scheme, generalized selection combining scheme, or selection combining scheme.

On the other hand, satellites sharing the same frequency may serve different devices within their overlapping coverage area, which leads to interference to the ground devices located in the same area that communicate with different satellites. Considering the target ground receiver located in the intersection area interfered by $N_I(t)$ satellites, the received SNR can be re-expressed as

$$\gamma(t) = \frac{\frac{P_S(t)}{d(t)^2}|h(t)|^2}{\sum_{i=1}^{N_I(t)} \frac{P_{S,i}(t)}{d_i(t)^2}|h_i(t)|^2 + \sigma^2}. \tag{2.118}$$

Since each satellite's serving time is different, the instantaneous system performance and throughput within different satellite visibility windows are also varying.

Another interesting problem that can be introduced is satellite handover. Based on the proposed satellite coordinate, the rotation routing of satellites can be predicted as well as the serving time. By acknowledging the users' position, the start point of connection time and the endpoint can be accessed. The next satellite can reserve enough resources to perform a more fluent satellite handover. For example, there are 12 LEO satellites uniformly distributed in the same orbit, with same $\beta_{\max} = \frac{2\pi}{9}$. Under the $(P, Q, W)$ coordinate, the positions of satellites are assumed to be $\left(H_S \cos\left(\frac{\pi}{6}i\right), H_S \cos\left(\frac{\pi}{6}i\right), 0\right)$, $i \in 1, \ldots, 12$. Based on the derived distance expression over time shown in (2.66) and maximal communication distance $d_{\max} = \frac{H_S - H_E}{\cos\left(\frac{2\pi}{9}\right)}$. The start point and endpoint for the connection between $i$th satellite and the ground user can be obtained by solving the equation $d(t) = d_{\max}$. It should be noted that there are some time overlaps between adjacent satellites because of the coverage overlap. Based on the derived time and satellite visibility window, the controller can schedule the active satellite for the ground user and implement satellite handover appropriately. It is just a simple example of satellite handover based on our derived results considering a single orbit. In a more realistic case, there are several orbits with different altitudes and directions that cross the target area simultaneously. They will act as a transmitter or an interferer. Further discussions and novel handover schemes will be left for our future works.

Moreover, an LEO satellite can only serve the devices within a limited satellite visibility window. The information bits may not be delivered by one satellite. When there are a large amount of data packets for delivery, they may be allocated to several satellites to deliver to the ground user at the same time or subsequently. The described satellite coordinate can predict the rotation routing of satellites and allocate the transmissions in advance.

#### 2.2.6.3 Mobile Devices

In practice, there are mobile users on the ground or flying over the air [4, 26], resulting in varying positions over time. By taking the moving speed and direction into consideration, the system performance can be analyzed in a similar way. Especially for fixed route moving devices, like trains, flights, and buses, the resources can be allocated properly in advance according to the fixed route and moving speed. For example, assuming a user moves on the ground with azimuthal angle and polar angle variations $\Delta\theta_u(t)$ and $\Delta\varphi_u(t)$ with ECEF coordinates

$$u_1 = H_E \sin(\theta_u + \Delta\theta_u(t)) \cos(\varphi_u + \Delta\varphi_u(t)), \tag{2.119}$$

$$u_2 = H_E \sin(\theta_u + \Delta\theta_u(t)) \sin(\varphi_u + \Delta\varphi_u(t)), \tag{2.120}$$

and

$$u_3 = H_E \cos(\theta_u + \Delta\theta_u(t)) \tag{2.121}$$

of the user at $t$. By replacing $\theta_u$ and $\varphi_u$ in (2.66) by $\theta_u + \Delta\theta_u(t)$ and $\varphi_u + \Delta\varphi_u(t)$, the instantaneous distance can also be derived. Accordingly, the instantaneous OP and capacity as well as the system throughput can also be derived. The flying devices over the air like flying cars, UAVs, HAPs also can be analyzed in the same way with different radial distance. Except for the fixed route case, the user may have random movements within a satellite visibility window. One way to model the randomness position is to assume $\theta_u$ and $\varphi_u$ are uniformly distributed within $[\theta_1, \theta_2]$ and $[\varphi_1, \varphi_2]$ respectively, where $0 \leq \theta_1 \leq \theta_2 \leq \pi$ and $0 \leq \varphi_1 \leq \varphi_2 \leq 2\pi$.

Furthermore, due to the high path loss over the propagation in the satellite communication systems, there will be cooperation devices like relays, RISs to assist the communications. Whether they are fixed or moving, and no matter whether they are in the air or on the ground or underwater, the system performance can always be analyzed through the results replacing different communication distances. The resource allocation and satellite handover scheme can also be further discussed accordingly.

### 2.2.7 *Extreme Short Communication Duration*

The position of U, the rotation rate of S, and the position of the orbit may cause a short communication time. In this case, the channel fading is time-varying rather than ergodic and time-fixed. In other words, the channel state cannot be ergodic as suggested by the channel distribution within a short satellite visibility window. The experienced channel within with satellite visibility window $t_e - t_s$ can be expressed as $h(t) \in h\left(t_s + \frac{t_e - t_s}{n}\right), n = 1, \ldots, N$, where $N$ is a small integer number. The worst occasion happens when $N = 1$, that is, $h(t) = h(t_e)$, which means that the channel condition is fixed within a short satellite visibility window $t_e - t_s$. Since the communication happens within a short satellite visibility window, the distance is approximated to the maximum communication distance, which is $d_{\max}$. Thus, the throughput could be expressed as

$$T_s = \frac{BT}{N} \sum_{n=1}^{N} \log_2 \left( 1 + \frac{P_S \xi^2 \left| h \left( t_s + \frac{t_e - t_s}{n} \right) \right|^2}{\sigma^2 C d_{\max}^2} \right). \tag{2.122}$$

In this case, the average OP could be counted as

$$P_{out} = \frac{1}{N} \sum_{n=1}^{N} \Pr \left\{ \frac{P_S \xi^2 \left| h \left( t_s + \frac{t_e - t_s}{n} \right) \right|^2}{\sigma^2 C d_{\max}^2} < \gamma_{th} \right\}. \tag{2.123}$$

It should be noted that, since the channel is not ergodic within $T$, the throughput and the OP would be different compared to the general case, which can be clearly seen from the numerical results presented in the next section.

### 2.2.8 *Numerical Results*

In this subsection, Monte-Carlo simulation and analytical results will be provided to study the performance of the considered system and verify the proposed analytical expression. In the following, we denote the case considering the Earth's rotation as ER, while the one ignoring the Earth's rotation is named as WER.

#### 2.2.8.1 Instantaneous OP

Firstly, we study the instantaneous OP under parameter settings: $f = 12\,\text{GHz}$, $P_S = 50$ dBW, $N_0 = -154$ dB, $b = 0.158$, $\Omega_h = 1.29$, $m = 19$, $\Omega = \pi/3$, $i = \pi/6$, $v = \pi/4$, $\theta_u = \pi/2$, $\varphi_u = \pi/8$, $\omega_S = 0.025$ deg/s, $\beta_{\max} = 35°$ and $H_S = 16000$ km. As shown in Fig. 2.12, the instantaneous OP mainly depends on the communication distance between S and U. There is an apparent difference in communication distance for WER and ER, especially when the satellite orbit is high, and the rotation speed is slow. As time goes by, the OP of WER will be periodic while the OP of ER changes in each rotation period, which is caused by the Earth's rotation. Moreover, the gaps between the instantaneous OP and the distance of ER and WER become large firstly, then decrease. After several periods, the gaps go back to the same as the first one, which implies that the distance difference between ER and WER is a periodic function. It also proved that we should not ignore the distance variation in the realistic scenario since it brings performance variation and outage. In Fig. 2.13, we could see that the asymptotic OP matches perfectly with the exact OP in high SNR regime with $b = \Omega_h = 15$ dB and $m = 2$. Moreover, the communication time could be greatly reduced by increasing the rotation speed of the satellite, because U stays more shortly in the coverage area. Moreover, we conduct the experiment on the optimal power allocation, which is shown in Fig. 2.14. Based on the derived closed-expression of OP, we use the function "vpasolve" in MATLAB to get the exact minimum transmit power to make the OP satisfy the system requirement,

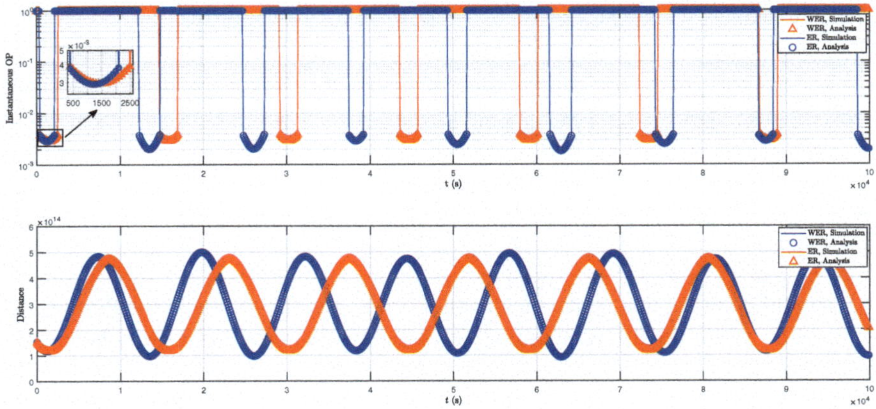


**Fig. 2.12** ER instantaneous OP versus WER instantaneous OP and ER instantaneous distance versus WER instantaneous distance

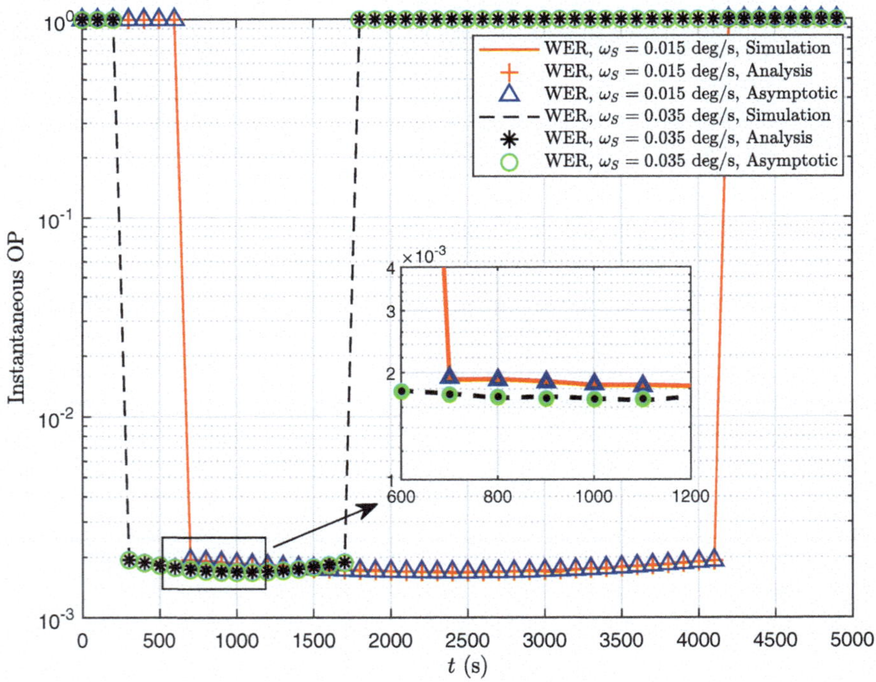


**Fig. 2.13** WER instantaneous OP for various $\omega_S$

that is, $P_o = 0.001$. We can see that the transmit power decreases firstly and then increases, which varies corresponding to the communication distance. We also adopt the closed-form solution shown in (2.112) derived by the asymptotic OP, which is shown to be accurate in the high SNR regime.

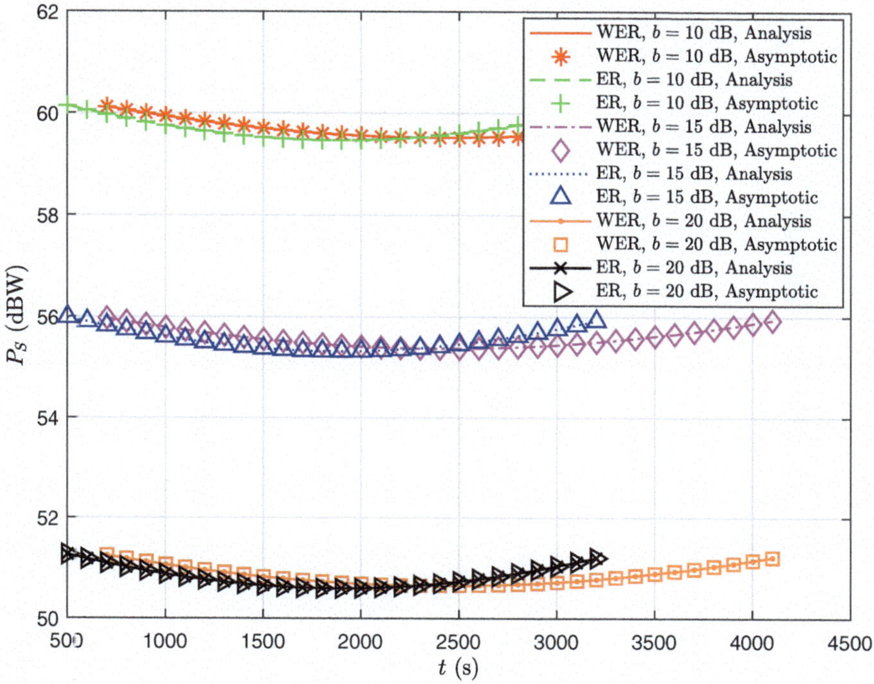


**Fig. 2.14** ER instantaneous transmit power versus WER instantaneous transmit power for various $b$

#### 2.2.8.2 Instantaneous Channel Capacity

By re-setting $b = 0.063$, $\Omega_h = 8.97 \times 10^{-4}$, $m = 1$, $\Omega = \pi/12$, $i = \pi/10$, $v = 0$, $\theta_u = \pi/2$, $\varphi_u = 0$ and $H_S = 7800$ km, the instantaneous channel capacity variation is shown in Fig. 2.15. We can see that the instantaneous channel capacities in the first satellite visibility window for different satellite rotation rates are very close. Moreover, the fast-rotating satellite could fall into the next satellite visibility window earlier. There are apparent transmission duration and communication distance variations among each satellite visibility window. In Fig. 2.16 with resetting parameters $b = \Omega_h = 15$ dB and $m = 2$, we could see that the instantaneous channel capacity in the first satellite visibility window under ER is the same as the one under WER case. In the following communication time, there are apparent differences between the performance under ER and the one under WER. We could conclude that even if the satellite has a high rotation speed, the Earth's rotation could only be ignored in the first satellite visibility window. Also, under WER, the instantaneous channel capacity could remain the same in every satellite visibility window, while the one under ER keeps varying.

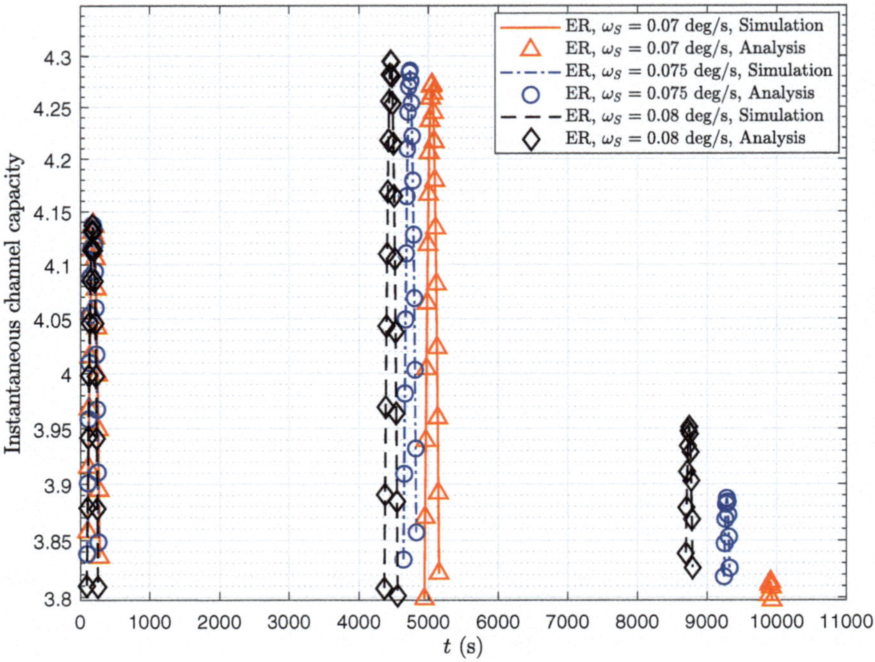


**Fig. 2.15** ER instantaneous channel capacity for various $\omega_S$

#### 2.2.8.3 Throughput

By setting $P_S = 50$ dBW, $b = 15$ dB, communication time $\Delta t = 5000$ s and keeping other parameters in subsection A, we verify the correctness of our proposed throughput expression. In Fig. 2.17, we could see that the throughput increases as the height of the satellite's orbit increases because high orbit leads to the large coverage area and long communication time. Moreover, WER gains higher throughput than ER caused by longer communication time and shorter communication distance, which could be treated as an upper bound for a realistic case. The asymptotic throughput matches the exact one closely. We also compare the throughput with the cases considering the random moving user, named WER-RU and ER-RU. The case is just a simple example, where the mobile user changes its position randomly within a limited area $\theta_u \in \left[\frac{\pi}{2}, \frac{\pi}{2} + \frac{\pi}{20}\right]$, $\varphi_u \in \left[\frac{\pi}{8}, \frac{\pi}{8} + \frac{\pi}{20}\right]$. It can be seen that the cases considering mobile users may lead to higher throughput within a limited time, because its random position may lead to shorter communication distance or longer communication time. It should be noted that the users' mobility can lead to better throughput or much worse throughput, which depends on practical movements. As seen in Fig. 2.18 with resetting parameters $P_S = 10$ dBW, the system throughput increases as the fading severity parameter increases due to stronger LoS component. However, it does not

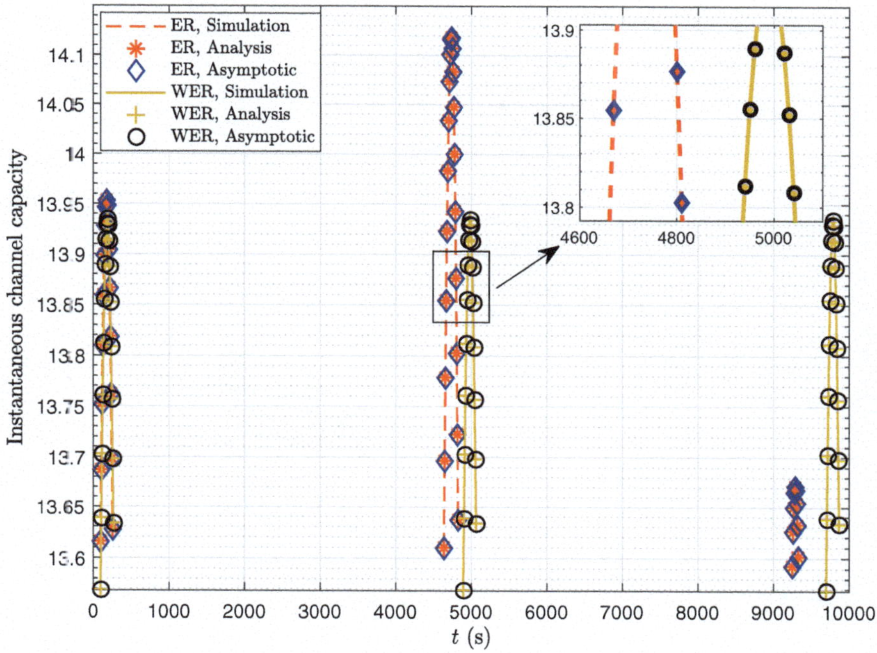


**Fig. 2.16** ER instantaneous channel capacity versus WER instantaneous channel capacity

increase apparently when $m$ is bigger than 5. From Figs. 2.17 and 2.18, it is obvious that better LoS channel conditions could greatly improve system performance.

### 2.2.8.4 Extreme Short Communication Duration

In this part, we compare the throughput and the OP under a non-ergodic channel with the ones under an ergodic channel with parameter setting: $b = \Omega_h = 5$ dB and the other same parameters in Part A of this section. As shown in Fig. 2.19, the throughput under non-ergodic cases could be higher or lower than the ergodic case since the instantaneous channel could be better or worse than the ergodic one. Similarly, in Fig. 2.20, the OP also depends on the instantaneous channel quality, although with the same fixed channel power gain. It should be noted that the OP could only take $N$ results, that is, $\frac{n}{N}$, $n = 0, \ldots, N$. We could conclude that the throughput and the OP within a short transmission time could not be represented by the ones of the general case anymore. Moreover, the throughput and the OP in both cases could be greatly improved by increasing the transmit power at the satellite.

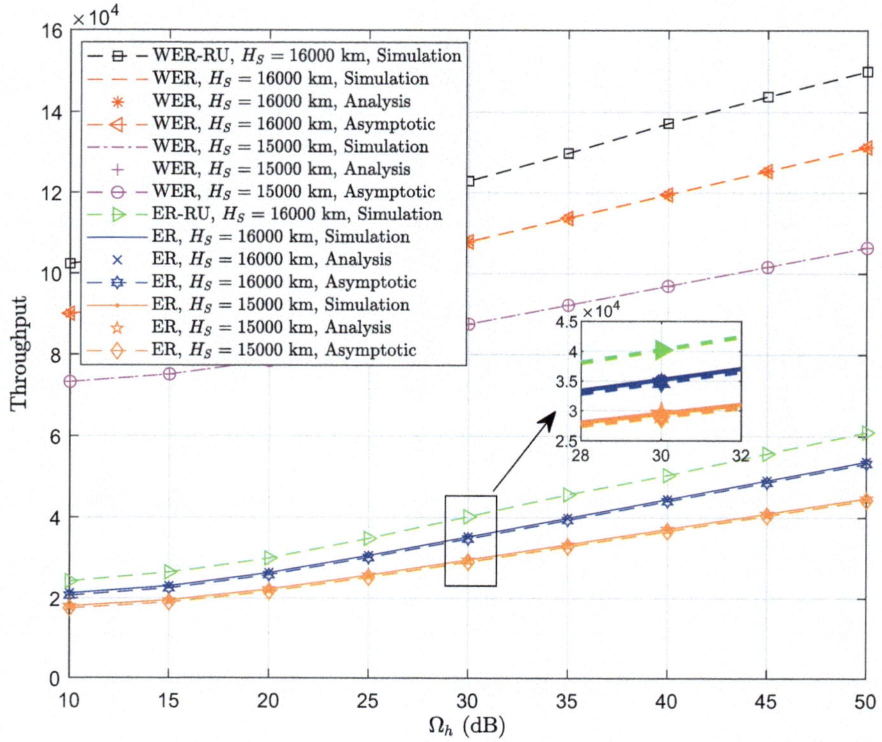


**Fig. 2.17** ER throughput versus WER throughput for various $H_S$

## 2.2.9 *Conclusion*

Taking the Earth's rotation and satellite rotation into account, a non-stationary satellite communication system is modeled. We study the instantaneous OP, instantaneous channel capacity, and throughput during the finite satellite visibility window of the proposed non-stationary satellite communication system. A simpler case without the Earth's rotation is also studied in this section, as well as asymptotic results for both cases. Applications involved with our derived results and future research direction are also discussed. Considering a short satellite visibility window, new definitions for throughput and OP are proposed.

Observing from the numerical results, some remarks can be achieved as follows:

(1) The distance variation shows a great influence on the system performance, because of the unavoidable path loss.

(2) There is an apparent difference in the communication distance for WER and ER, which leads to performance gaps. Also, the distance difference between ER and WER is a periodic function.

(3) A faster rotation speed of the satellite leads to shorter communication duration because a shorter satellite visibility window for U is available staying in the coverage

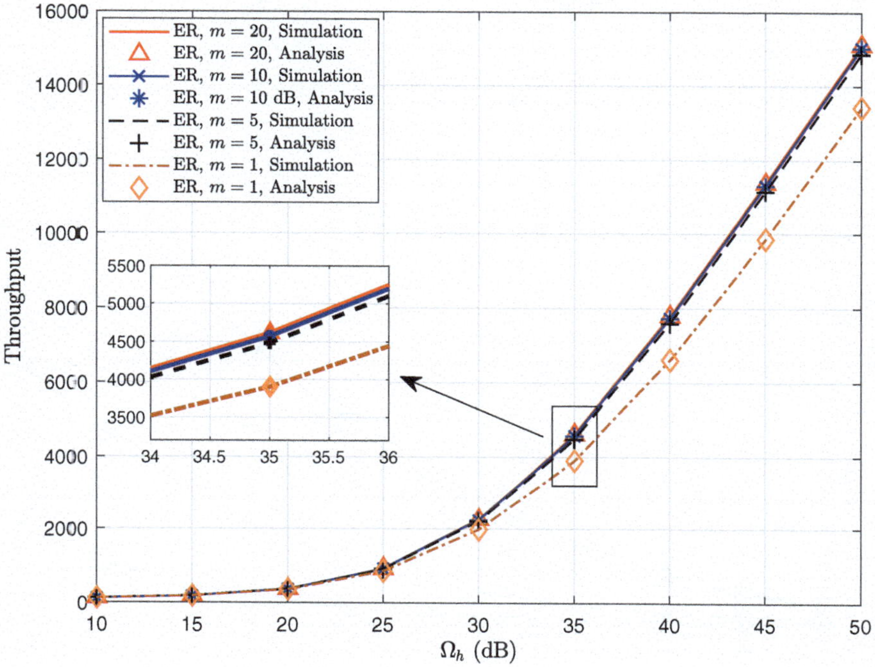


**Fig. 2.18** ER throughput for various $m$

area of the satellite. Moreover, the fast-rotating satellite could fall into the next satellite visibility window earlier.

(4) Even with a high satellite rotation speed, the Earth's rotation could only be ignored in the first satellite visibility window.

(5) A large radius of satellite orbit brings an improved throughput. However, it can be eliminated by high path loss.

(6) The throughput of WER could be treated as an upper bound for the throughput of ER.

(7) The throughput and the OP within short transmission time under non-ergodic channel fading could not be represented by the ones under ergodic channel fading.

(8) The asymptotic results can be used to solve the transmit power allocation problem in closed-form solutions.

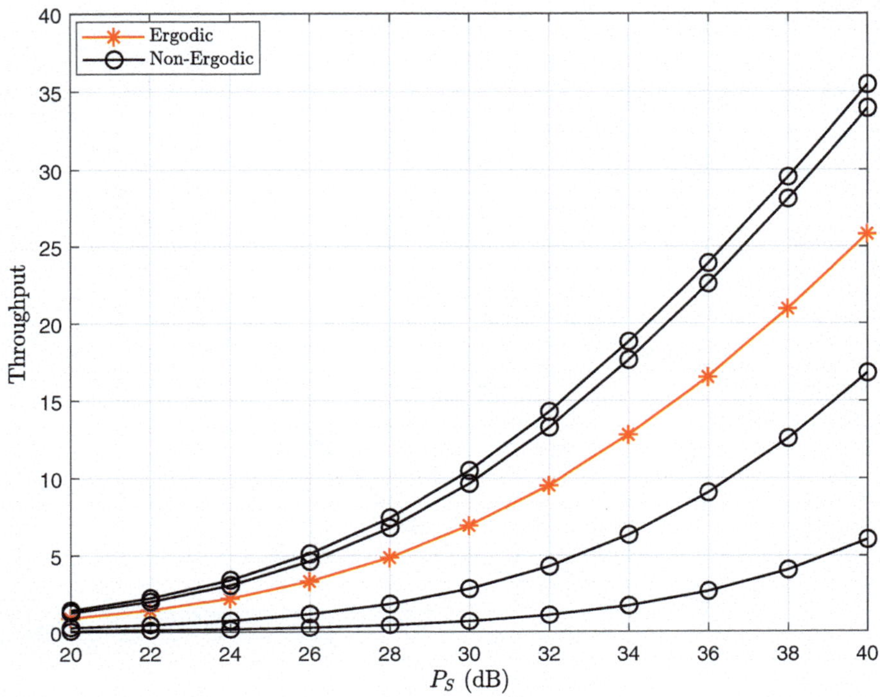


**Fig. 2.19** Ergodic throughput versus non-ergodic throughput

## 2.3 Secure Satellite-Terrestrial Downlink Transmissions

### *2.3.1 System Model*

As shown in Fig. 2.21, in this work, we consider a satellite-terrestrial transmission which consists of one satellite ($S$), one legitimate terrestrial receiver ($D$), and one terrestrial eavesdroppers ($E$). For the convenience of subsequent derivations, we treat the coverage area of the satellite as a circle with radius $L$. The projection of the satellite on the ground is the origin of this circle. We assume that $D$ and $E$ are equipped with $N_D \geq 2$ and $N_E \geq 2$ antennas, respectively.

Moreover, the locations of the two terrestrial receivers are modeled as a uniform distribution. According to [27], we can calculate the distance from the terrestrial receivers to the projection of $S$ by $\mathcal{M}$, the PDF of which can be written as $f_{\mathcal{M}}(x) = \frac{1}{\pi L^2}$.

Shadowed-Rician model proposed in [6] is adopted in this work to capture the statistical property of the satellite-terrestrial link. Then, the PDF of the channel gain for the link from $S$ to the $i$th antenna at receiver $j$ is given by

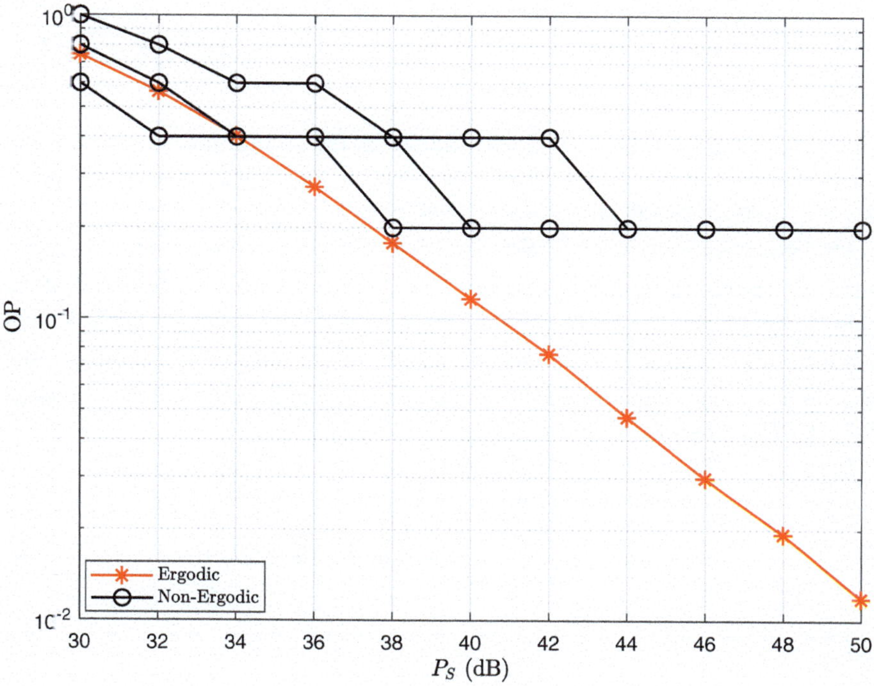


**Fig. 2.20** Ergodic OP versus non-ergodic OP with $N = 5$

**Fig. 2.21** System model

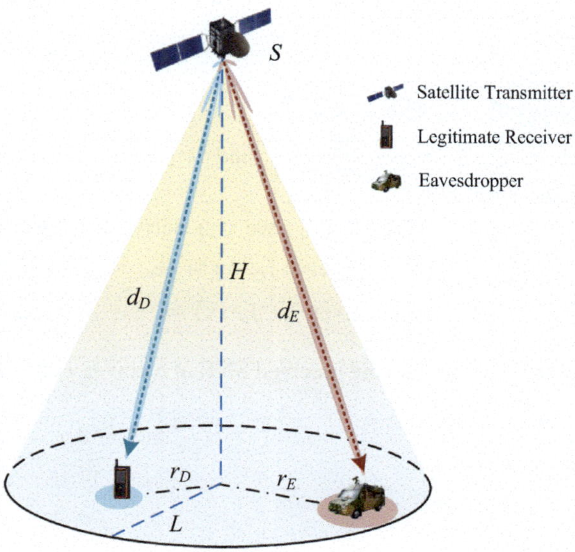

$$f_{|h_{ji}|^2}(x) = \left(\frac{2m_j b_j}{2m_j b_j + \Omega_j}\right)^{m_j} \frac{1}{2b_j} \exp\left(-\frac{x}{2b_j}\right) {}_1F_1\left(m_j, 1, \frac{\Omega_j x}{2b_j(2m_j b_j + \Omega_j)}\right), \tag{2.124}$$

where $j \in \{D, E\}$, $i \in \{1, ..., N_j\}$, $2b_j$ and $\Omega_j$ are the average power of scatter and line-of-sight components, respectively, $m_j$ denotes the fading severity parameter, ${}_1F_1(\cdot; \cdot; \cdot)$ represents the first kind confluent hypergeometric function.

Further, we can derive the SNR of the $i$th antenna at receiver $j$ as $\gamma_{ji} = \frac{P_s |h_{ji}|^2}{\sigma^2 d_j^n}$, where $P_s$ denotes the transmit power at $S$,[1] $\sigma^2$ represents the average power of the additive white Gaussian noise (AWGN), $n$ is the path-loss factor for the link from $S$ to terrestrial receiver $j$, $d_j = \sqrt{H^2 + r_j^2}$ is the distance between $S$ and terrestrial receiver $j$, $r_j$ denotes the distance from the receiver $j$ to the origin of the terrestrial disk, and $H$ denotes the height of the satellite.

Let $\lambda_{ji} = \frac{P_s |h_{ji}|^2}{\sigma^2} (0 \le j \le N_j)$, the PDF and CDF of $\lambda_{ji}$ can be given by

$$f_{\lambda_{ji}}(x) = \alpha_j \sum_{k=0}^{m_j - 1} \frac{\varsigma_j(k)}{\bar{\lambda}^{k+1}} x^k \exp\left(-\frac{\beta_j - \delta_j}{\bar{\lambda}} x\right) \tag{2.125}$$

and

$$F_{\lambda_{ji}}(x) = 1 - \alpha_j \sum_{k=0}^{m_j - 1} \frac{\varsigma_j(k)}{\bar{\lambda}^{k+1}} \sum_{p=0}^{k} \frac{k!}{p!} \left(\frac{\beta_j - \delta_j}{\bar{\lambda}}\right)^{-(k+1-p)} x^p \exp\left(-\frac{\beta_j - \delta_j}{\bar{\lambda}} x\right), \tag{2.126}$$

respectively, where $\bar{\lambda} = \frac{P_s}{\sigma^2}$, $\varsigma_j(k) = \frac{(-1)^k (1-m_j)_k \delta_j^k}{(k!)^2}$, $\alpha_j = \left(\frac{2_j b m_j}{2b_j m_j + \Omega_j}\right)^{m_j} / (2b_j)$, $\beta_j = \frac{1}{2b_j}$, $\delta_j = \frac{\Omega_j}{2b_j(2b_j m_j + \Omega_j)}$, and $(t)_k = t(t+1)\ldots(t+k-1)$ is the Pochhammer symbol [7].

Using the PDF of $\mathcal{M}$, we can derive the CDF and PDF of $r_j$ as $F_{r_j}(x) = \int_0^{2\pi} \int_0^x \frac{r}{\pi L^2} \mathrm{d}r\mathrm{d}\theta = \frac{x^2}{L^2}$ and $f_{r_j}(x) = \frac{2x}{L^2}$, $0 \le x \le L$, respectively. Further, the PDF of $d_j^n$ can be given as $f_{d_j^n}(x) = \frac{2x^{\frac{2}{n}-1}}{nL^2}$. By using maximal ratio combining (MRC) technique, the received SNR at receiver $j$ can be expressed as $\gamma_j = \sum_{i=0}^{N_j} \gamma_{ji} = \left(\sum_{i=0}^{N_j} \lambda_{ji}\right) / d_j^n$.

---

[1] Here, it is also assumed that there is no channel state information at the satellite, which meaning that the transmit power at the satellite will not be adjusted during the information transmission process. Then, secrecy outage events will happen.

### 2.3.2 Secrecy Outage Analysis

As suggested by [28], the PDF of $\lambda_j = \sum_{i=0}^{N_j} \lambda_{ji}$ is given by

$$f_{\lambda_j}(x) = \left(\frac{2b_j m_j}{\Omega_j + 2b_j m_j}\right)^{\tilde{m}_j} \cdot \frac{x^{N-1}\exp\left(-\frac{x}{2b_j\bar{\lambda}}\right)}{\Gamma(N_j)(2b_j)^{N_j}\bar{\lambda}^N} {}_1F_1\left(\tilde{m}_j, N_j; \frac{\Omega_j x}{\bar{\lambda}2b_j(2b_j m_j + \Omega_j)}\right), \tag{2.127}$$

where $\bar{\lambda} = \frac{P_S}{\sigma^2}$ and $\tilde{m}_j = \sum_{i=0}^{N_j} m_j$.

By using Kummer's transform method to represent the hypergeometric function [29], we can rewrite (2.127) as

$$f_{\lambda_j}(x) = \frac{\tilde{\alpha}_j}{\tilde{\beta}_j} \sum_{k=0}^{\tilde{m}_j - N_j} \theta_{jk}\exp\left(-\psi_j x\right) x^{k+N_j-1}, \tag{2.128}$$

where $\tilde{\alpha}_j = \left(\frac{2b_j\tilde{m}_j}{\bar{\Omega}_j+2b_j\tilde{m}_j}\right)^{\tilde{m}_j}$, $\tilde{\beta}_j = \Gamma(N_j)(2b_j)^{N_j}\bar{\lambda}^{N_j}$, $\tilde{\sigma}_j = \frac{\Omega_j}{\bar{\lambda}2b_j(2b_j m_j+\Omega_j)}$, $\theta_{jk} = \frac{(\tilde{m}_j-N)!\sigma_j^k}{k!(\tilde{m}_j-N-k)!(N)_k}$, and $\psi_j = \frac{1}{2b_j\bar{\lambda}} - \sigma_j$.

Then, the CDF of $\lambda_j = \sum_{i=0}^{N_j} \lambda_{ji}$ can be written as

$$F_{\lambda_{\_}}(x) = 1 - \frac{\tilde{\alpha}_j}{\tilde{\beta}_j} \sum_{k=0}^{\tilde{m}_j - N_j} \theta_{jk} \sum_{p=0}^{k+N_j-1} \tilde{\varsigma}_{kp}\left(\psi_j\right)^{-(k+N_j-p)} x^p \exp\left(-\psi_j x\right), \tag{2.129}$$

where $\tilde{\varsigma}_{kp} = (k + N_j - 1)!/p!$.

The instantaneous secrecy capacity of the considered system can be expressed as

$$C_s(\gamma_D, \gamma_E) = \log_2(1 + \gamma_D) - \log_2(1 + \gamma_E). \tag{2.130}$$

Then, the outage is defined as the event that the eavesdropping link's secrecy capacity is negative. Therefore, the secrecy outage probability (SOP) of the considered system is given as

$$P_{\text{SOP}} = \Pr\{C_s(\gamma_D, \gamma_E) \le 0\}. \tag{2.131}$$

**Corollary 2.1** *A closed-form analytical expression for SOP without considering the the randomness of the positions of D and E is given by*

$$P_{SOP|d_D,d_E} = 1 - \sum_{k=0}^{\tilde{m}_D-N_D} \sum_{p=0}^{k+N_D-1} \sum_{l=0}^{\tilde{m}_E-N_E} \Xi_{k,p,l} \left(\frac{d_D^n}{d_E^n}\right)^p \frac{\Gamma(l+p+N)}{\left(\psi_D \frac{d_D^n}{d_E^n} + \psi_E\right)^{l+p+N}}. \tag{2.132}$$

***Proof***

$$\begin{aligned} P_{\mathrm{SOP}|d_D,d_E} &= \Pr\{\gamma_D \leq \gamma_E | d_D, d_E\} \\ &= \Pr\left\{\lambda_D \leq \frac{\lambda_E d_D^n}{d_E^n} \middle| d_D, d_E\right\} \\ &= 1 - \int_0^\infty \sum_{k=0}^{\tilde{m}_D-N_D} \sum_{p=0}^{k+N_D-1} \sum_{l=0}^{\tilde{m}_E-N_E} \Xi_{k,p,l} \left(\frac{d_D^n}{d_E^n}\right)^p \\ &\quad \times \exp\left[-\left(\psi_D \frac{d_D^n}{d_E^n} + \psi_E\right) x\right] x^{l+p+N_E-1} \mathrm{d}x, \end{aligned} \tag{2.133}$$

where $\Xi_{k,p,l} = \tilde{\alpha}_D \tilde{\alpha}_E \theta_{jk} \theta_{El} \zeta_{kp} / (\tilde{\beta}_D \tilde{\beta}_E \psi_D^{k+N_D-p})$. In the following, we use $\sum_{k,p,l}$ instead of $\sum_{k=0}^{\tilde{m}_D-N_D} \sum_{p=0}^{k+N_D-1} \sum_{l=0}^{\tilde{m}_E-N_E}$ for simplification.

Let $I$ denote the integral part in (2.133), and using [7, Eq. (3.326.2.10)], we have

$$I = \sum_{k,p,l} \Xi_{k,p,l} \left(\frac{d_D^n}{d_E^n}\right)^p \frac{\Gamma(l+p+N)}{\left(\psi_D \frac{d_D^n}{d_E^n} + \psi_E\right)^{l+p+N}}. \tag{2.134}$$

This proof is concluded by substituting (2.134) into (2.133). □

Taking the randomness of $d_D$ and $d_E$ into account, and using Corollary 2.1, $SOP$ can be written as

$$\begin{aligned} \mathrm{SOP} &= \int_{d_{\min}^n}^{d_{\max}^n} \int_{d_{\min}^n}^{d_{\max}^n} \frac{2x_0^{\frac{2}{n}-1}}{nL^2} \frac{2x_e^{\frac{2}{n}-1}}{nL^2} \Bigg[1 - \\ &\quad \sum_{k,p,l} \Xi_{k,p,l} \left(\frac{x_0}{x_e}\right)^p \frac{\Gamma(l+p+N)}{\left(\psi_D \frac{x_0}{x_e} + \psi_E\right)^{l+p+N}}\Bigg] \mathrm{d}x_0 \mathrm{d}x_e \\ &= 1 - \sum_{k,p,l} \Xi_{k,p,l} \frac{4\Gamma(l+p+N)}{n^2 L^4} \int_{d_{\min}^n}^{d_{\max}^n} x_e^{l+N+\frac{2}{n}-1} \\ &\quad \int_{d_{\min}^n}^{d_{\max}^n} \frac{x_0^{\frac{2}{n}+p-1}}{(\psi_D x_0 + \psi_E x_e)^{l+p+N}} \mathrm{d}x_0 \mathrm{d}x_e, \end{aligned} \tag{2.135}$$

where $d_{\max} = \sqrt{L^2 + H^2}$ and $d_{\min} = H$, respectively.

$$\begin{aligned}\text{SOP} = 1 - \sum_{k,p,l} \Xi_{k,p,l} \frac{4\Gamma(l+p+N)}{n^2 L^4} \\ \times \left\{ \frac{(d_{\min}^n)^{-\mu_l}}{\mu_l \psi_D{}^{\nu_{pl}}} \left[ g\left(d_{\max}^n, \frac{\psi_E}{\psi_0 d_{\min}^n}, l, p\right) - g\left(d_{\min}^n, \frac{\psi_E}{\psi_D d_{\min}^n}, l, p\right) \right] \right. \\ \left. - \frac{(d_{\max}^n)^{-\mu_l}}{\mu_l \psi_D{}^{\nu_{pl}}} \left[ g\left(d_{\max}^n, \frac{\psi_E}{\psi_D d_{\max}^n}, l, p\right) - g\left(d_{\min}^n, \frac{\psi_E}{\psi_D d_{\max}^n}, l, p\right) \right] \right\}. \end{aligned} \tag{2.136}$$

$$\begin{aligned}\text{SOP} = 1 - \sum_{k,p,l} \Xi_{k,p,l} \frac{4\Gamma(l+p+N)}{n^2 L^4} \int_{d_{\min}^n}^{d_{\max}^n} x_e{}^{l+N+\frac{2}{n}-1} \\ \left[ \frac{(d_{\min}^n)^{-\mu_l}}{\mu_l \psi_D{}^{\nu_{pl}}} \cdot {}_2F_1\left(\nu_{pl}, \mu_l; 1+\mu_l, -\frac{\psi_E x_e}{\psi_D d_{\min}^n}\right) \right. \\ \left. - \frac{(d_{\max}^n)^{-\mu_l}}{\mu_l \psi_D{}^{\nu_{pl}}} \cdot {}_2F_1\left(\nu_{pl}, \mu_l; 1+\mu_l, -\frac{\psi_E x_e}{\psi_D d_{\max}^n}\right) \right] \mathrm{d}x_e. \end{aligned} \tag{2.137}$$

**Corollary 2.2** *A closed-form analytical expression for SOP is given in* (2.136), *as shown on the bottom of the previous page.*

***Proof*** By using [30, Eq. (1.2.4.4)], we can obtain the closed form of the inner integral in (2.135) as

$$\begin{aligned} I_1 &= \int_{d_{\min}^n}^{d_{\max}^n} \frac{x_0{}^{\frac{2}{n}+p-1}}{(\psi_D x_0 + \psi_E x_e)^{l+p+N}} \mathrm{d}x_0 \\ &= \frac{(d_{\min}^n)^{-\mu_l}}{\mu_l \psi_D{}^{\nu_{pl}}} \cdot {}_2F_1\left(\nu_{pl}, \mu_l; 1+\mu_l, -\frac{\psi_E x_e}{\psi_D d_{\min}^n}\right) \\ &\quad - \frac{(d_{\max}^n)^{-\mu_l}}{\mu_l \psi_D{}^{\nu_{pl}}} \cdot {}_2F_1\left(\nu_{pl}, \mu_l; 1+\mu_l, -\frac{\psi_E x_e}{\psi_D d_{\max}^n}\right), \end{aligned} \tag{2.138}$$

where $\nu_{pl} = l + p + N$, $\mu_l = N + l - 2/n$.

By employing (2.138), we can deduce (2.135) into (2.137), as shown on the bottom of the previous page, where ${}_2F_1(\cdot, \cdot; \cdot, \cdot)$ is the hypergeometric function.

To simplify the analysis, we use Meijer's G-function to represent the hypergeometric function and define a new function $g(\tau, \xi, l, p)$ as

$$\begin{aligned} g(\tau, \xi, l, p) &= \int_0^\tau x^{\eta_l} {}_2F_1\left(\nu_{pl}, \mu_l; 1+\mu_l, -\xi x\right) \mathrm{d}x \\ &= \int_0^{\xi\tau} \left(\frac{1}{\xi}\right)^{\eta_l+1} t^{\eta_l} {}_2F_1\left(\nu_{pl}, \mu_l; 1+\mu_l, -t\right) \mathrm{d}t \end{aligned}$$

$$= \int_0^{\xi\tau} \left(\frac{1}{\xi}\right)^{\eta_l+1} \frac{\Gamma(1+\mu_l)}{\Gamma(\nu_{pl})\Gamma(\mu_l)} t^{\eta_l} \mathrm{G}_{22}^{12}\left[t \middle| \begin{matrix} 1-\nu_{pl},\ 1-\mu_l \\ 0,\ -\mu_l \end{matrix}\right] \mathrm{d}t, \tag{2.139}$$

where $\eta_l = l + N + 2/n - 1$, and $\mathrm{G}_{p,q}^{m,n}[\cdot|\cdot]$ denotes the single variable Meijer's G-function.

Using [11, Eq. (26)], we can rewrite (2.139) as

$$g(\tau, \xi, l, p) = \left(\frac{1}{\xi}\right)^{\eta_l+1} \frac{\Gamma(1+\mu_l)}{\Gamma(\nu_{pl})\Gamma(\mu_l)} (\xi\tau)^{\eta_l+1} \mathrm{G}_{33}^{13}\left[\xi\tau \middle| \begin{matrix} 1-\nu_{pl}, 1-\mu_l, -\eta_l \\ 0, -1-\eta_l, -\mu_l \end{matrix}\right]. \tag{2.140}$$

This proof is concluded by substituting (2.140) and (2.138) into (2.133). □

### 2.3.3 Numerical Results

In this section, we present Monte Carlo simulation to validate our analytical expressions for SOP. In each case, we perform at least $10^5$ times of the realizations of the channel and the positions of terrestrial receivers. Besides, we set $n = 2$, $m_E = m_D = 2$, $b_E = 0.25$, $b_D = 0.36$, $\Omega_E = 1$ dB, $N_D = 2$, and $L = 15$ km.[2]

Figure 2.22 depicts that the changes in SOP versus $\Omega_D$. We can observe that SOP decreases when $\Omega_D$ improves. This is reasonable since that high $\Omega_D$ leads to improved legitimate channel quality. We can observe that for fixed $\Omega_D$, SOP increase with $N_D$ increase, as MRC diversity gain increases at the legitimate receiver. In the high $\Omega_D$ region, Fig. 2.22 indicates that the gap between asymptotic and the exact results is negligible. Figure 2.22 also shows that the SOP of MRC scheme outperforms the one of selection combining (SC) scheme. Besides, Fig. 2.23 indicates that the increment of $H$ shows a weak impact on SOP. The differences among the SOP for various $H$ from 250 to 350 km is very narrow (roughly on the orders of $10^{-6}$. Recalling that the expression for the received SNR at terrestrial receivers, adjusting the height of the satellite leads to an approximate identical influence to both legitimate and eavesdropping receivers.

### 2.3.4 Conclusion

In this letter, we investigated the secrecy outage performance in multi-antenna satellite-terrestrial transmissions by deriving the closed-form expressions for the exact and asymptotic SOP, while considering the randomness of the positions of

[2] In this section, we set the height and the coverage size of the footprint of $S$ by considering the case of LEO satellites.

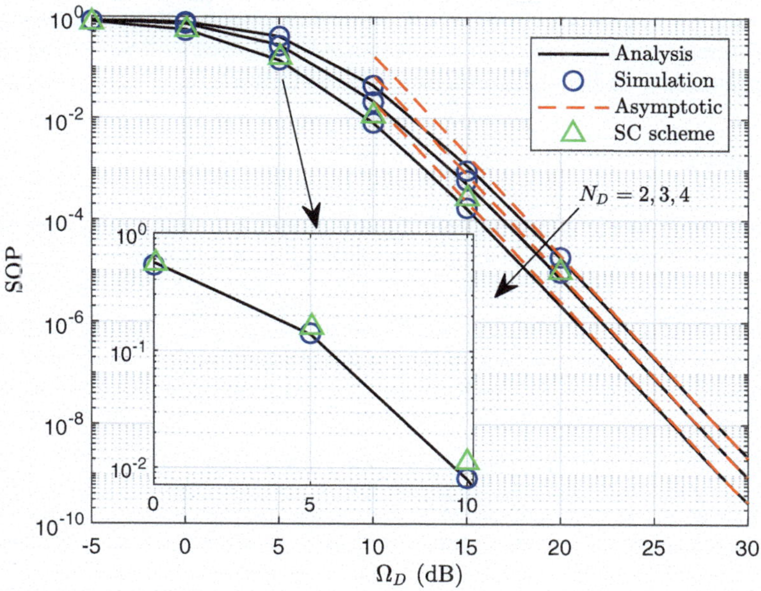


**Fig. 2.22** SOP versus $\Omega_D$ for various $N_D$

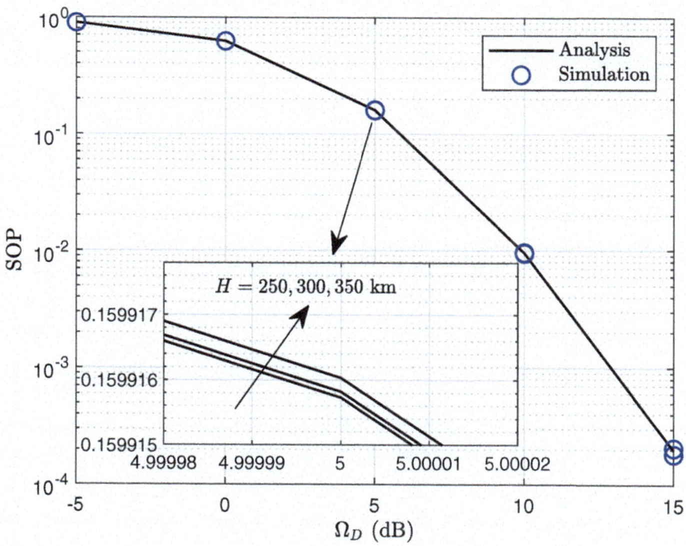


**Fig. 2.23** SOP versus $\Omega_D$ for various $H$

legitimate and eavesdropping receivers. Numerical results reveal that the height of the satellite exhibits a weak effect on SOP, while the number of antennas at the eavesdropper shows a positive impact on it. As expected, compared with SC scheme, MRC scheme can achieve a better performance in terms of SOP.

## References


1. J. Ye, C. Zhang, H. Lei, G. Pan, Z. Ding, "Secure UAV-to-UAV systems with spatially random UAVs," *IEEE Wireless Commun. Lett.*, vol. 8, no. 2, pp. 564–567, Apr. 2019.
2. G. Pan, H. Lei, J. An, et al., "On the secrecy of UAV systems with linear trajectory," *IEEE Trans. Wireless Commun.*, vol. 19, no. 10, pp. 6277–6288, Oct. 2020.
3. G. Pan, J. Ye and Z. Ding, "Secure hybrid VLC-RF systems with light energy harvesting," *IEEE Trans. Commun.*, vol. 65, no. 10, pp. 4348–4359, Oct. 2017.
4. G. Pan, J. Ye, Y. Zhang, et al., "Performance analysis and optimization of cooperative satellite-aerial-terrestrial systems," *IEEE Trans. Wireless Commun.*, vol. 19, no. 10, pp. 6693–6707, Oct. 2020.
5. P. Wu and N. Jindal, "Performance of hybrid-ARQ in block-fading channels: A fixed outage probability analysis," *IEEE Trans. Commun.*, vol. 58, no. 4, pp. 1129–1141, April 2010.
6. A. Abdi, W. Lau, M.-S. Alouini, and M. Kaveh, "A new simple model for land mobile satellite channels: First and second order statistics" *IEEE Trans. Wireless Commun.*, vol. 2, no. 3, pp. 519–528, May 2003.
7. I.S. Gradshteyn and I.M. Ryzhik, *Table of Integrals, Series and Products, 7 Ed.* San Diego: Academic Press, 2007.
8. H. Zhao, L. Yang, G. Pan, and M.-S. Alouini, "Secrecy outage analysis over fluctuating two-ray fading channels," *Electron. Lett.*, vol. 55, no. 15, pp. 866–868, Jul. 2019.
9. M. R. Bhatnagar and Arti M.K., "On the closed-form performance analysis of maximal ratio combining in shadowed-Rician fading LMS channels," *IEEE Commun. Lett.*, vol. 18, no. 1, pp. 54–57, Jan. 2014.
10. A. P. Prudnikov, Y. A. Brychkov, and O. I. Marichev, *Integrals and Series: More Special Functions, vol. 3, 3rd ed.*, New York: Gordon & Breach Sci. Publ., 1992.
11. V. S. Adamchik and O. I. Marichev, "The algorithm for calculating integrals of hypergeometric type functions and its realization in REDUCE system," in *Proceedings of the international symposium on Symbolic and algebraic computation (ISSAC' 90)*, Tokyo, Japan, Aug. 1990, pp. 212–224.
12. L. Ma, C. Hu, and J. Pei, "Polarization tracking study of Earth station in satellite communications," *Sci. Res.*, vol. 5, no. 1, pp. 10–19, 2016.
13. M. S. Grewal, L. R. Weill, and A. P. Andrews, *Global Positioning Systems, Inertial Navigation, and Integration*. Hoboken, NJ, USA: Wiley, 2007.
14. *Study on New Radio (NR) to Support Non-Terrestrial Networks*, Standard TR 38.811 (Release 15), 3GPP, 2020.
15. J. Lin, Z. Hou, Y. Zhou, L. Tian, and J. Shi, "Map estimation based on Doppler characterization in broadband and mobile LEO satellite communications," in *Proc. IEEE 83rd Veh. Technol. Conf. (VTC Spring)*, May 2016, pp. 1–5.
16. L. You, K.-X. Li, J. Wang, X. Gao, X.-G. Xia, and B. Ottersten, "Massive MIMO transmission for LEO satellite communications," *IEEE J. Sel. Areas Commun.*, vol. 38, no. 8, pp. 1851–1865, Aug. 2020.
17. S. K. S. Shin, K. L. Lim, K. C. Choi, and K. K. Kang, "Rain attenuation and Doppler shift compensation for satellite communications," *ETRI J.*, vol. 24, no. 1, pp. 31–42, Feb. 2002.
18. J. Li, Y. Zhang, Y. Zhang, W. Xiong, Y. Huang, and Z. Wang, "Fast tracking Doppler compensation for OFDM-based LEO satellite data transmission," in *Proc. 2nd IEEE Int. Conf. Comput. Commun. (ICCC)*, Oct. 2016, pp. 1814–1817.
19. U. Naeem, Z. Jawaid, and S. Sadruddin, "Doppler shift compensation techniques for LEO satellite on-board receivers," in *Proc. 9th Int. Bhurban Conf. Appl. Sci. Technol. (IBCAST)*, Islamabad, Pakistan, Jan. 2012, pp. 391–393.

20. M. Lin, Z. Lin, W.-P. Zhu, and J.-B. Wang, "oint beamforming for secure communication in cognitive satellite terrestrial networks," *IEEE J. Sel. Areas Commun.*, vol. 36, no. 5, pp. 1017–1029, May 2018.
21. Wolfram Research, MeijerG, Wolfram Functions Site. [Online]. Available: http://functions.wolfram.com/HypergeometricFunctions/MeijerG/07/01/.
22. M. K. Arti and V. Jain, "Relay selection-based hybrid satellite-terrestrial communication systems," *IET Commun.*, vol. 11, no. 17, pp. 2566–2574, Nov. 2017.
23. K. An, M. Lin, J. Ouyang, and W.P. Zhu, "Secure transmission in cognitive satellite terrestrial networks," *IEEE J. Sel. Areas Commun.*, vol. 34, no. 11, pp. 3025–3037, Nov. 2016.
24. A. Zappone, P. Cao, and E. A. Jorswieck, "Energy efficiency optimization in relay-assisted MIMO systems with perfect and statistical CSI," *IEEE Trans. Signal Process.*, vol. 62, no. 2, pp. 443–457, Jan. 2014.
25. M. Handley, "Using ground relays for low-latency wide-area routing in megaconstellations," in *Proc. 18th ACM Workshop Hot Topics Netw.*, Nov. 2019, pp. 125–132.
26. Y. Zhang, J. Ye, G. Pan, and M.-S. Alouini, "Secrecy outage analysis for satellite-terrestrial downlink transmissions," *IEEE Wireless Commun. Lett.*, vol. 9, no. 10, pp. 1643–1647, Oct. 2020.
27. G. Pan, J. Ye, and Z. Ding, "On Secure VLC Systems With Spatially Random Terminals," *IEEE Commun. Lett.*, vol. 21, no. 3, pp. 492–495, Mar. 2017.
28. Qingquan Huang, Min Lin, Wei-Ping Zhu, Symeon Chatzinotas, and Mohamed-Slim Alouini, "Performance Analysis of Integrated Satellite-Terrestrial Multiantenna Relay Networks with Multiuser Scheduling," *IEEE Trans. Aerosp. Electron. Syst.*, vol. 56, no. 4, pp. 2718–2731, Aug. 2020.
29. M. Abramowitz and I. A. Stegun, *Handbook of Mathematical Functions with Formulas, Graphs, and Mathematical Tables*. New York: Dover, 1972.
30. A. P. Prudnikov, Y. A. Brychkov, and O. I. Marichev, *Integrals and Series: Special Functions, 3rd ed.* New York: Gordon & Breach Sci. Publ., 1992.

# Chapter 3
# Satellite-Aerial Communications

**Haoxing Zhang, Shuai Wang, Gaofeng Pan, and Jianping An**

## 3.1 Secure Uplink Transmissions in Satellite-Aerial Systems

### *3.1.1 System Model*

In this paper, we consider a system, which consists of an aerial platform (S), a legitimate satellite (D), and an eavesdropping satellite (E), as shown in Fig. 3.1. Specifically, D receives data from S and E tries to eavesdrop the information transmitted between S and D simultaneously. As depicted in Fig. 3.1, the beam angle of S is denoted as $2\Psi$ ($0 < \Psi < \pi/2$).

Thus, for the S-D link, it is assumed that D is uniformly distributed on the surface of the spherical cone, which is determined by the sphere center, O, the radius $R_D$, and the apex angle $2\Theta$. From Fig. 3.1, one can observe that $\Theta$ is decided by $R_S$, $R_D$, and $\Psi$, which depends on the antenna at S. For the S-E link, we assume that E is uniformly distributed in the space, which is a part of the coverage space of S' beam, namely the spherical cone with radius $U_1$ minus the one with radius $U_2$ ($U_1 \geq U_2$) given Fig. 3.1. The two spherical cones are respectively with radius $U_1$ and $U_2$, as well as with sphere central, S, and the same apex angle, $2\Psi$.

Then, it is easy to have the received signal noise ratio (SNR) at D/E as

$$\gamma_i = \frac{P_S}{\sigma^2}\frac{|h_{Si}|^2}{d_i^{n_i}} = \frac{\lambda_i}{d_i^{n_i}}, \quad i \in \{D, E\}, \tag{3.1}$$

where $\lambda_i = \frac{P_S|h_{Si}|^2}{\sigma^2}$, $P_S$ and $\sigma^2$ are the transmit power at S and the average power of the additive white Gaussian noise, which is assumed to follow the same distribution

H. Zhang · S. Wang · G. Pan (✉) · J. An
Beijing Institute of Technology, Beijing, China
e-mail: gfpan@bit.edu.cn



S. Wang et al. (eds.), *Understanding Satellite Communications: The Stochastic Geometry Perspective*, https://doi.org/10.1007/978-981-97-7102-8_3

**Fig. 3.1** System model

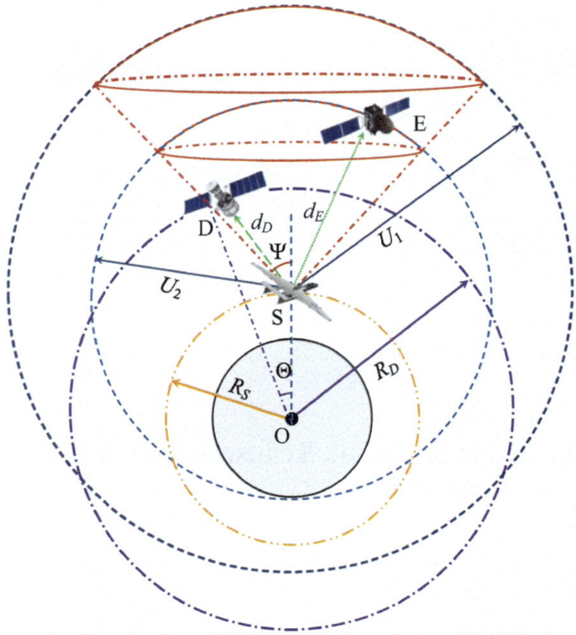


at both D and E, $d_i$ is the distance between S and D/E , $|h_{Si}|^2$ is the channel power gain, and $n_i$ is the path-loss factor over S-D/S-E link.

Moreover, though LoS propagation plays a leading role over air-to-air/space links, the path impairments in the aeronautical mobile-satellite communications include surface reflection (multi-path) effects [1]. Thus, in this work, we consider both S-D and S-E links are affected by independent and identically distributed Nakagami-$m$ distribution with cumulative distribution function (CDF) and probability density function (PDF) given by

$$F_{|h_{Si}|^2}(x) = \frac{1}{\Gamma(m_i)} \gamma\left(m_i, \frac{m_i x}{\Omega_i}\right) \tag{3.2}$$

and

$$f_{|h_{Si}|^2}(x) = \frac{x^{m_i - 1}}{\Gamma(m_i)} \left(\frac{m_i}{\Omega_i}\right)^{m_i} \exp\left(-\frac{m_i x}{\Omega_i}\right), \tag{3.3}$$

respectively, where $m_i$ is shape parameter, which is assumed to be an integer number, $\Omega_i$ represents the average channel power gains, $\gamma(a, x) = \int_0^x t^{a-1} e^{-t} dt$ is the lower incomplete Gamma function, and $\Gamma(x) = \int_0^{infty} t^{x-1} e^{-t} dt$ is the upper incomplete Gamma function.

Then, it is easy to obtain the received SNR at D/E as

$$\gamma_i = \frac{P_S}{\sigma^2}\frac{|h_{Si}|^2}{d_i^{n_i}} = \frac{\lambda_i}{d_i^{n_i}}, \tag{3.4}$$

where $\lambda_i = \frac{P_S|h_{Si}|^2}{\sigma^2}$, $P_S$ and $\sigma^2$ are the transmit power at S and the average power of the additive white Gaussian noise (AWGN), which is assumed to follow the same distribution at both D and E, $d_i$ (where $i \in \{D, E\}$) is the distance between S and i, and $n_i$ is the path-loss factor over S-i link.

By variable substitution, the CDF and PDF of $\lambda_i$ can be presented as

$$F_{\lambda_i}(x) = \frac{1}{\Gamma(m_i)}\gamma\left(m_i, \frac{m_i x}{\epsilon\Omega_i}\right) \tag{3.5}$$

and

$$f_{\lambda_i}(x) = \frac{x^{m_i-1}}{\Gamma(m_i)}\left(\frac{m_i}{\epsilon\Omega_i}\right)^{m_i}\exp\left(-\frac{m_i x}{\epsilon\Omega_i}\right), \tag{3.6}$$

respectively, where $\epsilon = \frac{P_S}{\sigma^2}$.

### 3.1.2 Secrecy Outage Analysis

In this section, we investigate the impact of the randomness of the satellites' position on the secrecy outage performance of the information transmissions over the up-link satellite-aerial link. To facilitate the following analysis, it is assumed that D and E operate in the circular orbit with the earth's center, O. Furthermore, the distance between S and D is denoted as $d_D$. Therefore, the instantaneous secrecy capacity of the S-D link can be given as

$$C(\gamma_D, \gamma_E) = \max\left\{\log_2(1+\gamma_D) - \log_2(1+\gamma_E), 0\right\}. \tag{3.7}$$

In this work, secrecy outage probability (SOP) is defined as the probability that the secrecy capacity is smaller than a threshold $C_{th}$, which can be written as

$$\begin{aligned} P_{\text{out}} &= \Pr\{C(\gamma_D, \gamma_E) \le C_{th}\} \\ &= \Pr\left\{\log_2(1+\gamma_D) - \log_2(1+\gamma_E) \le C_{th}\right\} \\ &= \Pr\{\gamma_D \le \Xi\gamma_E + \Xi - 1\} \\ &\ge \Pr\{\gamma_D \le \Xi\gamma_E\} \\ &= \Pr\left\{\frac{\lambda_D}{d_D^{n_D}} \le \Xi\frac{\lambda_E}{d_E^{n_E}}\right\}, \end{aligned} \tag{3.8}$$

where $\Xi = 2^{C_{th}}$.

Further, one can rewrite the lower bound of SOP as

$$P_{\text{out}} = \Pr\left\{\frac{\lambda_D}{\lambda_E} \le \Xi \frac{d_D^{n_D}}{d_E^{n_E}}\right\} = \Pr\{X \le \Xi Y\}$$
$$= \int_0^\infty F_X(\Xi y) f_Y(y)\, dy, \tag{3.9}$$

where $X = \frac{\lambda_D}{\lambda_E}$ and $Y = \frac{d_D^{n_D}}{d_E^{n_E}}$.

**Theorem 3.1** *The CDF of X can be derived as follows:*

$$F_X(x) = \frac{1}{B(m_D, m_E)} B\left(\frac{\alpha(x)}{\alpha(x)+\beta}; m_D, m_E\right). \tag{3.10}$$

*where* $\alpha(x) = \frac{m_D x}{\epsilon \Omega_D}$ *and* $\beta = \frac{m_E}{\epsilon \Omega_E}$.

***Proof*** We can obtain the CDF of X as

$$F_X(x) = \Pr\left\{\frac{\lambda_D}{\lambda_E} \le x\right\} = \Pr\{\lambda_D \le \lambda_E\} = \int_0^\infty F_{\lambda_D}(xy) f_{\lambda_E}(y)\, dy$$
$$= \frac{\left(\frac{m_E}{\epsilon\Omega_E}\right)^{m_E}}{\Gamma(m_D)\Gamma(m_E)} \int_0^\infty x^{m_E-1} \exp\left(-\frac{m_E y}{\epsilon\Omega_E}\right) s\gamma\left(m_D, \frac{m_D x y}{\epsilon\Omega_D}\right) dy. \tag{3.11}$$

Using the incomplete gamma function [2, Eq(6.455.2)], which is presented as $\int_0^\infty x^{\mu-1} e^{-\beta x} \gamma(\nu, \epsilon x)\, dx = \frac{\epsilon^\nu \Gamma(\mu+\nu)}{\nu(\epsilon+\beta)^{\mu+\nu}} {}_2F_1\left(1, \mu+\nu; \nu+1; \frac{\epsilon}{\epsilon+\beta}\right)$, where ${}_2F_1(a, b, c, d) = \sum_{n=0} \infty \frac{(a)_n (b)_n}{(c)_n} \frac{z^n}{n!}$ is defined as hypergeometric function, we can further rewritten the expression as

$$F_X(x) = \frac{\Gamma(m_D + m_E)}{\Gamma(m_D)\Gamma(m_E) m_D} \frac{\beta^{m_E} \alpha^{m_E}(x)}{[\alpha(x)+\beta]^{m_D+m_E}}$$
$$\times {}_2F_1\left[1, m_D + m_E; m_D + 1; \frac{\alpha(x)}{\alpha(x)+\beta}\right], \tag{3.12}$$

Utilizing the beta function [2, Eq(8.384.1)] as $B(x, y) = \frac{\Gamma(x)\Gamma(y)}{\Gamma(x+y)}$ and the incomplete beta function [2, Eq(8.391.1)] as $B(x; p, q) = \int_0^x t^{p-1}(1-t)^{q-1}\, dt = \frac{x^p}{p} {}_2F_1(p, 1-q; p+1; x)$, the CDF of X can be finally presented as (3.10).

Then the proof is completed. □

To characterize the statistical characteristics of $Y$, we first derive the CDF and PDF of $d_D$ and $d_E$ as follows.

**Theorem 3.2** *When the position of S is fixed and D is uniformly distributed on the surface of the spherical cone shown in Fig. 3.1, the CDF and PDF of $d_D$ can be derived as*

$$F_{d_D}(x) = \frac{1 - \frac{R_D^2 + R_S^2 - x^2}{2R_D R_S}}{1 - \cos\Theta}, \quad \Gamma_{D1} \le x \le \Gamma_{D2} \tag{3.13}$$

*and*

$$f_{d_D}(x) = \frac{x}{(1 - \cos\Theta) R_D R_S}, \tag{3.14}$$

*respectively, where* $\Gamma_{D1} = R_D - R_S$, $\Gamma_{D2} = \sqrt{R_D^2 + R_S^2 - 2R_D R_S \cos\Theta}$, $\Theta = \arccos\left\{\frac{1}{R_D}\left[R_S\left(1 - \cos^2\Psi\right), +\cos\Psi\sqrt{R_S^2\cos^2\Psi - R_S^2 + R_D^2}\right]\right\}$.

***Proof*** As shown in Fig. 3.1, the distance between S and D, $d_D$, can be given as $R_D^2 + R_S^2 - 2R_D R_S \cos\Theta = d_D^2$. Furthermore, we can also easily obtain $d_D^2 + R_S^2 - 2d_D R_S \cos(\pi - \Psi) = R_D^2$.

Thus, $\Theta$ can be presented as

$$\Theta = \arccos\left\{\frac{1}{R_D}\left[R_S\left(1 - \cos^2\Psi\right) + \cos\Psi\sqrt{R_S^2\cos^2\Psi - R_S^2 + R_D^2}\right]\right\}. \tag{3.15}$$

Furthermore, the CDF and PDF of $\theta$ can be easily given as

$$F_\theta(x) = \frac{1}{4\pi R_D^2}\int_0^{2x}\int_0^{2\pi} R_D^2 \sin\phi d\varphi d\phi = \frac{1 - \cos x}{1 - \cos\Theta} \tag{3.16}$$

and

$$f_\theta(x) = \begin{cases} \frac{\sin x}{1 - \cos\Theta}, & \text{if } 0 \le x \le \Theta \\ 0, & \text{else} \end{cases}, \tag{3.17}$$

respectively.

Thus, the CDF of $d_D$ can be given by

$$\begin{aligned} F_{d_D}(x) &= \Pr\{d_D \le x\} \\ &= \Pr\left\{\sqrt{R_D^2 + R_S^2 - 2R_D R_S \cos\theta} \le x\right\} \\ &= F_\theta\left(\arccos\frac{x^2 - R_D^2 - R_S^2}{2R_D R_S}\right). \end{aligned} \tag{3.18}$$

Using (3.16), one can obtain the CDF of $d_D$ as (3.18). Similar to the derivative of (3.18), one can also derive the PDF of $d_D$ as (3.14).

Then, the proof is completed. □

**Remark 3.1** $F_{d_D}$ is related to the distance between D and O, $R_D$, the distance between S and O, $R_S$, and the apex angle $\Theta$. Because we consider S operates in the circular orbit with the earth's center, thus, it can be seen as a random distribution.

**Lemma 3.1** *As E is uniformly distributed in the space shown in Fig. 3.1, the CDF and PDF of* $d_E$ *can be presented as*

$$F_{d_E}(x) = \frac{x^3 - U_2^3}{U_1^3 - U_2^3}, \quad U_2 \le x \le U_1 \tag{3.19}$$

*and*

$$f_{d_E}(x) = \begin{cases} \frac{3x^2}{U_1^3 - U_2^3}, & \text{if } U_2 \le x \le U_1; \\ 0, & \text{else} \end{cases}. \tag{3.20}$$

**Remark 3.2** $F_{d_E}$ is related to the radius of the spherical cone, $U_1$ and $U_2$. This is because in this work, we consider the eavesdropper is randomly distributed in a spherical cone.

**Corollary 3.1** *The CDF and PDF of Y can be obtained as*

$$F_Y(y) = \begin{cases} \frac{3[\Delta(U_1) - \Delta(\Gamma_{E1})]}{(U_1^3 - U_2^3)(1-\cos\Theta)}, \\ \qquad \text{if } \Gamma_{y1} \le y \le \min\{\Gamma_{y2}, \Gamma_{y3}\}; \\ \frac{3[\Delta(\Gamma_{E2}) - \Delta(\Gamma_{E1})]}{(U_1^3 - U_2^3)(1-\cos\Theta)} + \frac{U_1^3 - \Gamma_{E2}^3}{U_1^3 - U_2^3}, \\ \qquad \text{if } \Gamma_{y3} \le y \le \Gamma_{y2}; \\ \frac{3[\Delta(U_1) - \Delta(U_2)]}{(U_1^3 - U_2^3)(1-\cos\Theta)}, \\ \qquad \text{if } \Gamma_{y2} \le y \le \Gamma_{y3}; \\ \frac{3[\Delta(\Gamma_{E2}) - \Delta(U_2)]}{(U_1^3 - U_2^3)(1-\cos\Theta)} + \frac{U_1^3 - \Gamma_{E2}^3}{U_1^3 - U_2^3}, \\ \qquad \text{if } \max\{\Gamma_{y2}, \Gamma_{y3}\} \le y \le \Gamma_{y4} \end{cases} \tag{3.21}$$

*and*

$$f_Y(y) = \begin{cases} \frac{3\left(\frac{U_1^5}{10R_D R_S} + \frac{3}{2}Ay^{-\frac{5}{2}}\right)}{(U_1^3 - U_2^3)(1-\cos\Theta)}, \\ \qquad \text{if } \Gamma_{y1} \le y \le \min\{\Gamma_{y2}, \Gamma_{y3}\}; \\ \frac{3\left(-\frac{3}{2}By^{-\frac{5}{2}} + \frac{3}{2}Ay^{-\frac{5}{2}}\right)}{(U_1^3 - U_2^3)(1-\cos\Theta)} + \frac{3}{2(U_1^3 - U_2^3)}\Gamma_{D2}^3 y^{-\frac{5}{2}}, \\ \qquad \text{if } \Gamma_{y3} \le y \le \Gamma_{y2}; \\ \frac{3\left(\frac{U_1^5}{10R_D R_S} - \frac{U_2^5}{10R_D R_S}\right)}{(U_1^3 - U_2^3)(1-\cos\Theta)}, \\ \qquad \text{if } \Gamma_{y2} \le y \le \Gamma_{y3}; \\ \frac{3\left(-\frac{3}{2}By^{-\frac{5}{2}} - \frac{U_2^5}{10R_D R_S}\right)}{(U_1^3 - U_2^3)(1-\cos\Theta)} + \frac{3}{2(U_1^3 - U_2^3)}\Gamma_{D2}^3 y^{-\frac{5}{2}}, \\ \qquad \text{if } \max\{\Gamma_{y2}, \Gamma_{y3}\} \le y \le \Gamma_{y4} \end{cases}, \tag{3.22}$$

*where* $\Gamma_{y1} = \frac{(R_D-R_S)^2}{U_1^2}$, $\Gamma_{y2} = \frac{(R_D-R_S)^2}{U_2^2}$, $\Gamma_{y3} = \frac{(R_D^2+R_S^2-2R_DR_S\cos\Theta)}{U_1^2}$, $\Gamma_{y4} = \frac{(R_D^2+R_S^2-2R_DR_S\cos\Theta)}{U_2^2}$, $\Gamma_{E1} = \frac{R_D-R_S}{\sqrt{y}}$, $\Gamma_{E2} = \sqrt{\frac{R_D^2+R_S^2-2R_DR_S\cos\Theta}{y}}$, *and* $\Delta(x) = \frac{1}{3}x^3 + \frac{3yx^5-5(R_D^2+R_S^2)x^3}{30R_DR_S}$, $A = \frac{1}{3}(R_D-R_S)^3 + \frac{3(R_D-R_S)^5-5(R_D^2+R_S^2)(R_D-R_S)^3}{30R_DR_S}$, $B = \frac{1}{3}(R_D^2-R_S^2-2R_DR_S\cos\Theta)^{\frac{3}{2}} + \frac{3(R_D^2+R_S^2-2R_DR_S\cos\Theta)^{\frac{5}{2}}-5(R_D^2+R_S^2)(R_D^2+R_S^2-2R_DR_S\cos\Theta)^{\frac{3}{2}}}{30R_DR_S}$.

***Proof*** Using Eqs. (3.13) and (3.20), the CDF of $Y$ can be expressed as

$$F_Y(y) = \Pr\left\{\frac{d_D^{n_D}}{d_E^{n_E}} \le y\right\} = \Pr\left\{d_D \le \sqrt[n_D]{y d_E^{n_E}}\right\}$$
$$= \int_0^\infty F_{d_D}\left(\sqrt[n_D]{y d_E^{n_E}}\right) f_{d_E}(d_E)\, d(d_E). \tag{3.23}$$

In open space, we can have $n_D = n_E = 2$. Then, $F_Y(y)$ can be obtained as $F_Y(y) = \int_0^\infty F_{d_D}\left(\sqrt{y d_E^2}\right) f_{d_E}(d_E)\, d(d_E)$.

Then, it can be derived in the following cases:

(1) When $R_D - R_S \le \sqrt{yU_2^2} \le \sqrt{R_D^2 + R_S^2 - 2R_DR_S\cos\Theta} \le \sqrt{yU_1^2}$, the range of $y$ can be calculated as $\max\{\Gamma_{y2}, \Gamma_{y3}\} \le y \le \Gamma_{y4}$. Then, the CDF of $Y$ can be further obtained as $F_Y(y) = \frac{3}{(U_1^3-U_2^3)(1-\cos\Theta)}[\Delta(\Gamma_{E2}) - \Delta(U_2)] + \frac{U_1^3-\Gamma_{E2}^3}{U_1^3-U_2^3}$.

(2) When $R_D - R_S \le \sqrt{yU_2^2} \le \sqrt{yU_1^2} \le \sqrt{R_D^2 + R_S^2 - 2R_DR_S\cos\Theta}$, the range of $y$ can be derived as $\Gamma_{y2} \le y\Gamma_{y3}$. Plugging into (3.23), we can be further have $F_Y(y) = \frac{3}{(U_1^3-U_2^3)(1-\cos\Theta)}[\Delta(U_1) - \Delta(U_2)]$.

(3) When $\sqrt{yU_2^2} \le R_D - R_S \le \sqrt{R_D^2 + R_S^2 - 2R_DR_S\cos\Theta} \le \sqrt{yU_1^2}$, the range of $y$ can be given as $\Gamma_{y3} \le y \le \Gamma_{y2}$. Thus, the CDF of $Y$ can be further written as $F_Y(y) = \frac{3[\Delta(\Gamma_{E2})-\Delta(\Gamma_{E1})]}{(U_1^3-U_2^3)(1-\cos\Theta)} + \frac{U_1^3-\Gamma_{E2}^3}{U_1^3-U_2^3}$.

(4) When $\sqrt{yU_2^2} \le R_D - R_S \le \sqrt{yU_1^2} \le \sqrt{R_D^2 + R_S^2 - 2R_DR_S\cos\Theta}$, the range of $y$ can be calculated as $\Gamma_{y1} \le y \le \min\{\Gamma_{y2}, \Gamma_{y3}\}$. Similarly, it gives $F_Y(y) = \frac{3}{(U_1^3-U_2^3)(1-\cos\Theta)}[\Delta(U_1) - \Delta(\Gamma_{E1})]$.

Combining the four cases completes the proof. □

Finally, SOP can be derived by using Eqs. (3.10) and (3.21) into (3.9).

When $\Gamma_{y3} \le \Gamma_{y2}$, the SOP can be expressed as

$$P_{\text{out}} = \int_0^\infty F_X(\Xi y) f_Y(y)\, dy$$
$$= \frac{1}{B(m_D, m_E)} \frac{3}{U_1^3 - U_2^3} \left\{ \int_{\Gamma_{y1}}^{\Gamma_{y3}} \frac{1}{1-\cos\Theta} \right.$$
$$\times B\left(\frac{\alpha(\Xi y)}{\alpha(\Xi y)+\beta}; m_D, m_E\right)\left(\frac{U_1^5}{10R_DR_S} + \frac{3}{2}Ay^{-\frac{5}{2}}\right) dy$$

$$+\int_{\Gamma_{y3}}^{\Gamma_{y2}} B\left(\frac{\alpha(\Xi y)}{\alpha(\Xi y)+\beta}; m_D, m_E\right)\left[\frac{1}{1-\cos\Theta}\right.$$
$$\left.\times\left(-\frac{3}{2}By^{-\frac{5}{2}}+\frac{3}{2}Ay^{-\frac{5}{2}}\right)+\frac{1}{2}\Gamma_{D2}^{3}y^{-\frac{5}{2}}\right]dy$$
$$+\int_{\Gamma_{y2}}^{\Gamma_{y4}} B\left(\frac{\alpha(\Xi y)}{\alpha(\Xi y)+\beta}; m_D, m_E\right)\left[\frac{1}{1-\cos\Theta}\right.$$
$$\left.\left.\times\left(-\frac{3}{2}By^{-\frac{5}{2}}-\frac{U_2^5}{10R_DR_S}\right)+\frac{1}{2}\Gamma_{D2}^{3}y^{-\frac{5}{2}}\right]dy\right\}$$
$$=\frac{1}{B(m_D,m_E)}\frac{3}{U_1^3-U_2^3}\left[\delta_1+\delta_2+\delta_3\right]. \tag{3.24}$$

Due to the complexity of the integral, we use Chebyshev-Gauss quadrate to facilitate further analysis with an adequate accuracy, which is presented as $\int_{-1}^{1}\frac{f(x)}{\sqrt{1-x^2}}dx \approx \sum_{i=1}^{T}\omega_i f(x_i)$ with $x_i=\cos\left(\frac{2i-1}{2T}\pi\right)$ and the weight $\omega_i=\frac{\pi}{T}$. Thus, $P_{\text{out}}$ can be rewritten as

$$P_{\text{out}}=\frac{1}{B(m_D,m_E)}\frac{3}{U_1^3-U_2^3}\left[\delta_1+\delta_2+\delta_3\right]. \tag{3.25}$$

where $\delta_1=\frac{\pi}{Q}\frac{1}{1-\cos\Theta}\sum_{i=1}^{Q}\sqrt{1-{\chi_i}^2}B\left(\frac{\alpha(\Xi y1)}{\alpha(\Xi y1)+\beta}; m_D, m_E\right)\times\left(\frac{U_1^5}{10R_DR_S}+\frac{3}{2}Ay_1^{-\frac{5}{2}}\right)$, with $\chi_i=\cos\left(\frac{2i-1}{2Q}\pi\right)$, $y_1=\frac{\Gamma_{y3}-\Gamma_{y1}}{2}\chi_i+\frac{\Gamma_{y3}+\Gamma_{y1}}{2}$, $\delta_2=\frac{\pi}{Q}\frac{1}{1-\cos\Theta}\times\sum_{i=1}^{Q}B\left(\frac{\alpha(\Xi y_2)}{\alpha(\Xi y_2)+\beta}; m_D, m_E\right)\times\left[\frac{1}{1-\cos\Theta}\left(-\frac{3}{2}By_2^{-\frac{5}{2}}+\frac{3}{2}Ay_2^{-\frac{5}{2}}\right)+\frac{1}{2}\Gamma_{D2}^{3}y^{-\frac{5}{2}}\right]$ with $y_2=\frac{\Gamma_{y2}-\Gamma_{y3}}{2}\chi_i+\frac{\Gamma_{y2}+\Gamma_{y3}}{2}$, and $\delta_3=B\left(\frac{\alpha(\Xi y_3)}{\alpha(\Xi y_3)+\beta}; m_D, m_E\right)\left[\frac{1}{1-\cos\Theta}\left(-\frac{3}{2}By_3^{-\frac{5}{2}}-\frac{U_2^5}{10R_DR_S}\right)+\frac{1}{2}\Gamma_{D2}^{3}y_3^{-\frac{5}{2}}\right]$, with $y_3=\frac{\Gamma_{y4}-\Gamma_{y2}}{2}\chi_i+\frac{\Gamma_{y4}+\Gamma_{y2}}{2}$.

When $\Gamma_{y2}\le\Gamma_{y3}$, the SOP can be expressed as

$$P_{\text{out}}=\int_0^{\infty}F_X(\Xi y)f_Y(y)\,dy$$
$$=\frac{1}{B(m_D,m_E)}\frac{3}{U_1^3-U_2^3}\left\{\int_{\Gamma_{y1}}^{\Gamma_{y2}}\frac{1}{1-\cos\Theta}\right.$$
$$\times B\left(\frac{\alpha(\Xi y)}{\alpha(\Xi y)+\beta}; m_D, m_E\right)\left(\frac{U_1^5}{10R_DR_S}+\frac{3}{2}Ay^{-\frac{5}{2}}\right)dy$$
$$+\int_{\Gamma_{y2}}^{\Gamma_{y3}}\frac{1}{1-\cos\Theta}B\left(\frac{\alpha(\Xi y)}{\alpha(\Xi y)+\beta}; m_D, m_E\right)$$

$$
\begin{aligned}
&\times \left( \frac{U_1^5}{10 R_D R_S} - \frac{U_2^5}{10 R_D R_S} \right) dy \\
&+ \int_{\Gamma_{y3}}^{\Gamma_{y4}} B\left( \frac{\alpha(\Xi y)}{\alpha(\Xi y)+\beta}; m_D, m_E \right) \left[ \frac{1}{1-\cos\Theta} \right. \\
&\left. \times \left( -\frac{3}{2} B y^{-\frac{5}{2}} - \frac{U_2^5}{10 R_D R_S} \right) + \frac{1}{2} \Gamma_{D2}^3 y^{-\frac{5}{2}} \right] dy \Bigg\} \\
&= \frac{1}{B(m_D, m_E)} \frac{3}{U_1^3 - U_2^3} \left[ \delta_4 + \delta_5 + \delta_6 \right].
\end{aligned}
\tag{3.26}
$$

where $\delta_4 = \frac{\pi}{Q} \frac{1}{1-\cos\Theta} \sum_{i=1}^{Q} \sqrt{1-\chi_i{}^2} B\left( \frac{\alpha(\Xi y_4)}{\alpha(\Xi y_4)+\beta}; m_D, m_E \right) \times \left( \frac{U_1^5}{10 R_D R_S} + \frac{3}{2} A y_4^{-\frac{5}{2}} \right)$, with $y_4 = \frac{\Gamma_{y2}-\Gamma_{y1}}{2} \chi_i + \frac{\Gamma_{y2}+\Gamma_{y1}}{2}$, $\delta_5 = \frac{\pi}{Q} \frac{1}{1-\cos\Theta} \times \sum_{i=1}^{Q} \sqrt{1-\chi_i{}^2} B\left( \frac{\alpha(\Xi y_5)}{\alpha(\Xi y_5)+\beta}; m_D, m_E \right)$ $\left( \frac{U_1^5}{10 R_D R_S} - \frac{U_2^5}{10 R_D R_S} \right)$, $\delta_6 = B\left( \frac{\alpha(\Xi y_6)}{\alpha(\Xi y_6)+\beta}; m_D, m_E \right) \left[ \frac{1}{1-\cos\Theta} \left( -\frac{3}{2} B y_6^{-\frac{5}{2}} - \frac{U_2^5}{10 R_D R_S} \right) + \frac{1}{2} \Gamma_{D2}^3 y_6^{-\frac{5}{2}} \right]$, with $y_6 = \frac{\Gamma_{y4}-\Gamma_{y3}}{2} \chi_i + \frac{\Gamma_{y4}+\Gamma_{y3}}{2}$.

Moreover, as shown in Fig. 3.1, $R_D - R_S \le d_D \le \sqrt{R_D^2 + R_S^2 - 2 R_D R_S \cos\Theta}$, $U_2 \le d_E \le U_1$. Thus, the worst case of the eavesdropping can be obtained when $d_D = R_D - R_S$, $d_E = U_1$. In this case, the SOP can be derived as

$$
P_{\text{out}} = \Pr\left\{ \frac{\lambda_D}{\lambda_E} \le \Xi \frac{d_D^{n_D}}{d_E^{n_E}} \right\} = \Pr\{X \le \Xi Y\} = \int_0^\infty F_X(\Xi Y)\, dx, \tag{3.27}
$$

where $X = \frac{\lambda_D}{\lambda_E}$ and $Y = \frac{d_D^{n_D}}{d_E^{n_E}}$.

Besides, SOP is maximized when $d_D = \sqrt{R_D^2 + R_S^2 - 2 R_D R_S \cos\Theta}$, $d_E = U_2$.

In this work, the outage performance of the considered system with a single eavesdropper was studied. However, the proposed analysis method and the adopted technical road-map can serve as a meaningful reference to investigate the outage probability (OP) in multi-eavesdroppers scenarios, e.g., the PDF/CDF of the output SNR in multi-eavesdroppers scenarios with $N$ eavesdroppers can be derived as $\gamma_{E,\text{non-coop}} = \max\{\gamma_1, \gamma_2, \ldots, \gamma_N\}$ while these eavesdroppers work in non-cooperative mode, and $\gamma_{E,\text{coop}} = \gamma_1 + \gamma_2 + \cdots + \gamma_N$ while these eavesdroppers work in cooperative mode.

### 3.1.3 Numerical Results

In this subsection, we will provide numerical results to verify the proposed analytical models. Following the parameters adopted in [3–5], in this section, the main parameters are set as $U_1 = 2000$ km, $U_2 = 200$ km, $R_D = 1000$ km, $R_S = 6381$ km,

$\chi = \frac{\pi}{3}$ rad, $m_D = m_E = m_0 = 2, \Omega_D = \Omega_E = \Omega_0 = 0$ dB, $P_S = 30$ dBW, $C_{th} = 1$, and $n_D = n_E = 2$.

The influence of the transmit power at S, $P_S$, on the SOP of the considered system is investigated in Fig. 3.2. It can be easily observed that $P_S$ exhibits a weak impact on the SOP, since changing $P_S$ shows the same trend and scale on the received SNR at the D and E. Furthermore, Fig. 3.2 shows the impact of the distance between O and D, $R_D$, on the SOP of the system. As expected, $R_D$ shows a negative effect. This is because with the increase of $R_D$, the distance between S and D, $d_D$, increases, and therefore results in the degraded received signal power loss at D. Moreover, it can be seen that $d_D = R_D - R_S, d_E = U_1$ is the worst case of the eavesdropping and $d_D = \sqrt{R_D^2 + R_S^2 - 2R_D R_S \cos\Theta}, d_E = U_2$ are the optimal position of the destination and the eavesdropper, which further minimize or maximize the SOP.

In Fig. 3.3, the SOP of the considered system is studied for various Nakagami parameter $m$. We can observe that $m$ shows a positive impact on the secrecy outage performance. This is because when $m$ increases, the fading is less serious in the environment, and thus resulting in a lower SOP.

Figure 3.4 presents the SOP for various $\Psi$, while $R_D$ increasing. Obviously, $\Psi$ shows a positive impact on the secrecy outage performance of the considered system. Since $\Psi$ both affects $d_D$ and $d_E$, when $\Psi$ increases, the probability that $d_E$ is smaller than $d_D$ increases, which leads to the upgraded SOP.

The impact of the average channel power gain at the eavesdropper, $\Omega_E$, on the SOP was studied and shown in Fig. 3.5. We can see that SOP decreases as $\Omega_E$ increases,

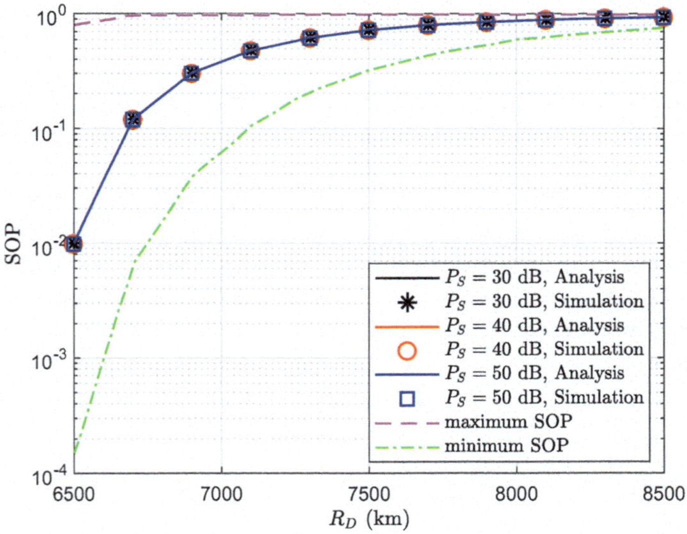


**Fig. 3.2** SOP for various $P_S$

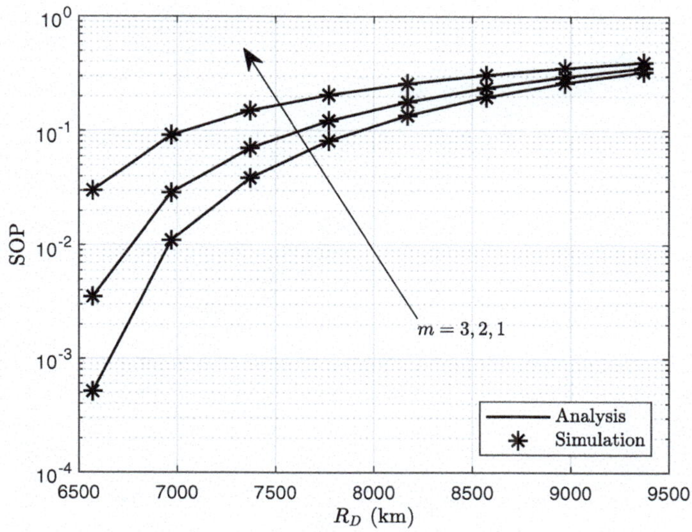


**Fig. 3.3** SOP for various $m$

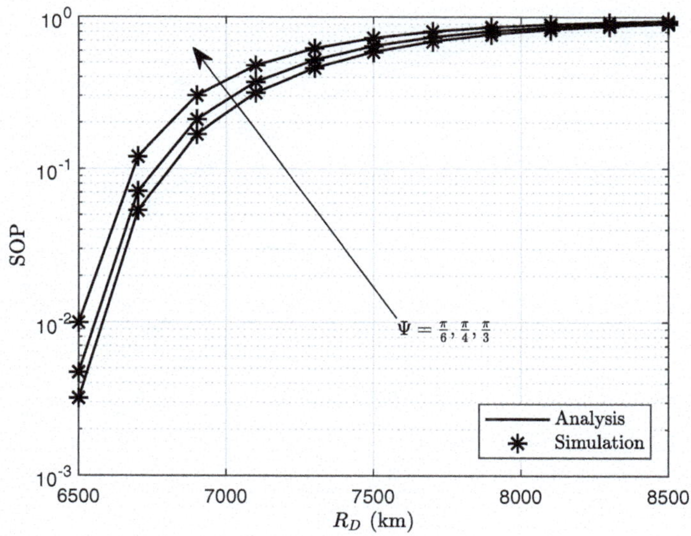


**Fig. 3.4** SOP for various $\Psi$

which shows $\Omega_E$ has a negative effect on the secrecy outage performance. The reason is that a larger $\Omega_E$ gives a better channel condition for the eavesdropping link between S and E.

In Figs. 3.6 and 3.7, numerical results are presented to address the influence of the maximum and minimum distances between S and E, respectively. It can be seen

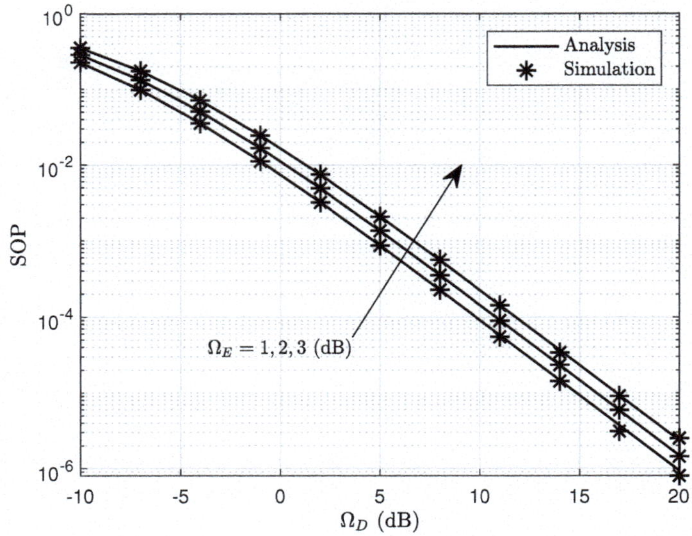


**Fig. 3.5** SOP for various $\Omega_E$

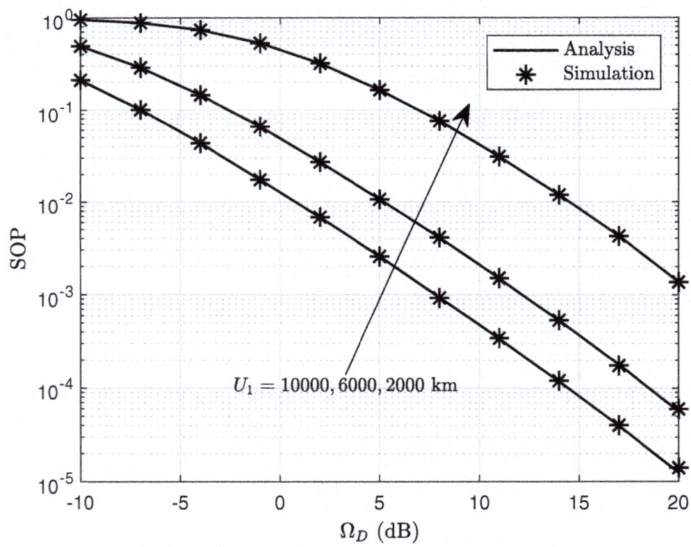


**Fig. 3.6** SOP for various $U_1$

that $U_1$ and $U_2$ exhibit a positive influence on SOP, which comes from the fact that a large $U_1/U_2$ leads to the farther distance between S and E, $d_E$, which results in the received signal power loss at E.

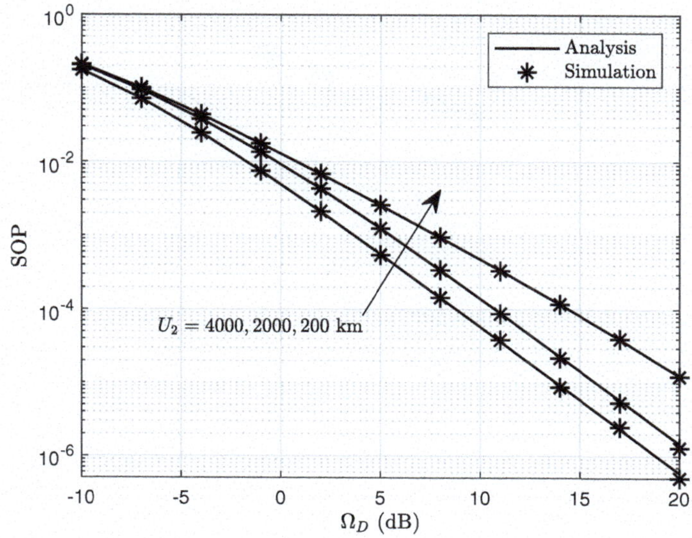


**Fig. 3.7** SOP for various $U_1$

## *3.1.4 Conclusion*

In this work, the secrecy outage performance of the considered system has been investigated by deriving the closed-form analytical expressions for the lower boundary of the SOP. We have considered the randomness of the positions of the legitimate satellite and the eavesdropping satellite to make our system more realistic.

(1) The maximum and minimum distances between S and E, $U_1$ and $U_2$, show a positive influence on the SOP of the considered system. Nevertheless, the distances between S and D, $R_D$ exhibit a negative effect on the SOP.

(2) The maximum SOP and the minimum SOP are determined by $U_1$, $U_2$, and $R_D$.

(3) The transmitting SNR at S provides a weak impact on the SOP over the uplink of the considered system.

(4) The apex angle $\Psi$ exhibits a positive effect on outage performance.

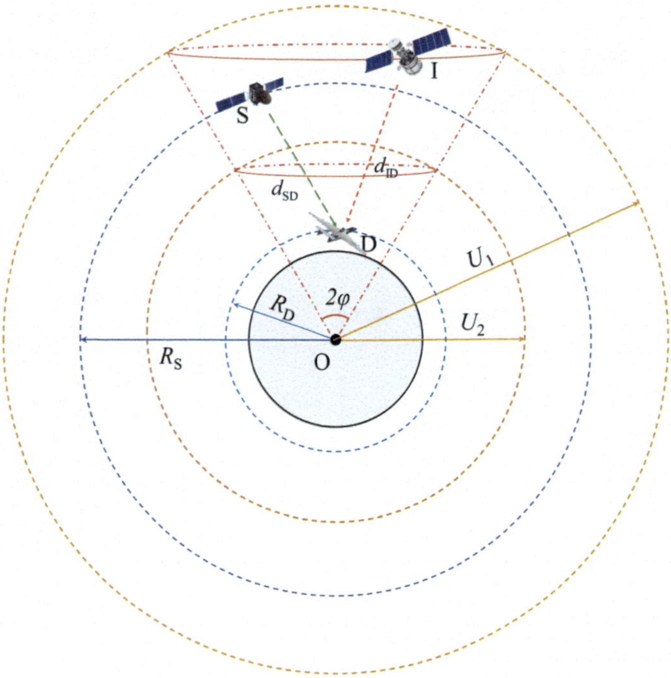


**Fig. 3.8** SAC system

## 3.2 Satellite-Aerial Downlink Deliveries in Presence of Spatially Random Space Interference

### *3.2.1 System Model*

In this paper, we consider a satellite-aerial communication (SAC) system, which consists of a satellite (S), an aerial platform (D), and an interference satellite (I), as shown in Fig. 3.8. Specifically, S operates in a circular orbit and transmits its data to D, which is decided by the Earth center, O, the distance between S and O, $R_S$, and the apex angle, $2\Psi$. Simultaneously, I generates jamming signals to interfere the information delivery between S and D. To reflect the practical scenarios, it is assume that I is uniformly distributed in the neighboring space of S, which is a part that the spherical cone with radius $U_1$ minus the one with radius $U_2$ ($U_1 \geq U_2$). The two spherical cones are respectively with radius $U_1$ and $U_2$, as well as with sphere central, O, and the same apex angle, $2\Psi$.[1]

[1] Here, we adopt a-single-interfering scenario to offer a more concise derivation process and further to facilitate the readers to understand our proposed analysis method.

According to the fading types experienced by the transmitted signal over the channel, in this work we consider both S-D and I-D links are affected by independent and identically distributed Nakagami-$m$ distribution as it can capture the fading characteristics of the line-of-sight propagation. The Nakagami-$m$ distribution can evolve into other distributions when the parameter $m$ takes some special values, for instance, when $m = 1$, it resorts to the Rayleigh distributions. Further, the signal fading condition can be spanned from severe to moderate by adjusting $m$. Moreover, the CDF and PDF of Nakagami-$m$ distribution can be given by

$$F_{|h_{iD}|^2}(x) = \frac{1}{\Gamma(m_i)} \gamma\left(m_i, \frac{m_i x}{\Omega_i}\right) \tag{3.28}$$

and

$$f_{|h_{iD}|^2}(x) = \frac{x^{m_i - 1}}{\Gamma(m_i)} \left(\frac{m_i}{\Omega_i}\right)^{m_i} \exp\left(-\frac{m_i x}{\Omega_i}\right), \tag{3.29}$$

respectively, where $i \in \{I, S\}$, $m_i$ is shape parameter, which is assumed to be an integer number, $\Omega_i$ represents the average channel power gains, $\gamma(a, x) = \int_0^x t^{a-1} e^{-t} dt$ is the lower incomplete Gamma function, and $\Gamma(x) = \int_0^\infty t^{x-1} e^{-t} dt$ is the upper incomplete Gamma function.

Thus, under this case, the received signal at D can be given as

$$y = \sqrt{\frac{P_S}{d_{SD}^{n_1}}} h_{SD} x_S + \sqrt{\frac{P_I}{d_{ID}^{n_2}}} h_{ID} x_I + n_0, \tag{3.30}$$

where $P_S$ and $P_I$ are the transmit power at S and I, respectively, $x_S$ is the transmitted signal from S, $x_I$ is the interfering signal from I, $d_{SD}$ is the distance between S and D, $d_{ID}$ is the distance between I and D, $n_1$ and $n_2$ are the path loss exponents for the channels over S-D link and I-D link, respectively, $n_0$ is the complex Gaussian noise at D with average power, $\sigma^2$. As the interfering power is normally much larger than that of the Gaussian noise, in the following we adopt signal-to-interference ratio (SIR) to address and study the effect of interfering signal on the outage performance of the considered SAC system.

So, the received SIR at D can be derived as

$$\gamma_{SD} = \frac{\frac{P_S |h_{SD}|^2}{d_{SD}^{n_1}}}{\frac{P_I |h_{ID}|^2}{d_{ID}^{n_2}}} = \frac{P_R |h_{SD}|^2 \, d_{ID}^{n_2}}{P_I |h_{ID}|^2 \, d_{SD}^{n_1}}, \tag{3.31}$$

where the path loss factor $n_1 = n_2 = 2$ for the open space considered in this work.

### 3.2.2 Outage Analysis

In this section, we will investigate the impacts of the randomness of the satellites' positions on the outage performance of the considered SAC system in the presence of space interference. It is assumed that S operates in the circular orbit with the earth's center, O.

In the following, we define the OP as $P_{\text{out}}$. Then, OP can be written as

$$\begin{aligned} P_{\text{out}} &= \Pr\left\{\gamma_{SD} \leq \gamma_{\text{out}}\right\} \\ &= \Pr\left\{\frac{P_R|h_{SD}|^2}{P_I|h_{ID}|^2}\frac{d_{ID}^2}{d_{SD}^2} \leq \gamma_{\text{out}}\right\} \\ &= \Pr\left\{Y \leq \epsilon X \gamma_{\text{out}}\right\}, \end{aligned} \tag{3.32}$$

where $X = \frac{|h_{ID}|^2}{|h_{SD}|^2}$, $Y = \frac{d_{ID}^2}{d_{SD}^2}$, and $\epsilon = \frac{P_I}{P_R}$.

Observing (3.32), in the following the statistical characteristics of $X$ and $Y$ will be respectively derived, to derive the expression of OP.

#### 3.2.2.1 The CDF and PDF of $X$

**Theorem 3.3** *The CDF and PDF of $X$ can be derived as*

$$F_X(x) = \frac{1}{B(m_S, m_I)} B\left(\frac{\alpha(x)}{\alpha(x)+\beta}; m_S, m_I\right) \tag{3.33}$$

*and*

$$f_X(x) = \frac{1}{B(m_S, m_I)} \frac{m_S m_I \Omega_S \Omega_I}{(\Omega_I m_S + \Omega_S m_I x)^2} \left[\frac{\alpha(x)}{\alpha(x)+\beta}\right]^{m_I-1} \left[1 - \frac{\alpha(x)}{\alpha(x)+\beta}\right]^{m_S-1}, \tag{3.34}$$

*respectively, where $B(x, y) = \frac{\Gamma(x)\Gamma(y)}{\Gamma(x+y)}$ is the beta function, $B(x; p, q) = \int_0^x t^{p-1}(1-t)^{q-1} dt$ is the incomplete beta function [2, Eq(8.391.1)], $\alpha(x) = \frac{m_I x}{\Omega_I}$, and $\beta = \frac{m_S}{\Omega_S}$.*

***Proof*** We can obtain the CDF of $X$ as

$$\begin{aligned} F_X(x) &= \Pr\left\{\frac{|h_{ID}|^2}{|h_{SD}|^2} \leq x\right\} \\ &= \Pr\left\{|h_{ID}|^2 \leq x|h_{SD}|^2\right\} \\ &= \int_0^\infty F_{|h_{ID}|^2}(xy) f_{|h_{SD}|^2}(y)\, dy \end{aligned}$$

$$= \frac{\left(\frac{m_S}{\Omega_S}\right)^{m_S}}{\Gamma(m_I)\Gamma(m_S)} \int_0^\infty x^{m_S-1} \exp\left(-\frac{m_S y}{\Omega_S}\right) \gamma\left(m_I, \frac{m_I x y}{\Omega_I}\right) dy. \quad (3.35)$$

Using the incomplete gamma function [2, Eq(6.455.2)], which is presented as $\int_0^\infty x^{\mu-1} e^{-\beta x} \gamma(\xi, \epsilon x)\, dx = \frac{\epsilon^\xi \Gamma(\mu+\xi)}{\xi(\epsilon+\beta)^{\mu+\xi}} {}_2F_1\left(1, \mu+\xi; \xi+1; \frac{\epsilon}{\epsilon+\beta}\right)$, where ${}_2F_1(x)$ is Gauss hypergeometric function, we can further rewritten (3.35) as

$$F_X(x) = \frac{\Gamma(m_S+m_I)}{\Gamma(m_S)\Gamma(m_I) m_I} \frac{\beta^{m_S} \alpha^{m_S}(x)}{[\alpha(x)+\beta]^{m_S+m_I}} {}_2F_1\left[1, m_S+m_I; m_I+1; \frac{\alpha(x)}{\alpha(x)+\beta}\right], \quad (3.36)$$

where $\alpha(x) = \frac{m_I x}{\Omega_I}$, $\beta = \frac{m_S}{\Omega_S}$.

Utilizing the beta function [2, Eq(8.384.1)] as $B(x, y) = \frac{\Gamma(x)\Gamma(y)}{\Gamma(x+y)}$ and the incomplete beta function [2, Eq(8.391.1)] as $B(x; p, q) = \int_0^x t^{p-1}(1-t)^{q-1} dt = \frac{x^p}{p} {}_2F_1(p, 1-q; p+1; x)$, the CDF of $X$ can be finally presented as (3.33).

Then the proof is completed. □

#### 3.2.2.2 The CDF and PDF of *Y*

We used stochastic geometry-based analysis method to calculate the CDF of $Y$ [6–8]. The CDF of $d_{SD}$ is first derived as follows.

In this work, it is assumed that S is uniformly distributed in a circular arc, which is decided by the Earth center, O, the distance between S and O, $R_S$, and the apex angle, $2\Psi$. Therefore, the distance between S and D, $d_{SD}$, can be written as

$$d_{SD} = \sqrt{R_D^2 + R_S^2 - 2R_D R_S \cos\varphi_S}, \quad (3.37)$$

where $\varphi_S$ is the polar angle of S.

**Theorem 3.4** *When the position of D is fixed and S is uniformly distributed in the circular arc shown in Fig. 3.8, the CDF and PDF of $d_{SD}$ can derived as*

$$F_{d_{SD}}(x) = \frac{1}{\Psi} \arccos \frac{R_D^2 + R_S^2 - x^2}{2R_D R_S} \quad (3.38)$$

*and*

$$f_{d_{SD}}(x) = \frac{1}{\Psi} \frac{2x}{\sqrt{(2R_D R_S)^2 - \left(R_D^2 + R_S^2 - x^2\right)^2}}, \quad (3.39)$$

*respectively.*

***Proof*** The CDF and PDF of $\varphi_S$ can be easily given as

$$F_{\varphi_S}(x) = \frac{x}{\Psi}, \quad 0 \le x \le \Psi \tag{3.40}$$

and

$$f_{\varphi_S}(x) = \begin{cases} \frac{1}{\Psi}, \text{ if } 0 \le x \le \Psi; \\ 0, \text{ else} \end{cases}, \tag{3.41}$$

respectively.

Then, one can write the CDF of $d_{SD}$ as

$$\begin{aligned} F_{d_{SD}}(x) &= \Pr\{d_{SD} \le x\} \\ &= \Pr\left\{\sqrt{R_D^2 + R_S^2 - 2R_D R_S \cos\varphi_S} \le x\right\} \\ &= \Pr\left\{\cos\varphi_S \ge \frac{x^2 - R_D^2 - R_S^2}{2R_D R_S}\right\} \\ &= \Pr\left\{\varphi_S \le \arccos\frac{x^2 - R_D^2 - R_S^2}{2R_D R_S}\right\} \\ &= F_{\varphi_S}\left(\arccos\frac{x^2 - R_D^2 - R_S^2}{2R_D R_S}\right). \end{aligned} \tag{3.42}$$

Using (3.40), one obtains the CDF of $d_{SD}$ as (3.38).

Finally, taking the derivative of (3.38), one can obtain the PDF of $d_{SD}$ as (3.39), which completes the proof. □

**Remark 3.3** $f_{d_{SD}^2}$ is related to the distance between D and O, $R_D$, the distance between S and O, $R_S$, and the apex angle $\Psi$. Because we consider S operates in the circular orbit with the earth's center, thus, it can be seen as a random distribution.

Next, we derive the PDF of $d_{ID}$ as follows.

To reflect the practical scenarios, it is assumed that I is uniformly distributed in the neighboring space of S, which is a part that the spherical cone with radius $U_1$ minus the one with radius $U_2$ ($U_1 \ge U_2$). The two spherical cones are respectively with radius $U_1$ and $U_2$, as well as with sphere central, O, and the same apex angle, $2\Psi$. Therefore, the distance between S and E, $d_{ID}$, can be presented as

$$d_{ID} = \sqrt{R_I^2 + R_S^2 - 2R_I R_S \cos\varphi_I}, \tag{3.43}$$

where $\varphi_I$ is the polar angle of I.

**Theorem 3.5** *As I is uniformly distributed in the space shown in Fig. 3.8, the PDF of $d_{ID}^2$ can be written as the following two cases:*

*(1) When $U_2^2 + R_D^2 - 2U_2 R_D \cos\Psi \le (U_1 - R_D)^2$, the PDF of $d_{ID}^2$ can be given as*

$$f_{d_{ID}^2}(x) = \begin{cases} \frac{1}{2}\tau\left[\left(\sqrt{x}+R_D\right)^2 - U_2^2\right], & \text{if } \eta^2 \le x \le \mu^2; \\ \frac{1}{2}\tau\left[\left(\sqrt{x}+R_D\right)^2 - \left(R_D\cos\Psi + \sqrt{x - R_S^2\sin^2\Psi}\right)^2\right], & \text{if } \mu^2 \le x \le \rho^2; \\ \frac{1}{2}\tau\left[U_1^2 - \left(R_D\cos\Psi + \sqrt{x - R_S^2\sin^2\Psi}\right)^2\right], & \text{else if } \rho^2 \le x \le \xi^2 \end{cases}, \tag{3.44}$$

*where $\eta = U_2 - R_D$, $\mu = \sqrt{U_2^2 + R_D^2 - 2U_2R_D\cos\Psi}$, $\rho = \sqrt{U_1^2 + R_D^2 - 2U_1R_D\cos\Psi}$, $\xi = U_1 - R_D$, $\tau = \frac{3}{2R_D(1-\cos\Psi)(U_1^3 - U_2^3)}$, and $x - R_D^2\sin^2\Psi \ge 0$.*

*(2) When $(U_1 - R_D)^2 \le U_2^2 + R_D^2 - 2U_2R_D\cos\Psi$, one can derive the PDF of $d_{ID}^2$ as*

$$f_{d_{ID}^2}(x) = \begin{cases} \frac{1}{2}\tau\left[\left(\sqrt{x}+R_D\right)^2 - U_2^2\right], & \text{if } \eta^2 \le x \le \rho^2; \\ \frac{1}{2}\tau\left(U_1^2 - U_2^2\right), & \text{if } \rho^2 \le x \le \mu^2; \\ \frac{1}{2}\tau\left[U_1^2 - \left(R_D\cos\Psi + \sqrt{x - R_S^2\sin^2\Psi}\right)^2\right], & \text{else if } \mu^2 \le x \le \xi^2 \end{cases}. \tag{3.45}$$

***Proof*** Please refer to Appendix I. □

**Remark 3.4** $f_{d_{ID}^2}$ is related to the distance between D and O, $R_D$ and the radius of the spherical cone, $U_1$ and $U_2$. This is because in this work, we consider the location of the interference is randomly distributed in a spherical cone.

Then, the CDF of $d_{ID}^2$ can be derived in the following two cases by taking the integration of (3.44) and (3.45).

(1) When $U_2^2 + R_D^2 - 2U_2R_D\cos\Psi \le (U_1 - R_D)^2$

$$F_{d_{ID}^2}(x) = \begin{cases} 0, & \text{if } 0 \le x \le \eta^2; \\ \frac{1}{2}\tau\left[A(x) - \mathbb{A}\right], & \text{if } \eta^2 \le x \le \mu^2; \\ \frac{1}{2}\tau\left[B(x) + \mathbb{B}\right], & \text{if } \mu^2 \le x \le \rho^2; \\ \frac{1}{2}\tau\left[C(x) + \mathbb{C}\right], & \text{if } \rho^2 \le x \le \xi^2; \\ 1, & \text{else if } x \ge \xi^2 \end{cases}, \tag{3.46}$$

where

$$A(x) = \left(R_D^2 - U_2^2\right)x + \frac{4}{3}R_D x^{\frac{3}{2}} + \frac{1}{2}x^2, \tag{3.47}$$

$$\mathbb{A} = A\left(\eta^2\right), \tag{3.48}$$

$$B(x) = 2R_D^2\sin^2\Psi x + \frac{4}{3}R_D x^{\frac{3}{2}} - \frac{4}{3}R_D\cos\Psi\left(x - R_D^2\sin^2\Psi\right)^{\frac{3}{2}}, \tag{3.49}$$

$$\mathbb{B} = A\left(\mu^2\right) - A\left(\eta^2\right) - B\left(\mu^2\right), \tag{3.50}$$

$$C(x) = \left(U_1^2 - R_D^2\cos^2\Psi + R_D^2\sin^2\Psi\right)x - \frac{1}{2}x^2 - \frac{4}{3}R_D\cos\Psi\left(x - R_D^2\sin^2\Psi\right)^{\frac{3}{2}}, \tag{3.51}$$

and

$$\mathbb{C} = A\left(\mu^2\right) - A\left(\eta^2\right) + B\left(\rho^2\right) - B\left(\mu^2\right) - C\left(\rho^2\right). \tag{3.52}$$

(2) When $(U_1 - R_D)^2 \leq U_2^2 + R_D^2 - 2U_2R_D\cos\Psi$

$$F_{d_{ID}^2}(x) = \begin{cases} 0, & \text{if } 0 \leq x \leq \eta^2; \\ \frac{1}{2}\tau\left[A(x) - \mathbb{A}\right], & \text{if } \eta^2 \leq x \leq \rho^2; \\ \frac{1}{2}\tau\left[D(x) + \mathbb{D}\right], & \text{if } \rho^2 \leq x \leq \mu^2; \\ \frac{1}{2}\tau\left[C(x) + \mathbb{E}\right], & \text{if } \mu^2 \leq x \leq \xi^2; \\ 1, & \text{else if } x \geq \xi^2 \end{cases}, \tag{3.53}$$

where

$$D(x) = \left(U_1^2 - U_2^2\right)x, \tag{3.54}$$

$$\mathbb{D} = A\left(\rho^2\right) - A\left(\eta^2\right) - D\left(\rho^2\right), \tag{3.55}$$

and

$$\mathbb{E} = A\left(\rho^2\right) - A\left(\eta^2\right) + D\left(\mu^2\right) - D\left(\rho^2\right) - C\left(\mu^2\right). \tag{3.56}$$

**Corollary 3.2** *The CDF of Y can be further expressed as*

$$\begin{aligned} F_Y(y) &= \Pr\{Y \leq y\} = \Pr\left\{\frac{d_{ID}^2}{d_{SD}^2} \leq y\right\} \\ &= \int_0^\infty F_{d_{ID}}\left(\sqrt{y}d_{SD}\right) f_{d_{SD}}\left(d_{SD}\right) d\left(d_{SD}\right). \end{aligned} \tag{3.57}$$

*Then it can be obtained in the following cases:*

*(1) When $U_2^2 + R_D^2 - 2U_2R_D\cos\Psi \le (U_1 - R_D)^2$, the CDF of Y can be given as*

$$
F_Y(y) = \begin{cases}
0, & \text{if } y < \left(\frac{\eta}{\upsilon}\right)^2 \\
\frac{\tau}{\Psi}\mathbb{F}\left(y;\upsilon,\frac{\eta}{\sqrt{y}}\right), & \text{if } \left(\frac{\eta}{\upsilon}\right)^2 < y < \left(\frac{\eta}{\varpi}\right)^2; \\
\frac{\tau}{\Psi}\mathbb{F}\left(y;\upsilon,\varpi\right), & \text{if } \left(\frac{\eta}{\varpi}\right)^2 < y < \left(\frac{\mu}{\upsilon}\right)^2; \\
\frac{\tau}{\Psi}\left[\mathbb{F}\left(y;\frac{\mu}{\sqrt{y}},\varpi\right) + \mathbb{G}\left(y;\upsilon,\frac{\mu}{\sqrt{y}}\right)\right], & \text{if } \left(\frac{\mu}{\upsilon}\right)^2 < y < \min\left\{\left(\frac{\mu}{\varpi}\right)^2,\left(\frac{\xi}{\upsilon}\right)^2\right\}; \\
\frac{\tau}{\Psi}\left[\mathbb{F}\left(y;\frac{\mu}{\sqrt{y}},\varpi\right) + \mathbb{G}\left(y;\frac{\xi}{\sqrt{y}},\frac{\mu}{\sqrt{y}}\right) + \mathbb{H}\left(y;\upsilon,\xi\right)\right], & \text{if } \left(\frac{\xi}{\upsilon}\right)^2 < y < \min\left\{\left(\frac{\mu}{\varpi}\right)^2,\left(\frac{\rho}{\upsilon}\right)^2\right\}; \\
\frac{\tau}{\Psi}\left[\mathbb{F}\left(y;\frac{\mu}{\sqrt{y}},\varpi\right) + \mathbb{G}\left(y;\frac{\xi}{\sqrt{y}},\frac{\mu}{\sqrt{y}}\right) + \mathbb{H}\left(y;\frac{\rho}{\sqrt{y}},\frac{\xi}{\sqrt{y}}\right) + \mathbb{I}\right], & \text{if } \left(\frac{\rho}{\upsilon}\right)^2 < y < \left(\frac{\mu}{\varpi}\right)^2; \\
\frac{\tau}{\Psi}\mathbb{G}\left(y;\upsilon,\varpi\right), & \text{if } \left(\frac{\mu}{\varpi}\right)^2 < y < \left(\frac{\xi}{\upsilon}\right)^2; \\
\frac{\tau}{\Psi}\left[\mathbb{G}\left(y;\frac{\xi}{\sqrt{y}},\varpi\right) + \mathbb{H}\left(y;\upsilon,\frac{\xi}{\sqrt{y}}\right)\right], & \text{if } \max\left\{\left(\frac{\mu}{\varpi}\right)^2,\left(\frac{\xi}{\upsilon}\right)^2\right\} < y < \left(\frac{\rho}{\upsilon}\right)^2; \\
\frac{\tau}{\Psi}\left[\mathbb{G}\left(y;\frac{\xi}{\sqrt{y}},\frac{\mu}{\sqrt{y}}\right) + \mathbb{H}\left(y;\frac{\rho}{\sqrt{y}},\frac{\xi}{\sqrt{y}}\right) + \mathbb{I}\right], & \text{if } \max\left\{\left(\frac{\rho}{\upsilon}\right)^2,\left(\frac{\mu}{\varpi}\right)^2\right\} < y < \left(\frac{\xi}{\varpi}\right)^2; \\
\frac{\tau}{\Psi}\left[\mathbb{H}\left(y;\frac{\rho}{\sqrt{y}},\varpi\right) + \mathbb{I}\right], & \text{if } \left(\frac{\xi}{\varpi}\right)^2 < y < \left(\frac{\rho}{\varpi}\right)^2; \\
1, & \text{else if } y > \left(\frac{\rho}{\varpi}\right)^2
\end{cases}, \tag{3.58}
$$

*where* $\varpi = R_S - R_D$, $\upsilon = \sqrt{R_S^2 + R_D^2 - 2R_SR_D\cos\Psi}$, $a = 2R_SR_D$, $b = R_S^2 + R_D^2$, $\vartheta_i = \cos\left(\frac{2i-1}{2Q}\pi\right)$, $x_0 = \frac{\lambda-\kappa}{2}\vartheta_i + \frac{\lambda+\kappa}{2}$, $\mathbb{I} = \frac{1}{\tau}\arccos\left(\frac{\upsilon^2-b}{\sqrt{a}}\right) - \frac{1}{\tau}\arccos\left(\frac{\rho^2-b}{\sqrt{a}}\right)$, $\vartheta_i = \cos\left(\frac{2i-1}{2Q_1}\pi\right)$, *and*

$$
\mathbb{F}\left(y;\lambda,\kappa\right) = g_1\left(y;\lambda\right) - g_1\left(y;\kappa\right) + h_1\left(y;\lambda,\kappa\right), \tag{3.59}
$$

*where*

$$
\begin{aligned}
g_1\left(y;x\right) =& \frac{1}{2}\mathbb{A}\arccos\left(\frac{x^2-b}{\sqrt{a}}\right) - \frac{1}{2}y\left(R_D^2 - U_2^2\right) \\
&\times\left[\sqrt{a-\left(b-x^2\right)^2} + b\arccos\left(\frac{x^2-b}{\sqrt{a}}\right)\right] \\
&-\frac{1}{8}y^2\left[\left(2b^2+a\right)\arccos\left(\frac{x^2-b}{\sqrt{a}}\right) + \left(x^2+3b\right)\right. \\
&\left.\times\sqrt{a-\left(b-x^2\right)^2}\right],
\end{aligned} \tag{3.60}
$$

$$
h_1\left(y;\lambda,\kappa\right) = \frac{2\pi}{3Q}y^{\frac{3}{2}}R_D(\lambda-\kappa)\sum_{i=1}^{Q}\frac{x_0^4\sqrt{1-\vartheta_i^2}}{\sqrt{a-\left(b-x_0^2\right)^2}}, \tag{3.61}
$$

$$
\mathbb{G}\left(y;\lambda,\kappa\right) = g_2\left(y;\lambda\right) - g_2\left(y;\kappa\right) + h_2\left(y;\lambda,\kappa\right), \tag{3.62}
$$

*where*

$$g_2(y;x) = -yR_D^2\sin^2\Psi\left[\sqrt{a-(b-x)^2}\right.$$
$$\left.+b\arccos\left(\frac{x^2-b}{\sqrt{a}}\right)\right] - \frac{1}{2}\mathbb{B}\arccos\left(\frac{x^2-b}{\sqrt{a}}\right), \quad (3.63)$$

$$h_2(y;\lambda,\kappa) = \frac{2\pi R_D(\lambda-\kappa)}{3Q}\left[y^{\frac{3}{2}}\sum_{i=1}^{Q}\frac{x_0^4\sqrt{1-\vartheta_i^2}}{\sqrt{a-\left(b-x_0^2\right)^2}}\right.$$
$$\left.-\cos\Psi\sum_{i=1}^{Q}\sqrt{1-\vartheta_i^2}\left(yx_0^2-R_D^2\sin^2\Psi\right)\right], \quad (3.64)$$

$$\mathbb{H}(y;\lambda,\kappa) = g_3(y;\lambda) - g_3(y;\kappa) + h_3(y;\lambda,\kappa), \quad (3.65)$$

*where*

$$g_3(y;x) = -\frac{1}{2}y\left(U_1^2 - R_D^2cos^2\Psi + R_D^2\sin^2\Psi\right)$$
$$\times\left[\sqrt{a-(b-x)^2} + b\arccos\left(\frac{x^2-b}{\sqrt{a}}\right)\right]$$
$$+\frac{1}{8}y^2\left[\left(2b^2+a\right)\arccos\left(\frac{x^2-b}{\sqrt{a}}\right) + \left(x^2+3b\right)\right.$$
$$\left.\times\sqrt{a-(b-x)^2}\right] - \frac{1}{2}\mathbb{C}\arccos\left(\frac{x^2-b}{\sqrt{a}}\right), \quad (3.66)$$

$$h_3(y;\lambda,\kappa) = -\frac{2}{3}R_D\cos\Psi(\lambda-\kappa)\sum_{i=1}^{Q}\frac{\pi}{Q}\sqrt{1-\vartheta_i^2}\left(yx_0^2-R_D^2\sin^2\Psi\right). \quad (3.67)$$

*(2) When $U_2^2 + R_D^2 - 2U_2R_D\cos\Psi \geq (U_1-R_D)^2$, one can derive the CDF of $Y$ as*

$$F_Y(y) = \begin{cases} 0, & \text{if } y < \left(\frac{\eta}{\upsilon}\right)^2; \\ \frac{\tau}{\Psi}\mathbb{F}\left(y; \upsilon, \frac{\eta}{\sqrt{y}}\right), & \text{if } \left(\frac{\eta}{\upsilon}\right)^2 < y < \min\left\{\left(\frac{\eta}{\varpi}\right)^2, \left(\frac{\xi}{\upsilon}\right)^2\right\}; \\ \frac{\tau}{\Psi}\left[\mathbb{F}\left(y; \frac{\xi}{\sqrt{y}}, \frac{\eta}{\sqrt{y}}\right) + \mathbb{J}\left(y; \upsilon, \frac{\xi}{\sqrt{y}}\right)\right], & \text{if } \left(\frac{\xi}{\upsilon}\right)^2 < y < \left(\frac{\eta}{\varpi}\right)^2; \\ \frac{\tau}{\Psi}\mathbb{F}\left(y; \upsilon, \varpi\right), & \text{if } \left(\frac{\eta}{\varpi}\right)^2 < y < \left(\frac{\xi}{\upsilon}\right)^2; \\ \frac{\tau}{\Psi}\left[\mathbb{F}\left(y; \frac{\xi}{\sqrt{y}}, \varpi\right) + \mathbb{J}\left(y; \upsilon, \frac{\xi}{\sqrt{y}}\right)\right], & \text{if } \max\left\{\left(\frac{\eta}{\varpi}\right)^2, \left(\frac{\xi}{\upsilon}\right)^2\right\} < y < \left(\frac{\mu}{\upsilon}\right)^2; \\ \frac{\tau}{\Psi}\left[\mathbb{F}\left(y; \frac{\xi}{\sqrt{y}}, \varpi\right) + \mathbb{J}\left(y; \frac{\mu}{\sqrt{y}}, \frac{\xi}{\sqrt{y}}\right) + \mathbb{K}\left(y; \upsilon, \frac{\mu}{\sqrt{y}}\right)\right], & \text{if } \left(\frac{\mu}{\upsilon}\right)^2 < y < \left(\frac{\rho}{\upsilon}\right)^2; \\ \frac{\tau}{\Psi}\left[\mathbb{F}\left(y; \frac{\xi}{\sqrt{y}}, \varpi\right) + \mathbb{J}\left(y; \frac{\mu}{\sqrt{y}}, \frac{\xi}{\sqrt{y}}\right) + \mathbb{K}\left(y; \frac{\rho}{\sqrt{y}}, \frac{\mu}{\sqrt{y}}\right) + \mathbb{I}\right], & \text{if } \left(\frac{\rho}{\upsilon}\right)^2 < y < \left(\frac{\xi}{\varpi}\right)^2; \\ \frac{\tau}{\Psi}\left[\mathbb{J}\left(y; \frac{\mu}{\sqrt{y}}, \varpi\right) + \mathbb{K}\left(y; \frac{\rho}{\sqrt{y}}, \frac{\mu}{\sqrt{y}}\right) + \mathbb{I}\right], & \text{if } \left(\frac{\xi}{\varpi}\right)^2 < y < \left(\frac{\mu}{\varpi}\right)^2; \\ \frac{\tau}{\Psi}\left[\mathbb{K}\left(y; \frac{\rho}{\sqrt{y}}, \varpi\right) + \mathbb{I}\right], & \text{if } \left(\frac{\mu}{\varpi}\right)^2 < y < \left(\frac{\rho}{\varpi}\right)^2; \\ 1, & \text{else if } y > \left(\frac{\rho}{\varpi}\right)^2 \end{cases} \tag{3.68}$$

*where*

$$\mathbb{J}(y; \lambda, \kappa) = g_4(y; \lambda) - g_4(y; \kappa), \tag{3.69}$$

*where*

$$g_4(y; x) = -\frac{1}{2}\left(U_1^2 - U_2^2\right)\left[\sqrt{a - (b - x)^2} + b \arccos\left(\frac{x^2 - b}{\sqrt{a}}\right)\right] - \frac{\mathbb{D}}{2}\arccos\left(\frac{x^2 - b}{\sqrt{a}}\right), \tag{3.70}$$

$$\mathbb{K}(y; \lambda, \kappa) = g_5(y; \lambda) - g_5(y; \kappa) + h_3(y; \lambda, \kappa), \tag{3.71}$$

*where*

$$\begin{aligned} g_5(y; x) = &-\frac{1}{2}\left(U_1^2 - R_D^2\cos^2\Psi + R_D^2\sin^2\Psi\right) \\ &\times\left[\sqrt{a - (b - x)^2} + b \arccos\left(\frac{x^2 - b}{\sqrt{a}}\right)\right] \\ &+ \frac{1}{8}\left[\left(2b^2 + a\right)\arccos\left(\frac{x^2 - b}{\sqrt{a}}\right) + \left(x^2 + 3b\right)\right. \\ &\left.\times\sqrt{a - (b - x)^2}\right] - \frac{\mathbb{E}}{2}\arccos\left(\frac{x^2 - b}{\sqrt{a}}\right). \end{aligned} \tag{3.72}$$

***Proof*** Please refer to Appendix II. □

### 3.2.2.3 Outage Probability

Finally, OP can be derived by substituting (3.58), (3.68) and (3.34) into (3.32).

1. When $\mu^2 \leq \xi^2$, $\left(\frac{\mu}{\varpi}\right)^2 < \left(\frac{\xi}{\upsilon}\right)^2$,

$$\begin{aligned}
P_{\text{out}} &= \int_0^\infty F_Y\left(x\gamma_{\text{out}}\epsilon\right) f_X\left(x\right) dx \\
&= \int_0^\infty F_Y\left(x\gamma_{\text{out}}\epsilon\right) \frac{1}{B\left(m_S, m_I\right)} \frac{m_S m_I \Omega_S \Omega_I}{\left(\Omega_I m_S x + \Omega_S m_I\right)^2} \\
&\quad \times \left[\frac{\alpha\left(x\right)}{\alpha\left(x\right)+\beta}\right]^{m_I-1} \left[1-\frac{\alpha\left(x\right)}{\alpha\left(x\right)+\beta}\right]^{m_S-1} dx \\
&= 1 - \frac{B\left(\frac{\alpha\left[\left(\frac{\rho}{\varpi}\right)^2\right]}{\alpha\left[\left(\frac{\rho}{\varpi}\right)^2\right]+\beta}; m_S, m_I\right)}{B\left(m_S, m_I\right)} + \frac{1}{B\left(m_S, m_I\right)} \frac{\tau}{\Psi} \\
&\quad \times \left\{\delta\left[\zeta\left(\frac{\eta}{\varpi}, \frac{\eta}{\upsilon}\right)\right] + \delta\left[\zeta\left(\frac{\mu}{\upsilon}, \frac{\eta}{\varpi}\right)\right] + \delta\left[\zeta\left(\frac{\mu}{\varpi}, \frac{\mu}{\upsilon}\right)\right]\right. \\
&\quad + \delta\left[\zeta\left(\frac{\xi}{\upsilon}, \frac{\mu}{\varpi}\right)\right] + \delta\left[\zeta\left(\frac{\rho}{\upsilon}, \frac{\xi}{\upsilon}\right)\right] \\
&\quad \left. + \delta\left[\zeta\left(\frac{\xi}{\varpi}, \frac{\rho}{\upsilon}\right)\right] + \delta\left[\zeta\left(\frac{\rho}{\varpi}, \frac{\xi}{\varpi}\right)\right]\right\},
\end{aligned} \tag{3.73}$$

where $\delta(y) = \frac{\pi}{S} \sum_{i=1}^{S} \sqrt{1-\chi_i{}^2} \frac{m_S m_I \Omega_S \Omega_I}{\left(\Omega_I m_S \zeta_1 + \Omega_S m_I\right)^2} F_Y\left(y\gamma_{\text{out}}\epsilon\right) \left[\frac{\alpha(y)}{\alpha(y)+\beta}\right]^{m_I-1} \times \left[1-\frac{\alpha(y)}{\alpha(y)+\beta}\right]^{m_S-1}$, $\chi_i = \cos\left(\frac{2i-1}{2S}\pi\right)$ and $\zeta(x, y) = \frac{\left(x^2-y^2\right)\chi_i + x^2 + y^2}{2\gamma_{\text{out}}\epsilon}$.

2. When $\mu^2 \leq \xi^2$, $\left(\frac{\xi}{\upsilon}\right)^2 < \left(\frac{\mu}{\varpi}\right)^2 < \left(\frac{\rho}{\upsilon}\right)^2$,

$$\begin{aligned}
P_{\text{out}} &= 1 - \frac{B\left(\frac{\alpha\left[\left(\frac{\rho}{\varpi}\right)^2\right]}{\alpha\left[\left(\frac{\rho}{\varpi}\right)^2\right]+\beta}; m_S, m_I\right)}{B\left(m_S, m_I\right)} + \frac{1}{B\left(m_S, m_I\right)} \frac{\tau}{\Psi} \\
&\quad \times \left\{\delta\left[\zeta\left(\frac{\eta}{\varpi}, \frac{\eta}{\upsilon}\right)\right] + \delta\left[\zeta\left(\frac{\mu}{\upsilon}, \frac{\eta}{\varpi}\right)\right] + \delta\left[\zeta\left(\frac{\xi}{\varpi}, \frac{\rho}{\upsilon}\right)\right]\right. \\
&\quad + \delta\left[\zeta\left(\frac{\rho}{\varpi}, \frac{\xi}{\varpi}\right)\right] + \delta\left[\zeta\left(\frac{\xi}{\upsilon}, \frac{\mu}{\upsilon}\right)\right] \\
&\quad \left. + \delta\left[\zeta\left(\frac{\mu}{\varpi}, \frac{\xi}{\upsilon}\right)\right] + \delta\left[\zeta\left(\frac{\rho}{\upsilon}, \frac{\mu}{\varpi}\right)\right]\right\}.
\end{aligned} \tag{3.74}$$

3. When $\mu^2 \leq \xi^2$, $\left(\frac{\xi}{\upsilon}\right)^2 < \left(\frac{\rho}{\upsilon}\right)^2 < \left(\frac{\mu}{\varpi}\right)^2$,

$$P_{\text{out}} = 1 - \frac{B\left(\frac{\alpha\left[\left(\frac{\rho}{\varpi}\right)^2\right]}{\alpha\left[\left(\frac{\rho}{\varpi}\right)^2\right]+\beta}; m_S, m_I\right)}{B\left(m_S, m_I\right)} + \frac{1}{B\left(m_S, m_I\right)}\frac{\tau}{\Psi}$$
$$\times\left\{\delta\left[\zeta\left(\frac{\eta}{\varpi},\frac{\eta}{\upsilon}\right)\right]+\delta\left[\zeta\left(\frac{\mu}{\upsilon},\frac{\eta}{\varpi}\right)\right]+\delta\left[\zeta\left(\frac{\rho}{\upsilon},\frac{\xi}{\upsilon}\right)\right]\right.$$
$$+\delta\left[\zeta\left(\frac{\rho}{\varpi},\frac{\xi}{\varpi}\right)\right]+\delta\left[\zeta\left(\frac{\xi}{\upsilon},\frac{\mu}{\upsilon}\right)\right]$$
$$\left.+\delta\left[\zeta\left(\frac{\mu}{\varpi},\frac{\rho}{\upsilon}\right)\right]+\delta\left[\zeta\left(\frac{\xi}{\varpi},\frac{\mu}{\varpi}\right)\right]\right\}. \tag{3.75}$$

4. When $\mu^2 > \xi^2$, $\left(\frac{\eta}{\varpi}\right)^2 < \left(\frac{\xi}{\upsilon}\right)^2$,

$$P_{\text{out}} = 1 - \frac{B\left(\frac{\alpha\left[\left(\frac{\rho}{\varpi}\right)^2\right]}{\alpha\left[\left(\frac{\rho}{\varpi}\right)^2\right]+\beta}; m_S, m_I\right)}{B\left(m_S, m_I\right)} + \frac{1}{B\left(m_S, m_I\right)}\frac{\tau}{\Psi}$$
$$\times\left\{\delta\left[\zeta\left(\frac{\eta}{\varpi},\frac{\eta}{\upsilon}\right)\right]+\delta\left[\zeta\left(\frac{\xi}{\varpi},\frac{\rho}{\upsilon}\right)\right]+\delta\left[\zeta\left(\frac{\xi}{\upsilon},\frac{\eta}{\varpi}\right)\right]\right.$$
$$+\delta\left[\zeta\left(\frac{\mu}{\upsilon},\frac{\xi}{\upsilon}\right)\right]+\delta\left[\zeta\left(\frac{\rho}{\upsilon},\frac{\mu}{\upsilon}\right)\right]$$
$$\left.+\delta\left[\zeta\left(\frac{\mu}{\varpi},\frac{\xi}{\varpi}\right)\right]+\delta\left[\zeta\left(\frac{\rho}{\varpi},\frac{\mu}{\varpi}\right)\right]\right\}. \tag{3.76}$$

5. When $\mu^2 > \xi^2$, $\left(\frac{\eta}{\varpi}\right)^2 > \left(\frac{\xi}{\upsilon}\right)^2$,

$$P_{\text{out}} = 1 - \frac{B\left(\frac{\alpha\left[\left(\frac{\rho}{\varpi}\right)^2\right]}{\alpha\left[\left(\frac{\rho}{\varpi}\right)^2\right]+\beta}; m_S, m_I\right)}{B\left(m_S, m_I\right)} + \frac{1}{B\left(m_S, m_I\right)}\frac{\tau}{\Psi}$$
$$\times\left\{\delta\left[\zeta\left(\frac{\mu}{\upsilon},\frac{\eta}{\varpi}\right)\right]+\delta\left[\zeta\left(\frac{\xi}{\varpi},\frac{\rho}{\upsilon}\right)\right]+\delta\left[\zeta\left(\frac{\rho}{\upsilon},\frac{\mu}{\upsilon}\right)\right]\right.$$
$$+\delta\left[\zeta\left(\frac{\mu}{\varpi},\frac{\xi}{\varpi}\right)\right]+\delta\left[\zeta\left(\frac{\rho}{\varpi},\frac{\mu}{\varpi}\right)\right]$$
$$\left.+\delta\left[\zeta\left(\frac{\xi}{\upsilon},\frac{\eta}{\upsilon}\right)\right]+\delta\left[\zeta\left(\frac{\eta}{\varpi},\frac{\xi}{\upsilon}\right)\right]\right\}. \tag{3.77}$$

#### 3.2.2.4 Asymptotic Analysis

To illustrate the relationship between outage performance and a set of system parameters and thereby provide profound insights, in this part, we will present the asymptotic analysis of the outage performance in the system. In the following, we derive the approximate expressions assuming $\lambda_S = \frac{P_S}{N_0} \to \infty$.

Adopting the series representations of the exponential function, $\exp\left(-\frac{m_S x}{\Omega_S}\right) = \sum_{n=0}^{\infty} \frac{\left(-\frac{m_S x}{\Omega_S}\right)^n}{n!}$, and keeping the first two terms while ignoring the higher order term, the asymptotic PDF of $\lambda_S |h_{SD}|^2$ can be obtained as

$$f^{\infty}_{\lambda_S |h_{SD}|^2}(x) = \frac{x^{m_S - 1}}{\Gamma(m_S)} \left(\frac{m_S}{\Omega_S \lambda_S}\right)^{m_S} \left(1 - \frac{m_S x}{\Omega_S \lambda_S}\right), \tag{3.78}$$

Further, we can derived the asymptotic CDF and PDF of $X$ as (3.79) and (3.80) shown on the top of next page.

$$\begin{aligned}
F_X^{\infty}(x) &= \Pr\left\{\frac{P_I |h_{ID}|^2}{P_S |h_{SD}|^2} \le x\right\} = \int_0^{\infty} F_{\lambda_I |h_{ID}|^2}(xy)\, f_{\lambda_S |h_{SD}|^2}(y)\, dy \\
&= \frac{\left(\frac{m_S}{\Omega_S \lambda_S}\right)^{m_S}}{\Gamma(m_I)\Gamma(m_S)} \int_0^{\infty} y^{m_S - 1} \left(1 - \frac{m_S y}{\Omega_S \lambda_S}\right) \gamma\left(m_I, \frac{m_I x y}{\Omega_I \lambda_I}\right) dy \\
&= \frac{\left(\frac{m_S}{\Omega_S \lambda_S}\right)^{m_S}}{\Gamma(m_I)\Gamma(m_S)} \left\{\frac{1}{m_S}\left[y^{m_S} \gamma\left(m_I, \frac{m_I x y}{\Omega_I \lambda_I}\right) - \left(\frac{m_I x}{\Omega_I \lambda_I}\right)^{-m_S} \gamma\left(m_I + m_S, \frac{m_I x y}{\Omega_I \lambda_I}\right)\right]\right. \\
&\quad \left. - \frac{1}{m_S + 1}\left[y^{m_S + 1} \gamma\left(m_I, \frac{m_I x y}{\Omega_I \lambda_I}\right) - \left(\frac{m_I x}{\Omega_I \lambda_I}\right)^{-m_S + 1} \gamma\left(m_I + m_S + 1, \frac{m_I x y}{\Omega_I \lambda_I}\right)\right]\right\}
\end{aligned} \tag{3.79}$$

$$\begin{aligned}
f_X^{\infty}(x) = {}& \frac{\left(\frac{m_S}{\Omega_S \lambda_S}\right)^{m_S}}{\Gamma(m_I)\Gamma(m_S)} \left\{\frac{1}{m_S}\left[y^{m_S}\left(-\frac{m_I y}{\Omega_I \lambda_I}\left(\frac{m_I x y}{\Omega_I \lambda_I}\right)^{m_I - 1} e^{-\frac{m_I x y}{\Omega_I \lambda_I}}\right) + m_S \left(\frac{m_I}{\Omega_I \lambda_I}\right)^{-m_S}\right.\right. \\
& \left. \times x^{-m_S - 1} \gamma\left(m_I + m_S, \frac{m_I x y}{\Omega_I \lambda_I}\right) - \left(\frac{m_I x}{\Omega_I \lambda_I}\right)^{-m_S}\left(-\frac{m_I y}{\Omega_I \lambda_I}\left(\frac{m_I x y}{\Omega_I \lambda_I}\right)^{m_I + m_S - 1} e^{-\frac{m_I x y}{\Omega_I \lambda_I}}\right)\right] \\
& - \frac{1}{m_S + 1}\left[y^{m_S + 1}\left(-\frac{m_I y}{\Omega_I \lambda_I}\left(\frac{m_I x y}{\Omega_I \lambda_I}\right)^{m_I - 1} e^{-\frac{m_I x y}{\Omega_I \lambda_I}}\right) - (-m_S + 1)\left(\frac{m_I}{\Omega_I \lambda_I}\right)^{-m_S + 1} x^{-m_S}\right. \\
& \left.\left. \times \gamma\left(m_I + m_S + 1, \frac{m_I x y}{\Omega_I \lambda_I}\right) - \left(\frac{m_I x}{\Omega_I \lambda_I}\right)^{-m_S + 1}\left(-\frac{m_I y}{\Omega_I \lambda_I}\left(\frac{m_I x y}{\Omega_I \lambda_I}\right)^{m_I + m_S} e^{-\frac{m_I x y}{\Omega_I \lambda_I}}\right)\right]\right\}
\end{aligned} \tag{3.80}$$

Finally, the asymptotic OP can be expressed as follows.

1. When $U_2^2 + R_D^2 - 2U_2 R_D \cos\Psi \le (U_1 - R_D)^2$, $\left(\frac{\mu}{\varpi}\right)^2 < \left(\frac{\xi}{\upsilon}\right)^2$

$$\begin{aligned}
P_{\text{out}}^{\infty} &= \int_0^{\infty} F_Y(x \gamma_{\text{out}})\, f_X^{\infty}(x)\, dx \\
&= \delta^{\infty}(\zeta_1) + \delta^{\infty}(\zeta_2) + \delta^{\infty}(\zeta_3) + \delta^{\infty}(\zeta_4) + \delta^{\infty}(\zeta_5) + \delta^{\infty}(\zeta_6) + \delta^{\infty}(\zeta_7),
\end{aligned} \tag{3.81}$$

where $\delta^{\infty}(y) = \frac{\pi}{S}\sum_{i=1}^{S}\sqrt{1-\chi_i{}^2}F_Y(y\gamma_{\text{out}})\,f_X^{\infty}(y)$. 2. When $U_2^2 + R_D^2 - 2U_2R_D\cos\Psi \le (U_1 - R_D)^2$, $\left(\frac{\xi}{\upsilon}\right)^2 < \left(\frac{\mu}{\varpi}\right)^2 < \left(\frac{\rho}{\upsilon}\right)^2$

$$P_{\text{out}} = \delta^{\infty}(\zeta_1) + \delta^{\infty}(\zeta_2) + \delta^{\infty}(\zeta_8) + \delta^{\infty}(\zeta_9) + \delta^{\infty}(\zeta_{10}) + \delta^{\infty}(\zeta_{11}) + \delta^{\infty}(\zeta_7). \tag{3.82}$$

3. When $U_2^2 + R_D^2 - 2U_2R_D\cos\Psi \le (U_1 - R_D)^2$, $\left(\frac{\xi}{\upsilon}\right)^2 < \left(\frac{\rho}{\upsilon}\right)^2 < \left(\frac{\mu}{\varpi}\right)^2$

$$P_{\text{out}} = \delta^{\infty}(\zeta_1) + \delta^{\infty}(\zeta_2) + \delta^{\infty}(\zeta_8) + \delta^{\infty}(\zeta_{12}) + \delta^{\infty}(\zeta_{13}) + \delta^{\infty}(\zeta_{14}) + \delta^{\infty}(\zeta_7). \tag{3.83}$$

4. When $U_2^2 + R_D^2 - 2U_2R_D\cos\Psi > (U_1 - R_D)^2$, $\left(\frac{\eta}{\varpi}\right)^2 < \left(\frac{\xi}{\upsilon}\right)^2$

$$P_{\text{out}} = \delta^{\infty}(\zeta_1) + \delta^{\infty}(\zeta_{15}) + \delta^{\infty}(\zeta_{16}) + \delta^{\infty}(\zeta_{17}) + \delta^{\infty}(\zeta_{18}) + \delta^{\infty}(\zeta_{19}) + \delta^{\infty}(\zeta_{20}). \tag{3.84}$$

5. When $U_2^2 + R_D^2 - 2U_2R_D\cos\Psi > (U_1 - R_D)^2$, $\left(\frac{\eta}{\varpi}\right)^2 > \left(\frac{\xi}{\upsilon}\right)^2$

$$P_{\text{out}} = \delta^{\infty}(\zeta_{21}) + \delta^{\infty}(\zeta_{22}) + \delta^{\infty}(\zeta_{23}) + \delta^{\infty}(\zeta_{17}) + \delta^{\infty}(\zeta_{18}) + \delta^{\infty}(\zeta_{19}) + \delta^{\infty}(\zeta_{20}). \tag{3.85}$$

### *3.2.3 Numerical Results*

In this section, we will provide numerical results to verify the proposed analytical model. Following the parameters adopted in [3, 5, 9], here, the main parameters are set as $m_S = 2$, $m_I = 2$, $\Omega_S = 0$ dB, $\Omega_I = 0$ dB, $\gamma_{\text{out}} = 0$ dB, $R_S = 7531$ km, $R_D = 6371$ km, $U_1 = 30371$ km, $U_2 = 6531$ km, $\Psi = \pi/4$ rad, $P_S = 30$ dB, and $P_I = 30$ dB.

The influence of the transmit SNR at S, $P_S$, on the OP of the considered SAC system is investigated in Fig. 3.9. It can be observed that $P_S$ exhibits a positive effect, because the increasing of $P_S$ can definitely increase the transmit SNR at D. We can also see that $R_S$ shows a positive impact on the outage performance of the considered SAC system. This observation comes from the fact that a small $R_S$ denotes a low path-loss for the signal transmitted over S-D link.

Figure 3.10 presents the outage performance of the considered system for various $\Psi$. It can be observed that $\Psi$ shows a positive impact on the OP when $R_S$ is small, but $\Psi$ shows a negative impact on the OP when $R_S$ is large. This is because when $R_S$ is small, the probability that $d_{ID}$ is smaller than $d_{SD}$ is decided by the PDF of $d_{SD}$, $f_{d_{SD}}(x) = \frac{1}{\Psi}\frac{2x}{\sqrt{(2R_DR_S)^2 - \left(R_D^2 + R_S^2 - x^2\right)^2}}$ given in (3.39). Then, it

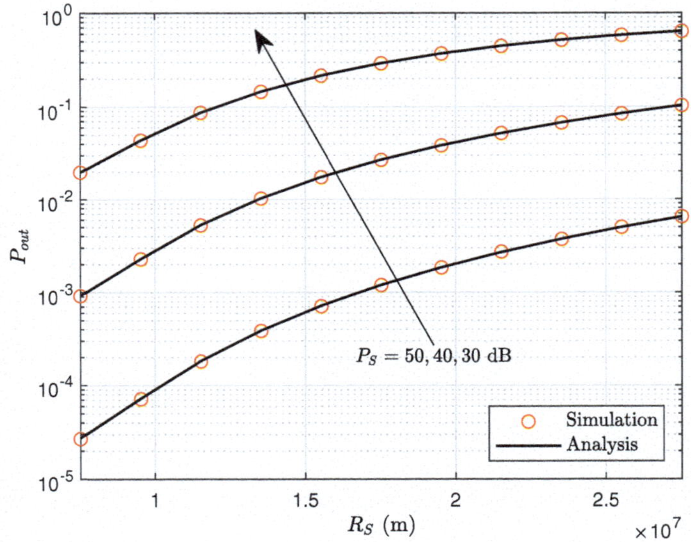


Fig. 3.9 OP for various $P_S$

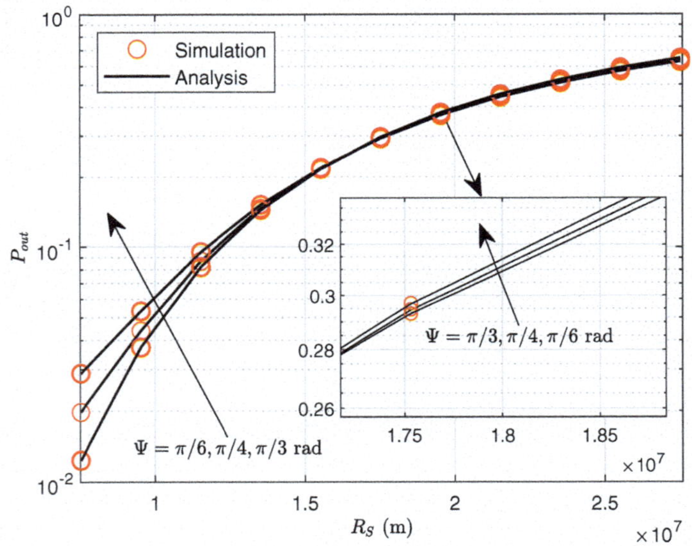


Fig. 3.10 OP for various $\Psi$

can be obtained that when $\Psi$ increases, the probability that $d_{ID}$ is smaller than $d_{SD}$ increases, which leads to the degraded OP. Furthermore, when $R_S$ is large, the probability that $d_{ID}$ is smaller than $d_{SD}$ also depends on the PDF of $d_{ID}$ given in

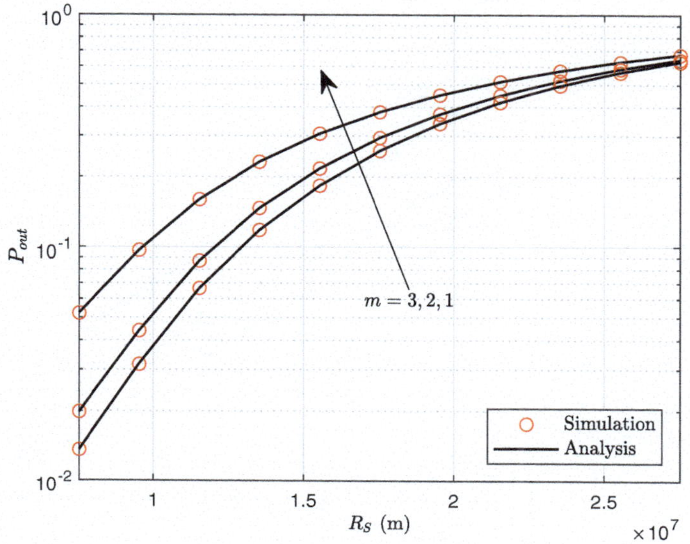


**Fig. 3.11** OP for various $m_S$

(3.39) as $f_{d_{ID}}(x) = \frac{1}{2}\tau\left[\left(\sqrt{x} + R_D\right)^2 - \left(R_D \cos\Psi + \sqrt{x - R_S^2\sin^2\Psi}\right)^2\right]$, $\mu^2 \leq x \leq \rho^2$ and $f_{d_{ID}}(x) = \frac{1}{2}\tau\left[U_1^2 - \left(R_D \cos\Psi + \sqrt{x - R_S^2\sin^2\Psi}\right)^2\right]$, $\rho^2 \leq x \leq \xi^2$. Then, it can be obtained that when $\Psi$ increases, the rate of the increase of $d_{ID}$ is faster than that of $d_{SD}$, and then causes the results that OP decreases while $\Psi$ increases in large $R_S$ region.

In Fig. 3.11, the OP of the considered SAC system is studied for various $m_S$. We can observe that $m_S$ shows a positive impact on the OP. This is because when $m_S$ increases, the fading is less serious over S-D link, and thus resulting in the improved OP.

In Fig. 3.12, the impact of the maximum height of I on the OP of the considered SAC system is presented. It can be seen that $U_1$ exhibits a negative influence on OP. This is because a large $U_1$ represents a large distribution space for I, leading to large path-loss over I-D link. Moreover, the asymptotic OP of the considered system gets close to the simulation and analysis ones as the average channel gain increases.

The OP for various $R_D$ is shown in Fig. 3.13. One can observe that the OP with various $R_D$ is similar, which indicates that adjusting $R_D$ is not a feasible way to improve the OP of the considered SAC system. It can be explained by fact that $R_D$ will influence both S-D and I-D links simultaneously, and then the outage performance cannot be improved. It is also easy to observe that the obtained asymptotic results for the OP can accurately match the simulation and analysis ones in the high SNR regime.

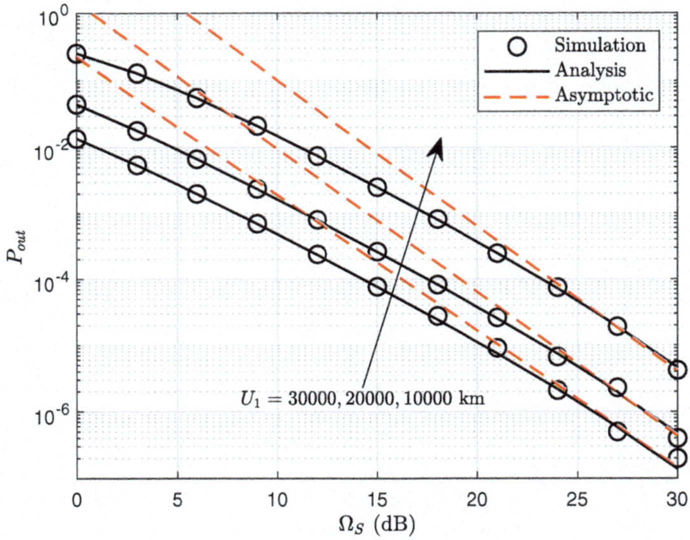


**Fig. 3.12** OP for various $U_1$

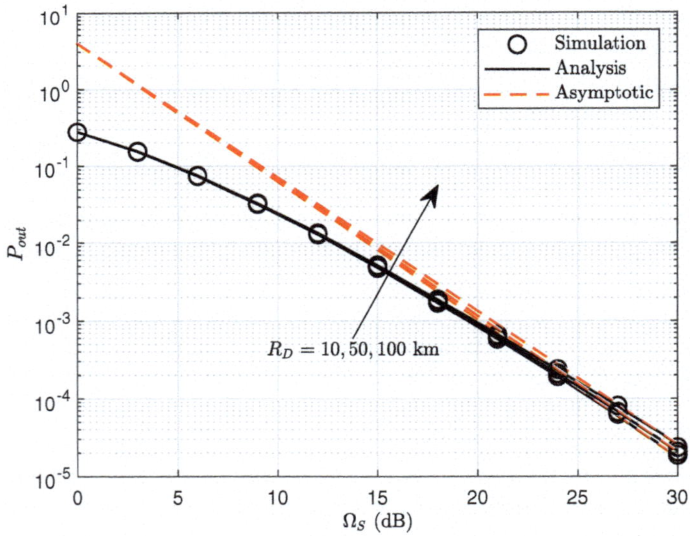


**Fig. 3.13** OP for various $R_D$

The impact of the number of interference nodes on the OP of the considered system is shown in Fig. 3.14. In this simulation scenario, we suppose that all interference nodes share the same transmit power $P_I$ for simplification purpose. It is obvious that, as expected, the outage performance degrades as the number of interference satellites increases.

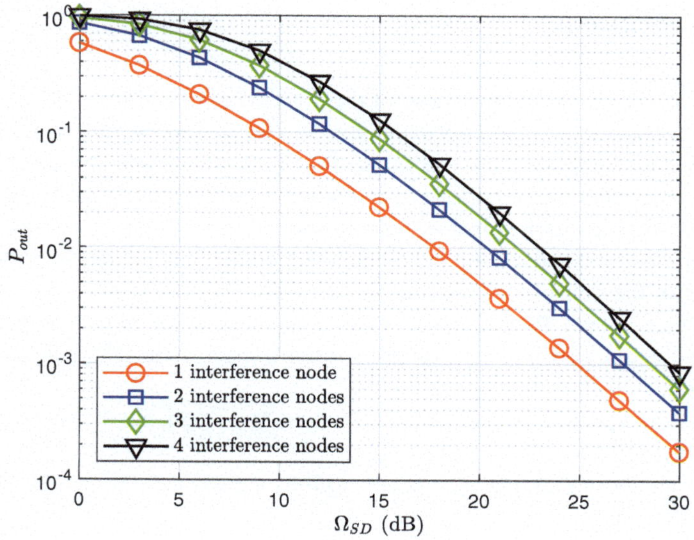


**Fig. 3.14** OP for various number of interference nodes

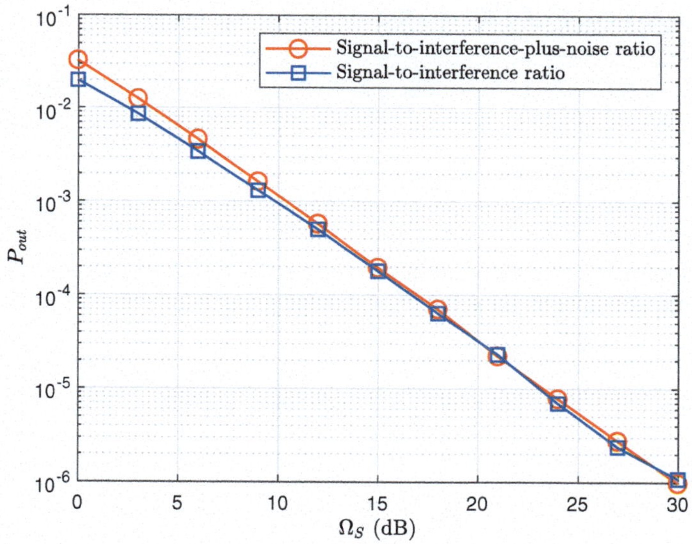


**Fig. 3.15** Comparison between $P_{out}$ considering SIR or SINR

As shown in Fig. 3.15, it can be seen that, signal-to-interference-plus-noise ratio (SINR) can be approximately but accurately evaluated by SIR, especially when the $\Omega_S$ is large, that is when the fading is less serious.

### 3.2.4 Conclusion

In this work, the outage performance of the considered downlink SAC system was investigated by deriving the approximation analytical expressions of the OP. We consider the randomness of the positions of the satellite and the space interference to make our system more realistic. Observing from the numerical results, some meaningful remarks can be obtained as follows:

(1) The maximum and minimum heights of I, $U_1$ and $U_2$, show a negative influence on the OP of the considered system. Nevertheless, the height of S, $R_S$ exhibits a positive effect on the OP. The height of D, $R_D$ shows a weak influence on the OP.

(2) The transmit power at S provides a positive impact on the OP of the considered SAC system.

(3) $\Psi$ shows a positive impact on the OP when $R_S$ is small, but $\Psi$ shows a negative impact on the OP when $R_S$ is large.

(4) The number of interference nodes exhibits a positive effect on the OP of the considered system.

## Appendix I: Proof of Theorem 3.2

The volume of the space between the spherical cones with radius $U_1$ and $U_2$ can be expressed as

$$V = \frac{2\pi}{3}(1 - \cos\Psi)\left(U_1^3 - U_2^3\right). \tag{3.86}$$

Thus, employing the method adopted in Appendix A of [10], and considering that I is uniformly distributed, the CDF of $\varphi_I$ can be given as

$$\begin{aligned} F_{\varphi_I}(x) &= \int_{U_2}^{U_1}\int_0^{2\pi}\int_0^x \frac{1}{V}\rho^2 \sin\varphi d\varphi d\theta d\rho \\ &= \frac{1}{V}\frac{2\pi}{3}\left(U_1^3 - U_2^3\right)(1 - \cos x) \\ &= \begin{cases} \frac{1-\cos x}{1-\cos\Psi}, & \text{if } 0 \le x \le \Psi; \\ 0, & \text{else} \end{cases}. \end{aligned} \tag{3.87}$$

So, it is easy to have the PDF of $\varphi_I$ as

$$f_{\varphi_I}(x) = \frac{\sin x}{1 - \cos\Psi}. \tag{3.88}$$

Similarly, the PDF of $R_I$ can be written as

$$f_{R_I}(x) = \begin{cases} \frac{3x^2}{U_1^3 - U_2^3}, & \text{if } U_2 \le x \le U_1; \\ 0, & \text{else} \end{cases}. \tag{3.89}$$

Thus, the joint PDF of $R_I$ and $\varphi_I$ can be presented as

$$f_{R_I,\varphi_I}(x, y) = \frac{3x^2}{U_1^3 - U_2^3} \frac{\sin y}{1 - \cos \Psi} \times \mathbb{I}. \tag{3.90}$$

Since $d_{ID} = \sqrt{R_I^2 + R_D^2 - 2R_I R_D \cos \varphi_I}$, one can have the Jacobian matrix as

$$\begin{aligned} \left| \frac{\partial \left(d_{ID}^2, R_I\right)}{\partial \left(R_I, \varphi_I\right)} \right| &= \begin{vmatrix} 2R_I - 2R_D \cos \varphi_I & 2R_I R_D \sin \varphi_I \\ 1 & 0 \end{vmatrix} \\ &= 2R_I R_D \sin \varphi_I. \end{aligned} \tag{3.91}$$

Then, the joint PDF of $d_{ID}^2$ and $R_I$ can be written as

$$f_{d_{ID}^2, R_I}(x, y) = \frac{f_{R_I,\varphi_I}(x, y)}{\left| \frac{\partial (d_{ID}^2, R_I)}{\partial (R_I, \varphi_I)} \right|} = \frac{1}{2R_D} \frac{1}{1 - \cos \Psi} \frac{3y}{U_1^3 - U_2^3}, \tag{3.92}$$

where $U_2 \le y \le U_1$ and $\cos \Psi \le \frac{y^2 + R_D^2 - x}{2yR_D} \le 1$.

So, we can further derive the PDF of $d_{ID}^2$ as

$$f_{d_{ID}^2}(x) = \int f_{d_{ID}^2, R_I}(x, y)\, dy. \tag{3.93}$$

To deal with the condition in (3.92), we discuss the following four cases:

**Case 1**: $(U_2 - R_D)^2 \le x \le \min \left\{U_2^2 + R_D^2 - 2U_2 R_D \cos \Psi, (U_1 - R_D)^2\right\}$. Under this case, we can obtain $U_2 \le y \le R_D + \sqrt{x}$ and

$$\begin{aligned} f_{d_{ID}^2}(x) &= \int_{U_2}^{R_D + \sqrt{x}} f_{d_{ID}^2, R_I}(x, y)\, dy \\ &= \frac{3}{4R_D} \frac{1}{1 - \cos \Psi} \frac{1}{U_1^3 - U_2^3} \left[\left(R_D + \sqrt{x}\right)^2 - U_2^2\right] \\ &= \frac{\tau}{2} \left[\left(R_D + \sqrt{x}\right)^2 - U_2^2\right], \end{aligned} \tag{3.94}$$

where $\tau = \frac{3}{2R_D} \frac{1}{1 - \cos \Psi} \frac{1}{U_1^3 - U_2^3}$ and $x - R_D^2 \cos^2 \Psi \ge 0$.

**Case 2**: $U_2^2 + R_D^2 - 2U_2 R_D \cos \Psi \le x < (U_1 - R_D)^2$.

Under this case, we can derive the range of $y$ as $R_D \cos\Psi + \sqrt{x - {R_D}^2 \sin^2\Psi} < y < R_D + \sqrt{x}$ and further derive (3.93) as

$$\begin{aligned} f_{d_{ID}^2}(x) &= \int_{R_D \cos\Psi + \sqrt{x - {R_D}^2 \sin^2\Psi}}^{R_D + \sqrt{x}} f_{d_{ID}^2, R_I}(x, y)\, dy \\ &= \frac{\tau}{2}\left[\left(R_D + \sqrt{x}\right)^2 - \left(R_D \cos\Psi + \sqrt{x - {R_D}^2 \sin^2\Psi}\right)^2\right]. \end{aligned} \tag{3.95}$$

**Case 3**: $(U_1 - R_D)^2 \leq x < U_2^2 + R_D^2 - 2U_2 R_D \cos\Psi$.
Under this case, we can obtain $U_2 < y < U_1$. and derive (3.93) as

$$f_{d_{ID}^2}(x) = \int_{U_2}^{U_1} f_{d_{ID}^2, R_I}(x, y)\, dy = \frac{\tau}{2}\left(U_1^2 - U_2^2\right). \tag{3.96}$$

**Case 4**:
$\max\left\{U_2^2 + R_D^2 - 2U_2 R_D \cos\Psi, (U_1 - R_D)^2\right\} < x < U_1^2 + R_D^2 - 2U_1 R_D \cos\Psi$.
Under this case, it gives $R_S \cos\Psi + \sqrt{x - {R_S}^2 \sin^2\Psi} < y < U_1$ and

$$\begin{aligned} f_{d_{ID}^2}(x) &= \int_{R_S \cos\Psi + \sqrt{x - {R_S}^2 \sin^2\Psi}}^{U_1} f_{d_{ID}^2, R_I}(x, y)\, dy \\ &= \frac{\tau}{2}\left[U_1^2 - \left(R_S \cos\Psi + \sqrt{x - {R_S}^2 \sin^2\Psi}\right)^2\right]. \end{aligned} \tag{3.97}$$

Thus, combining (3.94), (3.95), (3.96), and (3.97) completes the proof.

## *Appendix II: Proof of Theorem 3.3*

The CDF of $d_{ID}$ can be given by

$$\begin{aligned} F_{d_{ID}}(x) &= \Pr\{d_{ID} \leq x\} \\ &= \Pr\left\{d_{ID}^2 \leq x^2\right\} \\ &= F_{d_{ID}}\left(x^2\right) \end{aligned} \tag{3.98}$$

Since $F_{d_{ID}^2}$ is piece-wise with different range of integration as $\eta$, $\mu$, $\xi$, and $\rho$, the derivation of the OP changes when the values of $U_2$, $U_1$, $R_D$, $R_D$, and $\Psi$ are set differently. Therefore, the derivation of the OP is given with different settings of the parameters in the following.

1. When $\mu \leq \xi$, the CDF of $d_{ID}$ can be expressed as

$$F_{d_{ID}}(x) = \begin{cases} 0, & \text{if } 0 \leq x \leq \eta \\ \frac{1}{2}\tau\left[A\left(x^2\right) - \mathbb{A}\right], & \text{if } \eta \leq x \leq \mu \\ \frac{1}{2}\tau\left[B\left(x^2\right) + \mathbb{B}\right], & \text{if } \mu \leq x \leq \xi \\ \frac{1}{2}\tau\left[C\left(x^2\right) + \mathbb{C}\right], & \text{if } \xi \leq x \leq \rho \\ 1, & \text{if } x \geq \rho \end{cases}, \tag{3.99}$$

Then we can further obtain the following situations as:
(1) $\varpi\sqrt{y} < \upsilon\sqrt{y} < \eta$, $y < \left(\frac{\eta}{\upsilon}\right)^2$, where $\varpi = R_S - R_D$ and

$$\upsilon = \sqrt{R_S^2 + R_D^2 - 2R_S R_D \cos\Psi},$$

then the OP can be obtained as

$$F_Y(y) = \int_{\varpi}^{\upsilon} F_{d_{ID}}\left(\sqrt{y}x\right) f_{d_{SD}}(x)\, dx = 0. \tag{3.100}$$

(2) $\varpi\sqrt{y} < \eta < \upsilon\sqrt{y} < \mu$, in this case, $\left(\frac{\eta}{\upsilon}\right)^2 < y < \min\left\{\left(\frac{\eta}{\varpi}\right)^2, \left(\frac{\mu}{\upsilon}\right)^2\right\}$, the OP can be further derived as

$$\begin{aligned} F_Y(y) &= \int_{\varpi}^{\upsilon} F_{d_{ID}}\left(\sqrt{y}x\right) f_{d_{SD}}(x)\, dx \\ &= \int_{\frac{\eta}{\sqrt{y}}}^{\upsilon} \frac{\tau}{\Psi} \frac{x\left[A\left(yx^2\right) - \mathbb{A}\right]}{\sqrt{a^2 - \left(b - x^2\right)^2}} dx, \end{aligned} \tag{3.101}$$

where $a = 2R_S R_D$ and $b = R_S^2 + R_D^2$.

Since the integral is hard to solve, we use Chebyshev-Gauss quadrate to facilitate further analysis with an adequate accuracy, which is presented as $\int_{-1}^{1} \frac{f(x)}{\sqrt{1-x^2}} dx \approx \sum_{i=1}^{T} \omega_i f(x_i)$ with $x_i = \cos\left(\frac{2i-1}{2T}\pi\right)$ and the weight $\omega_i = \frac{\pi}{T}$. Thus, the OP can be rewritten as

$$\begin{aligned} F_Y(y) = \frac{\tau}{\Psi} &\left[ g_1(\upsilon) - g_1\left(\frac{\eta}{\sqrt{y}}\right) + \frac{2}{3} R_D \left(\Gamma_{S2} - \frac{\eta}{\sqrt{y}}\right) y^{\frac{3}{2}} \right. \\ &\left. \times \sum_{i=1}^{Q} \frac{\pi}{Q} \sqrt{1 - \vartheta_i^2} \frac{x_0^4}{\sqrt{a - \left(b - x_0^2\right)^2}} \right] \\ = \frac{\tau}{\Psi} &\left[ g_1(\upsilon) - g_1\left(\frac{\eta}{\sqrt{y}}\right) + h_1\left(\Gamma_{S2}, \frac{\eta}{\sqrt{y}}\right) \right] \end{aligned}$$

$$= \mathbb{F}\left(y; \upsilon, \frac{\eta}{\sqrt{y}}\right), \tag{3.102}$$

where $\mathbb{F}(y; \lambda, \kappa) = g_1(y; \lambda) - g_1(y; \kappa) + h_1(y; \lambda, \kappa)$, $\vartheta_i = \cos\left(\frac{2i-1}{2Q}\pi\right)$, $x_0 = \frac{\Gamma_{S2}-\Gamma_{E1}}{2}\vartheta_i + \frac{\Gamma_{S2}+\Gamma_{E1}}{2}$,

$$h_1(y; \lambda, \kappa) = \frac{2}{3} y^{\frac{3}{2}} R_D(\lambda - \kappa) \sum_{i=1}^{Q} \frac{\pi}{Q} \sqrt{1-\vartheta_i^2} \frac{x_0^4}{\sqrt{a - \left(b - x_0^2\right)^2}}, \tag{3.103}$$

and

$$\begin{aligned} g_1(y; x) =& \frac{1}{2} A \arccos\left(\frac{x^2 - b}{\sqrt{a}}\right) - \frac{1}{2} y \left(R_D^2 - U_2^2\right) \Bigg[\sqrt{a - \left(b - x^2\right)^2} \\ &+ b \arccos\left(\frac{x^2 - b}{\sqrt{a}}\right) - \frac{1}{8} y^2 \Bigg[\left(2b^2 + a\right) \arccos\left(\frac{x^2 - b}{\sqrt{a}}\right) \\ &+ \left(x^2 + 3b\right) \sqrt{a - \left(b - x^2\right)^2}\Bigg]. \end{aligned} \tag{3.104}$$

(3) $\eta < \varpi\sqrt{y} < \upsilon\sqrt{y} < \mu$, in this case, $\left(\frac{\eta}{\varpi}\right)^2 < y < \left(\frac{\mu}{\upsilon}\right)^2$, we can obtain

$$\begin{aligned} F_Y(y) &= \int_{\varpi}^{\upsilon} \frac{\tau}{\Psi} \frac{x\left[A\left(yx^2\right) - \mathbb{A}\right]}{\sqrt{a^2 - \left(b - x^2\right)^2}} dx \\ &= \frac{\tau}{\Psi} \left[\mathbb{F}(y; \upsilon, \varpi)\right], \end{aligned} \tag{3.105}$$

(4) $\eta < \varpi\sqrt{y} < \mu < \upsilon\sqrt{y} < \xi$, in this situation, $\max\left\{\left(\frac{\eta}{\varpi}\right)^2, \left(\frac{\mu}{\upsilon}\right)^2\right\} < y < \min\left\{\left(\frac{\mu}{\varpi}\right)^2, \left(\frac{\xi}{\upsilon}\right)^2\right\}$, then the OP can be written as

$$\begin{aligned} F_Y(y) &= \int_{\varpi}^{\frac{\mu}{\sqrt{y}}} \frac{\tau}{\Psi} \frac{x\left[A\left(yx^2\right) - \mathbb{A}\right]}{\sqrt{a^2 - \left(b - x^2\right)^2}} dx + \int_{\frac{\mu}{\sqrt{y}}}^{\upsilon} \frac{\tau}{\Psi} \frac{x\left[B\left(yx^2\right) + \mathbb{B}\right]}{\sqrt{a^2 - \left(b - x^2\right)^2}} dx \\ &= \frac{\tau}{\Psi} \left[\mathbb{F}\left(y; \frac{\mu}{\sqrt{y}}, \varpi\right) + \mathbb{G}\left(y; \upsilon, \frac{\mu}{\sqrt{y}}\right)\right], \end{aligned} \tag{3.106}$$

where

$$\mathbb{G}(y; \lambda, \kappa) = g_2(y; \lambda) - g_2(y; \kappa) + h_2(y; \lambda, \kappa), \tag{3.107}$$

$$g_2(y;x) = -\,yR_D^2\sin^2\Psi\left[\sqrt{a-(b-x)^2} + b\arccos\left(\frac{x^2-b}{\sqrt{a}}\right)\right] - \frac{\mathbb{B}}{2}\arccos\left(\frac{x^2-b}{\sqrt{a}}\right), \tag{3.108}$$

and

$$h_2(y;\lambda,\kappa) = \frac{2}{3}R_D(\lambda-\kappa)\left[y^{\frac{3}{2}}\sum_{i=1}^{Q}\frac{\pi}{Q}\sqrt{1-\vartheta_i^2}\frac{x_0^4}{\sqrt{a-\left(b-x_0^2\right)^2}} - \cos\Psi\sum_{i=1}^{Q}\frac{\pi}{Q}\sqrt{1-\vartheta_i^2}\left(yx_0^2-R_D^2\sin^2\Psi\right)\right]. \tag{3.109}$$

(5) $\eta < \varpi\sqrt{y} < \mu < \xi < \upsilon\sqrt{y} < \rho$, in this case, $\max\left\{\left(\frac{\eta}{\varpi}\right)^2, \left(\frac{\xi}{\upsilon}\right)^2\right\} < y < \min\left\{\left(\frac{\mu}{\varpi}\right)^2, \left(\frac{\rho}{\upsilon}\right)^2\right\}$,

$$\begin{aligned} F_Y(y) &= \int_{\varpi}^{\frac{\mu}{\sqrt{y}}}\frac{\tau}{\Psi}\frac{x\left[A\left(yx^2\right)-\mathbb{A}\right]}{\sqrt{a^2-\left(b-x^2\right)^2}}dx + \int_{\frac{\mu}{\sqrt{y}}}^{\frac{\xi}{\sqrt{y}}}\frac{\tau}{\Psi}\frac{x\left[B\left(yx^2\right)+\mathbb{B}\right]}{\sqrt{a^2-\left(b-x^2\right)^2}}dx \\ &\quad + \int_{\frac{\xi}{\sqrt{y}}}^{\upsilon}\frac{\tau}{\Psi}\frac{x\left[C\left(yx^2\right)+\mathbb{C}\right]}{\sqrt{a^2-\left(b-x^2\right)^2}}dx \\ &= \frac{\tau}{\Psi}\left[\mathbb{F}\left(y;\frac{\mu}{\sqrt{y}},\varpi\right) + \mathbb{G}\left(y;\frac{\xi}{\sqrt{y}},\frac{\mu}{\sqrt{y}}\right) + \mathbb{H}\left(y;\upsilon,\frac{\xi}{\sqrt{y}}\right)\right], \end{aligned} \tag{3.110}$$

where

$$\mathbb{H}(y;\lambda,\kappa) = g_3(y;\lambda) - g_3(y;\kappa) + h_3(y;\lambda,\kappa), \tag{3.111}$$

$$\begin{aligned} g_3(y;x) &= -\frac{1}{2}y\left(U_1^2 - R_D^2 cos^2\Psi + R_D^2\sin^2\Psi\right)\left[\sqrt{a-(b-x)^2} + b\arccos\left(\frac{x^2-b}{\sqrt{a}}\right)\right] \\ &\quad + \frac{1}{8}y^2\left[\left(2b^2+a\right)\arccos\left(\frac{x^2-b}{\sqrt{a}}\right) + \left(x^2+3b\right)\sqrt{a-(b-x)^2}\right] \\ &\quad - \frac{\mathbb{C}}{2}\arccos\left(\frac{x^2-b}{\sqrt{a}}\right), \end{aligned} \tag{3.112}$$

and

$$h_3(y;\lambda,\kappa) = -\frac{2}{3}R_D\cos\Psi(\lambda-\kappa)\sum_{i=1}^{Q}\frac{\pi}{Q}\sqrt{1-\vartheta_i^2}\left(yx_0^2 - R_D^2\sin^2\Psi\right). \tag{3.113}$$

(6) $\eta < \varpi\sqrt{y} < \mu < \xi < \rho < \upsilon\sqrt{y}$, in this situation, $\max\left\{\left(\frac{\eta}{\varpi}\right)^2, \left(\frac{\rho}{\upsilon}\right)^2\right\} < y < \left(\frac{\mu}{\varpi}\right)^2$, the OP is given by

$$\begin{aligned} F_Y(y) &= \int_{\varpi}^{\frac{\mu}{\sqrt{y}}} \frac{\tau}{\Psi} \frac{x\left[A\left(yx^2\right) - \mathbb{A}\right]}{\sqrt{a^2 - \left(b - x^2\right)^2}} dx + \int_{\frac{\mu}{\sqrt{y}}}^{\frac{\xi}{\sqrt{y}}} \frac{\tau}{\Psi} \frac{x\left[B\left(yx^2\right) + \mathbb{B}\right]}{\sqrt{a^2 - \left(b - x^2\right)^2}} dx \\ &\quad + \int_{\frac{\xi}{\sqrt{y}}}^{\rho} \frac{\tau}{\Psi} \frac{x\left[C\left(yx^2\right) + \mathbb{C}\right]}{\sqrt{a^2 - \left(b - x^2\right)^2}} dx + \int_{\rho}^{\upsilon} \frac{2}{\Psi} \frac{x}{\sqrt{a^2 - \left(b - x^2\right)^2}} dx \\ &= \frac{\tau}{\Psi}\left[\mathbb{F}\left(y; \frac{\mu}{\sqrt{y}}, \varpi\right) + \mathbb{G}\left(y; \frac{\xi}{\sqrt{y}}, \frac{\mu}{\sqrt{y}}\right) + \mathbb{H}\left(y; \rho, \frac{\xi}{\sqrt{y}}\right) + \mathbb{I}\right], \end{aligned} \tag{3.114}$$

where $\mathbb{I} = \frac{1}{\tau}\arccos\left(\frac{\upsilon^2 - b}{\sqrt{a}}\right) - \frac{1}{\tau}\arccos\left(\frac{\rho^2 - b}{\sqrt{a}}\right)$.

(7) $\mu < \varpi\sqrt{y} < \upsilon\sqrt{y} < \xi$, in this case, $\left(\frac{\mu}{\varpi}\right)^2 < y < \left(\frac{\xi}{\upsilon}\right)^2$, then the OP can be obtained as

$$\begin{aligned} F_Y(y) &= \int_{\varpi}^{\upsilon} \frac{\tau}{\Psi} \frac{x\left[B\left(yx^2\right) + \mathbb{B}\right]}{\sqrt{a^2 - \left(b - x^2\right)^2}} dx \\ &= \frac{\tau}{\Psi}\left[\mathbb{G}\left(y; \upsilon, \varpi\right)\right]. \end{aligned} \tag{3.115}$$

(8) $\mu < \varpi\sqrt{y} < \xi < \upsilon\sqrt{y} < \rho$, in this case, $\max\left\{\left(\frac{\mu}{\varpi}\right)^2, \left(\frac{\xi}{\upsilon}\right)^2\right\} < y < \min\left\{\left(\frac{\xi}{\varpi}\right)^2, \left(\frac{\rho}{\upsilon}\right)^2\right\}$, then we can derive the OP as

$$\begin{aligned} F_Y(y) &= \int_{\varpi}^{\frac{\xi}{\sqrt{y}}} \frac{\tau}{\Psi} \frac{x\left[B\left(yx^2\right) + \mathbb{B}\right]}{\sqrt{a^2 - \left(b - x^2\right)^2}} dx + \int_{\frac{\xi}{\sqrt{y}}}^{\upsilon} \frac{\tau}{\Psi} \frac{x\left[C\left(yx^2\right) + \mathbb{C}\right]}{\sqrt{a^2 - \left(b - x^2\right)^2}} dx \\ &= \frac{\tau}{\Psi}\left[\mathbb{G}\left(y; \frac{\xi}{\sqrt{y}}, \varpi\right) + \mathbb{H}\left(y; \upsilon, \frac{\xi}{\sqrt{y}}\right)\right]. \end{aligned} \tag{3.116}$$

(9) $\mu < \varpi\sqrt{y} < \xi < \rho < \upsilon\sqrt{y}$, in this case, $\max\left\{\left(\frac{\rho}{\upsilon}\right)^2, \left(\frac{\mu}{\varpi}\right)^2\right\} < y < \left(\frac{\xi}{\varpi}\right)^2$, the OP can be obtained as

$$\begin{aligned} F_Y(y) &= \int_{\varpi}^{\frac{\xi}{\sqrt{y}}} \frac{\tau}{\Psi} \frac{x\left[B\left(yx^2\right) + \mathbb{B}\right]}{\sqrt{a^2 - \left(b - x^2\right)^2}} dx + \int_{\frac{\xi}{\sqrt{y}}}^{\rho} \frac{\tau}{\Psi} \frac{x\left[C\left(yx^2\right) + \mathbb{C}\right]}{\sqrt{a^2 - \left(b - x^2\right)^2}} dx \\ &\quad + \int_{\rho}^{\upsilon} \frac{1}{\Psi} \frac{2x}{\sqrt{a^2 - \left(b - x^2\right)^2}} dx \end{aligned}$$

$$= \frac{\tau}{\Psi}\left[\mathbb{G}\left(y; \frac{\xi}{\sqrt{y}}, \varpi\right) + \mathbb{H}\left(y; \rho, \frac{\xi}{\sqrt{y}}\right) + \mathbb{I}\right]. \tag{3.117}$$

(10) $\xi < \varpi\sqrt{y} < \rho < \upsilon\sqrt{y}$, in this case, $\max\left\{\left(\frac{\xi}{\varpi}\right)^2, \left(\frac{\rho}{\upsilon}\right)^2\right\} < y < \left(\frac{\rho}{\varpi}\right)^2$,

$$\begin{aligned} F_Y(y) &= \int_{\varpi}^{\rho} \frac{\tau}{\Psi} \frac{x\left[C\left(yx^2\right) + \mathbb{C}\right]}{\sqrt{a^2 - \left(b - x^2\right)^2}} dx + \int_{\rho}^{\upsilon} \frac{2}{\Psi} \frac{x}{\sqrt{a^2 - \left(b - x^2\right)^2}} dx \\ &= \frac{\tau}{\Psi}\left[\mathbb{H}(y; \rho, \varpi) + \mathbb{I}\right]. \end{aligned} \tag{3.118}$$

(11) $\rho < \varpi\sqrt{y} < \upsilon\sqrt{y}$, in this case, $y > \left(\frac{\rho}{\varpi}\right)^2$,

$$F_Y(y) = \int_{\varpi}^{\upsilon} F_{d_{ID}}\left(\sqrt{y} d_{SD}\right) f_{d_{SD}}\left(d_{SD}\right) = 1. \tag{3.119}$$

2. When $\mu \geq \xi$, the CDF of $d_{ID}$ can be expressed as

$$F_{d_{ID}}(x) = \begin{cases} 0, & \text{if } 0 \leq x \leq \eta \\ \frac{\tau}{\Psi}\left[A\left(x^2\right) - \mathbb{A}\right], & \text{else if } \eta \leq x \leq \xi \\ \frac{\tau}{\Psi}\left[D\left(x^2\right) + \mathbb{D}\right], & \text{else if } \xi \leq x \leq \mu \\ \frac{\tau}{\Psi}\left[C\left(x^2\right) + \mathbb{E}\right], & \text{else if } \mu \leq x \leq \rho \\ 1, & \text{else } x \geq \rho \end{cases}, \tag{3.120}$$

Then we can further obtain the following situations as:

(1) $\varpi\sqrt{y} < \upsilon\sqrt{y} < \eta$, in this situation, $y < \left(\frac{\eta}{\upsilon}\right)^2$, then the OP is given by

$$F_Y(y) = \int_{\varpi}^{\upsilon} F_{d_{ID}}\left(\sqrt{y} d_{SD}\right) f_{d_{SD}}\left(d_{SD}\right) d\left(d_{SD}\right) = 0. \tag{3.121}$$

(2) $\varpi\sqrt{y} < \eta < \upsilon\sqrt{y} < \xi$, in this case, $\left(\frac{\eta}{\upsilon}\right)^2 < y < \min\left\{\left(\frac{\eta}{\varpi}\right)^2, \left(\frac{\xi}{\upsilon}\right)^2\right\}$, then the OP can be obtained the same as (3.102).

(3) $\varpi\sqrt{y} < \eta < \xi < \upsilon\sqrt{y} < \mu$, in this situation, $\left(\frac{\xi}{\upsilon}\right)^2 < y < \left(\frac{\eta}{\varpi}\right)^2$, the OP can be obtained as

$$\begin{aligned} F_Y(y) &= \int_{\eta}^{\frac{\xi}{\sqrt{y}}} \frac{\tau}{\Psi} \frac{x\left[A\left(yx^2\right) - \mathbb{A}\right]}{\sqrt{a^2 - \left(b - x^2\right)^2}} dx + \int_{\frac{\xi}{\sqrt{y}}}^{\upsilon} \frac{\tau}{\Psi} \frac{x\left[D\left(x^2\right) + \mathbb{D}\right]}{\sqrt{a^2 - \left(b - x^2\right)^2}} dx \\ &= \frac{\tau}{\Psi}\left[\mathbb{F}\left(y; \frac{\xi}{\sqrt{y}}, \eta\right) + g_4(\upsilon) - g_4\left(\frac{\xi}{\sqrt{y}}\right)\right] \\ &= \frac{\tau}{\Psi}\left[\mathbb{F}\left(y; \frac{\xi}{\sqrt{y}}, \eta\right) + \mathbb{J}\left(y; \upsilon, \frac{\xi}{\sqrt{y}}\right)\right], \end{aligned} \tag{3.122}$$

where

$$g_4(x) = -\frac{1}{2}y\left(U_1^2 - U_2^2\right)\left[\sqrt{a-(b-x)^2} + b\arccos\left(\frac{x^2-b}{\sqrt{a}}\right)\right] - \frac{1}{2}\mathbb{D}\arccos\left(\frac{x^2-b}{\sqrt{a}}\right) \tag{3.123}$$

and $\mathbb{J}(y; \lambda, \kappa) = g_4(\lambda) - g_4(\kappa)$.

(4) $\eta < \varpi\sqrt{y} < \upsilon\sqrt{y} < \xi$, in this situation, $\left(\frac{\eta}{\varpi}\right)^2 < y < \left(\frac{\xi}{\upsilon}\right)^2$, then the OP is the same as (3.114).

(5) $\eta < \varpi\sqrt{y} < \xi < \upsilon\sqrt{y} < \mu$, in this case, $\max\left\{\left(\frac{\eta}{\varpi}\right)^2, \left(\frac{\xi}{\upsilon}\right)^2\right\} < y < \min\left\{\left(\frac{\xi}{\varpi}\right)^2, \left(\frac{\mu}{\upsilon}\right)^2\right\}$, then the OP can be obtained as

$$\begin{aligned} F_Y(y) &= \int_{\varpi}^{\frac{\xi}{\sqrt{y}}} \frac{\tau}{\Psi}\frac{x\left[A\left(yx^2\right) - \mathbb{A}\right]}{\sqrt{a^2-\left(b-x^2\right)^2}}dx + \int_{\frac{\xi}{\sqrt{y}}}^{\upsilon} \frac{\tau}{\Psi}\frac{x\left[D\left(x^2\right) + \mathbb{D}\right]}{\sqrt{a^2-\left(b-x^2\right)^2}}dx \\ &= \frac{\tau}{\Psi}\left[\mathbb{F}\left(y; \frac{\xi}{\sqrt{y}}, \varpi\right) + \mathbb{J}\left(y; \upsilon, \frac{\xi}{\sqrt{y}}\right)\right]. \end{aligned} \tag{3.124}$$

(6) $\eta < \varpi\sqrt{y} < \xi < \mu < \upsilon\sqrt{y} < \rho$, in this situation, $\max\left\{\left(\frac{\eta}{\varpi}\right)^2, \left(\frac{\mu}{\upsilon}\right)^2\right\} < y < \min\left\{\left(\frac{\xi}{\varpi}\right)^2, \left(\frac{\rho}{\upsilon}\right)^2\right\}$, we can obtain that

$$\begin{aligned} F_Y(y) &= \int_{\varpi}^{\frac{\xi}{\sqrt{y}}} \frac{\tau}{\Psi}\frac{x\left[A\left(yx^2\right) - \mathbb{A}\right]}{\sqrt{a^2-\left(b-x^2\right)^2}}dx + \int_{\frac{\xi}{\sqrt{y}}}^{\frac{\mu}{\sqrt{y}}} \frac{\tau}{\Psi}\frac{x\left[D\left(yx^2\right) + \mathbb{D}\right]}{\sqrt{a^2-\left(b-x^2\right)^2}}dx \\ &\quad + \int_{\frac{\mu}{\sqrt{y}}}^{\upsilon} \frac{\tau}{\Psi}\frac{x\left[C\left(yx^2\right) + \mathbb{E}\right]}{\sqrt{a^2-\left(b-x^2\right)^2}}dx \\ &= \frac{\tau}{\Psi}\left[\mathbb{F}\left(y; \frac{\xi}{\sqrt{y}}, \varpi\right) + \mathbb{J}\left(y; \frac{\mu}{\sqrt{y}}, \frac{\xi}{\sqrt{y}}\right) + \mathbb{K}\left(y; \upsilon, \frac{\mu}{\sqrt{y}}\right)\right], \end{aligned} \tag{3.125}$$

where

$$\mathbb{K}(y; \lambda, \kappa) = g_5(\lambda) - g_5(\kappa) + h_3(\lambda, \kappa) \tag{3.126}$$

and

$$\begin{aligned} g_5(x) = &-\frac{1}{2}y\left(U_1^2 - R_D^2\cos^2\Psi + R_D^2\sin^2\Psi\right)\left[\sqrt{a-(b-x)^2} + b\arccos\left(\frac{x^2-b}{\sqrt{a}}\right)\right] \\ &+ \frac{1}{8}y^2\left[\left(2b^2+a\right)\arccos\left(\frac{x^2-b}{\sqrt{a}}\right) + \left(x^2+3b\right)\sqrt{a-(b-x)^2}\right] \end{aligned}$$

$$-\frac{1}{2}\mathbb{E}\arccos\left(\frac{x^2-b}{\sqrt{a}}\right). \tag{3.127}$$

(7) $\eta < \varpi\sqrt{y} < \xi < \mu < \rho < \upsilon\sqrt{y}$, in this situation, $\max\left\{\left(\frac{\eta}{\varpi}\right)^2, \left(\frac{\rho}{\upsilon}\right)^2\right\} < y < \left(\frac{\xi}{\varpi}\right)^2$, the OP can be derived as

$$\begin{aligned}F_Y(y) &= \int_{\varpi}^{\frac{\xi}{\sqrt{y}}} \frac{\tau}{\Psi}\frac{x\left[A\left(yx^2\right)-\mathbb{A}\right]}{\sqrt{a^2-\left(b-x^2\right)^2}}dx + \int_{\frac{\xi}{\sqrt{y}}}^{\frac{\mu}{\sqrt{y}}} \frac{\tau}{\Psi}\frac{x\left[D\left(yx^2\right)+\mathbb{D}\right]}{\sqrt{a^2-\left(b-x^2\right)^2}}dx \\ &+ \int_{\frac{\mu}{\sqrt{y}}}^{\rho} \frac{\tau}{\Psi}\frac{x\left[C\left(yx^2\right)+\mathbb{E}\right]}{\sqrt{a^2-\left(b-x^2\right)^2}}dx + \int_{\rho}^{\upsilon} \frac{2}{\Psi}\frac{x}{\sqrt{a^2-\left(b-x^2\right)^2}}dx \\ &= \frac{\tau}{\Psi}\left[\mathbb{F}\left(y;\frac{\xi}{\sqrt{y}},\varpi\right)+\mathbb{J}\left(y;\frac{\mu}{\sqrt{y}},\frac{\xi}{\sqrt{y}}\right)+\mathbb{K}\left(y;\rho,\frac{\mu}{\sqrt{y}}\right)+\mathbb{I}\right].\end{aligned} \tag{3.128}$$

(8) $\xi < \varpi\sqrt{y} < \mu < \rho < \upsilon\sqrt{y}$, in this case, $\max\left\{\left(\frac{\xi}{\varpi}\right)^2, \left(\frac{\rho}{\upsilon}\right)^2\right\} < y < \left(\frac{\mu}{\varpi}\right)^2$,

$$\begin{aligned}F_Y(y) &= \int_{\varpi}^{\frac{\mu}{\sqrt{y}}} \frac{\tau}{\Psi}\frac{x\left[D\left(yx^2\right)+\mathbb{D}\right]}{\sqrt{a^2-\left(b-x^2\right)^2}}dx + \int_{\frac{\mu}{\sqrt{y}}}^{\rho} \frac{\tau}{\Psi}\frac{x\left[C\left(yx^2\right)+\mathbb{E}\right]}{\sqrt{a^2-\left(b-x^2\right)^2}}dx \\ &+ \int_{\rho}^{\upsilon} \frac{2}{\Psi}\frac{x}{\sqrt{a^2-\left(b-x^2\right)^2}}dx \\ &= \frac{\tau}{\Psi}\left[\mathbb{J}\left(y;\frac{\mu}{\sqrt{y}},\varpi\right)+\mathbb{K}\left(y;\rho,\frac{\mu}{\sqrt{y}}\right)+\mathbb{I}\right].\end{aligned} \tag{3.129}$$

(9) $\mu < \varpi\sqrt{y} < \rho < \upsilon\sqrt{y}$, in this situation, $\max\left\{\left(\frac{\mu}{\varpi}\right)^2, \left(\frac{\rho}{\upsilon}\right)^2\right\} < y < \left(\frac{\rho}{\varpi}\right)^2$

$$\begin{aligned}F_Y(y) &= \int_{\varpi}^{\rho} \frac{\tau}{\Psi}\frac{x\left[C\left(yx^2\right)+\mathbb{E}\right]}{\sqrt{a^2-\left(b-x^2\right)^2}}dx + \int_{\rho}^{\upsilon} \frac{2}{\Psi}\frac{x}{\sqrt{a^2-\left(b-x^2\right)^2}}dx \\ &= \frac{\tau}{\Psi}\left[\mathbb{K}\left(y;\rho,\varpi\right)+\mathbb{I}\right].\end{aligned} \tag{3.130}$$

(10) $\rho < \varpi\sqrt{y} < \upsilon\sqrt{y}$, in this situation, $y > \left(\frac{\rho}{\varpi}\right)^2$,

$$F_Y(y) = \int_{\varpi}^{\upsilon} F_{d_{ID}^2}\left(yx^2\right) f_{d_{SD}}\left(d_{SD}\right)dx = 1. \tag{3.131}$$

## References


1. G. Pan, J. Ye, J. An, and S. Alouini, "Latency versus reliability in LEO mega-constellations: Terrestrial, aerial, or space relay," *IEEE Trans. Mob. Comput.*, vol. 22, no. 9, pp. 5330–5345, Sept. 2023.
2. I.S. Gradshteyn and I. M. Ryzhik, *Table of Integrals, Products*. New York, NY, USA: Academic, 2014.
3. A. Abdi, W. C. Lau, M. Alouini, and M. Kaveh, et al. "A new simple model for land mobile satellite channels: First- and second-order statistics," *IEEE Trans. Wireless Commun.*, vol. 2, no. 3, pp. 519–528, May 2003.
4. X. Yan, H. Xiao, K. An, G. Zheng, and W. Tao, "Hybrid satellite terrestrial relay networks with cooperative non-orthogonal multiple access," *IEEE Commun. Lett.*, vol. 22, no. 5, pp. 978–981, May 2018.
5. K. Guo, M. Lin, B. Zhang, W. Zhu, J. Wang, and T. A. Tsiftsis, "On the performance of LMS communication with hardware impairments and interference," *IEEE Trans. Commun.*, vol. 67, no. 2, pp. 1490–1505, Feb. 2019.
6. M. Haenggi, *Stochastic Geometry for Wireless Networks*. Cambridge, U.K.: Cambridge, Univ. Press, 2012.
7. S. M. Azimi-Abarghouyi, B. Makki, M. Haenggi, M. Nasiri-Kenari, and T. Svensson, "Stochastic geometry modeling and analysis of single- and multi-cluster wireless networks," *IEEE Trans. Commun.*, vol. 66, no. 10, pp. 4981–4996, Oct. 2018.
8. M. Afshang and H. S. Dhillon, "Fundamentals of modeling finite wireless networks using binomial point process," *IEEE Trans. Wireless Commun.*, vol. 16, no. 5, pp. 3355–3370, May 2017.
9. X. Yan, H. Xiao, K. An, G. Zheng, and W. Tao, "Hybrid satellite terrestrial relay networks with cooperative non-orthogonal multiple access," *IEEE Commun. Lett.*, vol. 22, no. 5, pp. 978–981, May 2018.
10. G. Pan, H. Lei, Z. Ding, and Q. Ni, "3-D hybrid VLC-RF indoor IoT systems with light energy harvesting," *IEEE Trans. Green Commun. Netw.*, vol. 3, no. 3, pp. 853–865, Sep. 2019.

# Chapter 4
# Satellite-Aerial-Terrestrial Communications

**Haoxing Zhang, Gaofeng Pan, Shuai Wang, and Mohamed-Slim Alouini**

## 4.1 Modeling and Optimization of Cooperative Satellite-Aerial-Terrestrial Systems

### *4.1.1 System Model*

In this work, a cooperative satellite-aerial-terrestrial communication (CSATC) system, which consists of a satellite transmitter (S), an aerial relay (R), and a group of terrestrial receivers ($D_i$, $1 \leq i \leq N$), is considered, as shown in Fig. 4.1. Specifically, S delivers its information to terrestrial receivers via the help of R, due to the deep fading between S and D, and the hardware limitations of terrestrial receivers (e.g., wireless sensors and common mobile phones) on realizing large-size radio-frequency receiving front end. Decode-and-forward (DF) relay scheme is adopted at R to process and forward the received signals from S.

For tractability purposes, in this work we also treat the coverage area of R on the ground as a circle, $\mathcal{S}$, with radius, $L$, where the projection of R, O, is located at the centre of the circle. The location of terrestrial receivers is modeled as a homogeneous Poisson Point process (PPP) with density $\vartheta$. The number of terrestrial receivers, which denoted by $N$ ($N \leq 1$), is Poisson distributed i.e., $P\{N = k\} = \left(\mu_S^k / k!\right) \exp\left(-\mu_S\right)$, where $\mu_S = \pi L^2 \vartheta$ is the mean measure. Then, the $N$ ($N \leq 1$) terrestrial receivers can be modeled as a set of independently and identically distributed points in the circle, $\mathcal{S}$, denoted by $\mathcal{W}$. Therefore, the distance

H. Zhang · G. Pan (✉) · S. Wang
Beijing Institute of Technology, Beijing, China
e-mail: gfpan@bit.edu.cn

M.-S. Alouini
King Abdullah University of Science and Technology (KAUST), Thuwal, Saudi Arabia



S. Wang et al. (eds.), *Understanding Satellite Communications: The Stochastic Geometry Perspective*, https://doi.org/10.1007/978-981-97-7102-8_4

**Fig. 4.1** System model

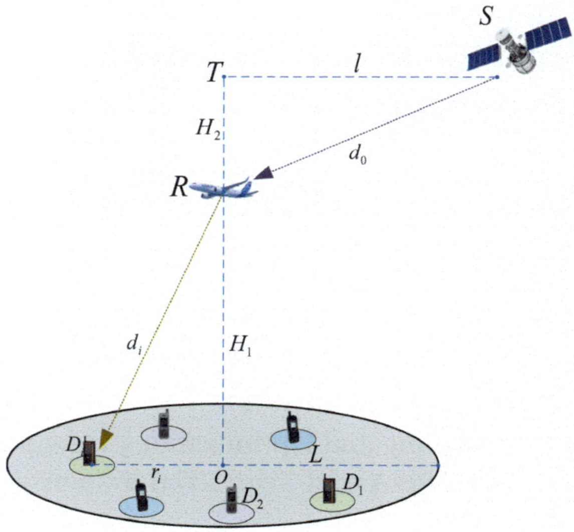


between terrestrial receivers and the projection of R (namely, O) can be calculated from $\mathcal{W}$, the probability density function (PDF) of which can be given by [1, 2]

$$f_{\mathcal{W}}(w) = \frac{\vartheta}{\mu(\mathcal{S})} = \frac{1}{\pi L^2}. \tag{4.1}$$

Furthermore, in this work shadowed-Rician SR model [3] is adopted to describe the statistical distribution of the satellite link from S to R, which has been proved to be an accurate, practical and applicable tool to evaluate the performance of the satellite propagation environments in various frequency bands, e.g., the UHF-band, L-band, S-band, Ku-band and Ka-band.[1]

Without loss of generality, the PDF of the power gain, $|h_{\mathrm{SR}}|^2$, over S-R link in SR fading is given as [3]

$$f_{|h_{\mathrm{SR}}|^2}(x) = \alpha \exp(-\beta x)\, {}_1F_1(m; 1; \delta x), \quad x \geq 0, \tag{4.2}$$

[1] In this work, a low-altitude platform (LAP) serves as the aerial relay, the height of which ranges from a few dozen meters to a few kilometers. Then, when the height of the LAP is low, the received signal will suffer both multi-path fading and large-scale fading/shadowing simultaneously; When the height of the LAP is high, the received signal may only experience line-of-sight (LOS) shadow fading. SR channel model is adopted here to address the channel fading of S-R link, as it is capable of describing the scenarios of pure LOS fading and mixed fading of multi-path fading and shadow fading [3].

where $\alpha = \left(\frac{2bm}{2bm+\Omega}\right)^m/(2b)$, $\beta = \frac{1}{2b}$, and $\delta = \frac{\Omega}{2b(2bm+\Omega)}$, $\Omega$ and $2b$ are the average power of the LOS and multi-path components, respectively, $m$ is the fading severity parameter,[2] and ${}_1F_1(\cdot;\cdot;\cdot)$ is the confluent hypergeometric function of the first kind.

Then, it is easy to obtain the received signal-to-noise ratio (SNR) at R as

$$\gamma_{\mathrm{SR}} = \frac{P_{\mathrm{S}}}{\sigma^2}\frac{|h_{\mathrm{SR}}|^2}{d_0^{n_1}} = \frac{\lambda_{\mathrm{S}}}{d_0^{n_1}}, \tag{4.3}$$

where $\lambda_{\mathrm{S}} = P_{\mathrm{S}}|h_{\mathrm{SR}}|^2/\sigma^2$, $P_{\mathrm{S}}$ and $\sigma^2$ are the transmit power at S and the average power of the additive white Gaussian noise (AWGN) at R,[3] $d_0$ is the distance between S and R, $n_1$ is the path-loss factor over S-R link).

By variable substitution, the PDF and cumulative distribution function (CDF) of $\lambda_{\mathrm{S}}$ can be presented as

$$f_{\lambda_{\mathrm{S}}}(x) = \alpha \sum_{k=0}^{m-1} \frac{\varsigma(k)}{\bar{\lambda}^{k+1}} x^k \exp\left(-\frac{\beta-\delta}{\bar{\lambda}}x\right) \tag{4.4}$$

and

$$F_{\lambda_{\mathrm{S}}}(x) = 1 - \alpha \sum_{k=0}^{m-1} \frac{\varsigma(k)}{\bar{\lambda}^{k+1}} \sum_{p=0}^{k} \frac{k!}{p!} \left(\frac{\beta-\delta}{\bar{\lambda}}\right)^{-(k+1-p)} x^p \exp\left(-\frac{\beta-\delta}{\bar{\lambda}}x\right), \tag{4.5}$$

respectively, where $\bar{\lambda} = P_{\mathrm{S}}/\sigma^2$, $\varsigma(k) = \frac{(-1)^k(1-m)_k\delta^k}{(k!)^2}$ and $(t)_k = t(t+1)\cdots(t+k-1)$ is the Pochhammer symbol [4].

Similarly, the instantaneous received SNR at the $i$th terrestrial receiver ($\mathrm{D}_i$, $1 \le i \le N$), can be written as[4]

$$\gamma_{\mathrm{RD}} = \frac{P_{\mathrm{R}}}{\sigma^2}\frac{|h_{\mathrm{RD}}|^2}{d_i^{n_2}} = \frac{\lambda_{\mathrm{R}}}{d_i^{n_2}}, \tag{4.6}$$

where $\lambda_{\mathrm{R}} = P_{\mathrm{R}}|h_{\mathrm{RD}}|^2/\sigma^2$, $P_{\mathrm{R}}$ is the transmit power at R, $d_i$ is the distance between R and the $i$th terrestrial receiver, $n_2$ is the path-loss factor over R-$\mathrm{D}_i$ link, and $h_{\mathrm{RD}}$ is the channel gain of R-$\mathrm{D}_i$ link.

In this work, we assume that the channel among R and the $i$th terrestrial receiver follows Rician distribution. Then, the PDF and CDF of $\lambda_{\mathrm{R}}$ can be given as [5]

---

[2] When $m = 0$, it represents the LOS case; when $0 < m < \infty$, it stands for the case of both small-scale fading and LOS; when $m = \infty$, it denotes the case without LOS.

[3] In this work, for simplification purposes, it is assumed that the average power of the AWGN at all terminals is same, namely, $\sigma^2$.

[4] In this following parts, we adopt D instead of $\mathrm{D}_i$ in equations by ignoring the subscript for simplification.

$$f_{\lambda_{\mathrm{R}}}(x) = \frac{(1+K)}{\Omega_{\mathrm{R}}} \exp\left(-K - \frac{x(1+K)}{\Omega_{\mathrm{R}}}\right) I_0\left(2\sqrt{\frac{K(1+K)x}{\Omega_{\mathrm{R}}}}\right), \quad x \geq 0 \tag{4.7}$$

and

$$F_{\lambda_{\mathrm{R}}}(x) = 1 - Q_1\left(\sqrt{2K}, \sqrt{\frac{2(1+K)x}{\Omega_{\mathrm{R}}}}\right), \tag{4.8}$$

respectively, where $\Omega_{\mathrm{R}}$ is the variance of the signal, $K$ is the Ricci factor, which corresponds to the ratio of the power of the LOS (specular) component to the average power of the scattered component, $I_0(\cdot)$ is modified Bessel function of order 0, $Q_1(a, b) = \int_b^{\infty} x \exp\left(-\frac{x^2+a^2}{2}\right) I_0(ax)\, dx$ is Marcum-$Q$ function.

### 4.1.2 *Coverage Analysis*

In this subsection, coverage analysis will be carried out under two cases, namely, the ones with/without interference. Also, without loss of generality, we take the $i$th terrestrial receiver, $\mathrm{D}_i$, as the target to study the coverage performance.

#### 4.1.2.1 Non-interference Scenario

As one of the most important metrics to evaluate the performance of wireless networks (especially for some application scenarios, e.g., battlefield and hot-spots), coverage probability (CP) is defined as the probability that a typical user can achieve some threshold of SNR/SINR. In this part, we will analyze the coverage performance of R-$\mathrm{D}_i$ link, while considering the randomness of the positions of terrestrial receivers.

In this work, without loss of generality, the $i$th terrestrial receiver, $\mathrm{D}_i$ $(1 \leq i \leq N)$, is considered as the target for the following analysis.

Then, CP can be written as

$$\begin{aligned} p_c &= \Pr\{\gamma_{\mathrm{RD}} \geq \gamma_{\mathrm{th}}\} \\ &= 1 - \Pr\left\{\lambda_{\mathrm{R}} \leq \gamma_{\mathrm{th}} d_i^{n_2}\right\} \\ &= Q_1\left(\sqrt{2K}, \sqrt{\frac{2(1+K)\gamma_{\mathrm{th}}}{\Omega_{\mathrm{R}}} d_i^{n_2}}\right), \end{aligned} \tag{4.9}$$

where $\gamma_{\mathrm{th}}$ is the predefined threshold.

Using (4.1), we can derive the CDF of $r_i$ $(1 \leq i \leq N)$ as

$$F_{r_i}(x) = \int_0^{2\pi}\int_0^{x} \frac{1}{\pi L^2} r dr d\theta = \frac{x^2}{L^2}. \tag{4.10}$$

Therefore, the PDF of $r_i$ $(1 \leq i \leq N)$ can be obtained as $f_{r_i}(x) = \frac{2x}{L^2}, 0 \leq x \leq L$. Further we can achieve the PDF of $d_i = \sqrt{{H_1}^2 + r_i^2}$ $(1 \leq i \leq N)$ as $f_{d_i}(x) = \frac{2}{L^2}x$, $H_1 \leq x \leq \sqrt{{H_1}^2 + L^2}$.

Considering the randomness of the position of $\mathrm{D}_i$, CP can be further calculated as

$$\begin{aligned} p_c &= \int_{H_1}^{\sqrt{{H_1}^2+L^2}} Q_1\left(\sqrt{2K}, \sqrt{\frac{2(1+K)\gamma_{\text{th}}}{\Omega_{\text{R}}} x^{n_2}}\right) f_{d_i}(x)\, dx \\ &= \frac{2}{L^2}\int_{H_1}^{\sqrt{{H_1}^2+L^2}} x Q_1\left(\sqrt{2K}, \sqrt{\frac{2(1+K)\gamma_{\text{th}}}{\Omega_{\text{R}}} x^{n_2}}\right) dx. \end{aligned} \tag{4.11}$$

As suggested in [6], Marcum $Q$-function can be approximated as[5]

$$Q_1(a, b) \approx \exp\left(-e^{\upsilon(a)} b^{\mu(a)}\right), \tag{4.12}$$

where $\mu(a) = 2.174 - 0.592a + 0.593a^2 - 0.092a^3 + 0.005a^4$ and $\upsilon(a) = -0.840 + 0.327a - 0.740a^2 + 0.083a^3 - 0.004a^4$.

Then, we can obtain

$$Q_1\left(\sqrt{2K}, \sqrt{\frac{2(1+K)\gamma_{\text{th}}}{\Omega_{\text{R}}} x^{n_2}}\right) \approx \exp\left(-e^{\upsilon}\Theta^{\frac{\mu}{2}} x^{\frac{n_2\mu}{2}}\right), \tag{4.13}$$

where $\Theta = \frac{2(1+K)\gamma_{\text{th}}}{\Omega_{\text{R}}}$, $\mu = 2.174 - 0.937\sqrt{K} + 1.186K - 0.260K^{\frac{3}{2}} + 0.010K^2$ and $\upsilon = -0.840 + 0.809\sqrt{K} - 1.480K + 0.235K^{\frac{3}{2}} - 0.008K^2$.

[5] As suggested by Fig. 3 in [6], the approximation of Marcum $Q$-function given by (4.12) is robust with respect to changes in $a$ from 1 to 5 and can adequately represent the mass of the Marcum $Q$-function while $b$ ranging from 0 to 10.

Using (4.13) into (4.11), the CP for non-interference scenario can be derived as

$$
\begin{aligned}
p_c &\approx \frac{2}{L^2} \int_{H_1}^{\sqrt{H_1{}^2+L^2}} x \exp\left(-e^{\upsilon}\Theta^{\frac{\mu}{2}} x^{\frac{n_2\mu}{2}}\right) dx \\
&= \frac{4}{L^2 n_2 \mu} e^{-\frac{4\upsilon}{n_2\mu}} \Theta^{-\frac{2}{n_2}} \int_{y_{\min}}^{y_{\max}} y^{\frac{4}{n_2\mu}-1} \exp(-y) dy \\
&= \frac{4}{L^2 n_2 \mu} e^{-\frac{4\upsilon}{n_2\mu}} \Theta^{-\frac{2}{n_2}} \left[\Gamma\left(\frac{4}{n_2\mu}, y_{\min}\right) - \Gamma\left(\frac{4}{n_2\mu}, y_{\max}\right)\right],
\end{aligned}
\tag{4.14}
$$

where in the second line we let $y = e^{\upsilon}\Theta^{\frac{\mu}{2}} x^{\frac{n_2\mu}{2}}$, $y_{\max} = e^{\upsilon}\Theta^{\frac{\mu}{2}} \left(H_1{}^2 + L^2\right)^{\frac{n_2\mu}{4}}$, $y_{\min} = e^{\upsilon}\Theta^{\frac{\mu}{2}} H_1{}^{\frac{n_2\mu}{2}}$, and $\Gamma(a, x) = \int_x^{\infty} t^{a-1} \exp(-t)\, dt$ is the upper incomplete gamma function.

#### 4.1.2.2 Interference Scenario

In this part, we will study the coverage performance of the target system in the presence of a grounded interfering source.

In this work, we assume there is an interfering node existing in the neighbor area of the target terminal, $D_i$, while R delivering information bits to $D_i$. This assumption is reasonable and practical, as satellite-terrestrial communications are mainly designed for the scenarios with or even without limited traditional terrestrial communication facilities, e.g., rural areas and battlefields. For example, a macro-cellular typically spans a few kilometers to tens of kilometers to provide radio coverage for rural areas. Then, it is meaningful to assume that there is only one interfering node for our considered CSATC systems.

As shown in Fig. 4.2, it is assumed that the interfering node, I, is uniformly distributed in the circle with original O and radius $T$ ($T > L$). In order to facilitate the following analysis, we denote the polar coordinates of $D_i$ and I as $(0, r_i)$ and $(\theta, r_I)$, respectively. Then, the distance between $D_i$ and I can be written as $u_i = \sqrt{r_i^2 + r_I^i - 2 r_i r_I \cos\theta}$.

Furthermore, the PDF of $r_I$ and $\theta$ can be easily given as

$$
f_{r_I}(x) = \begin{cases} \frac{1}{T}, & \text{if } 0 \le x \le T; \\ 0, & \text{else} \end{cases}
\tag{4.15}
$$

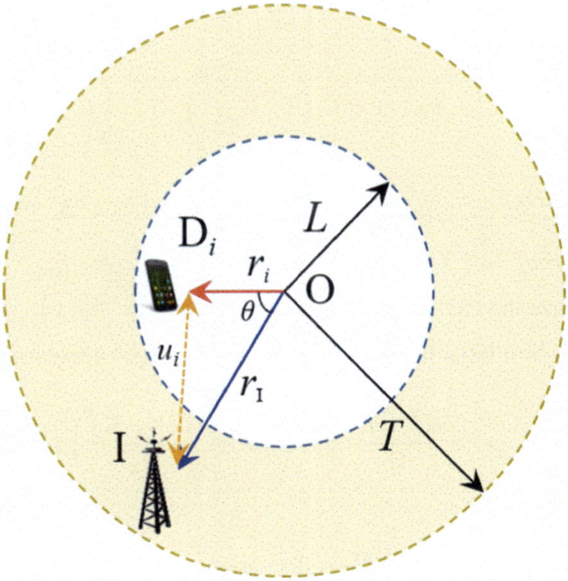


**Fig. 4.2** Interference scenario

and

$$f_\theta(x) = \frac{1}{2\pi}, \tag{4.16}$$

respectively.

Thus, under this case the received signal at $\mathrm{D}_i$ can be given as

$$y_{\mathrm{RD}} = \sqrt{\frac{P_{\mathrm{R}}}{d_i^{n_2}}} h_{\mathrm{RD}} x_{\mathrm{RD}} + \sqrt{\frac{P_{\mathrm{I}}}{u_i^{n_2}}} h_{\mathrm{ID}} x_{\mathrm{I}} + z, \tag{4.17}$$

where $P_{\mathrm{I}}$ is the transmit power at I, $h_{\mathrm{ID}}$ is the channel gain of the interfering I-$\mathrm{D}_i$ link (in this work Rayleigh fading is considered for terrestrial communications, namely, $f_{|h_{\mathrm{ID}}|^2}(x) = \lambda_{\mathrm{I}} \exp(-\lambda_{\mathrm{I}} x)$, $1/\lambda_{\mathrm{I}}$ is the mean of $|h_{\mathrm{ID}}|^2$), $x_i$ is the transmitted bits from I, $z$ is the complex Gaussian noise at $\mathrm{D}_i$ with average power, $\sigma^2$. As the interfering power is normally much larger than the one of AWGN, in the following we adopt signal-to-interference ratio (SIR) to address and study the effect of interfering signal on the coverage performance of the considered system.

So, the received SIR at $\mathrm{D}_i$ can be derived as

$$\gamma_{\mathrm{RD}} = \frac{\frac{P_{\mathrm{R}}|h_{\mathrm{RD}}|^2}{d_i^{n_2}}}{\frac{P_{\mathrm{I}}|h_{\mathrm{ID}}|^2}{u_i^{n_2}}} = \frac{P_{\mathrm{R}}|h_{\mathrm{RD}}|^2}{P_{\mathrm{I}}|h_{\mathrm{ID}}|^2} \frac{u_i^{n_2}}{d_i^{n_2}} = \frac{P_{\mathrm{R}}|h_{\mathrm{RD}}|^2}{P_{\mathrm{I}}|h_{\mathrm{ID}}|^2} \frac{r_i^2 + r_I^i - 2 r_i r_I \cos\theta}{r_i^2 + H_1^2}, \tag{4.18}$$

where the path loss factor $n_2 = 2$ is only considered in this work for tractability.

**Lemma 4.1** *Let* $Z = \frac{P_R |h_{RD}|^2}{P_I |h_{ID}|^2}$, *the CDF of Z can be written as*

$$F_Z(z) = \Gamma_1 \left(\Gamma_2 + \frac{\Gamma_3}{z}\right)^{-1} {}_1F_1\left(1; 1; \Gamma_4\left(\Gamma_2 + \frac{\Gamma_3}{z}\right)^{-1}\right), \tag{4.19}$$

*where* $\Gamma_1 = \frac{1+K}{\Omega_X} \exp(-K)$, $\Gamma_2 = \frac{1+K}{\Omega_X}$, $\Gamma_3 = \frac{\lambda_I P_R}{P_I}$, *and* $\Gamma_4 = \frac{K(1+K)}{\Omega_X}$.

***Proof*** Let $X = |h_{RD}|^2$ and $Y = \frac{P_I}{P_R} |h_{ID}|^2$. As $f_{|h_{ID}|^2}(x) = \lambda_I \exp(-\lambda_I x)$, one can have the CDF of $Y$ as $F_Y(y) = 1 - \exp\left(-\frac{\lambda_I P_R}{P_I} y\right)$.

Similarly, the PDF of $X = |h_{RD}|^2$ can be given as

$$f_X(x) = \frac{(1+K)}{\Omega_X} \exp\left(-K - \frac{x(1+K)}{\Omega_X}\right) I_0\left(2\sqrt{\frac{K(1+K)x}{\Omega_X}}\right), x \geq 0, \tag{4.20}$$

where $\Omega_X = \Omega\sigma^2 / P_R$ is the variance of $X$.

It is easy to write the CDF of $Z$ as

$$F_Z(z) = \Pr\left\{Z = \frac{X}{Y} \leq z\right\} = 1 - \Pr\left\{Y \leq \frac{X}{z}\right\} = \exp\left(-\frac{\lambda_I P_R}{P_I} \frac{X}{z}\right). \tag{4.21}$$

Then, using [7, Eq. (2.15.5.4)], the CDF of $Z = \frac{X}{Y}$ can be further derived as

$$\begin{aligned} F_Z(z) &= \int_0^\infty \exp\left(-\frac{\lambda_I P_R}{P_I} \frac{x}{z}\right) f_X(x)\, dx \\ &= \frac{(1+K)}{\Omega_X} \int_0^\infty \exp\left(-K - \left(\frac{1+K}{\Omega_X} + \frac{1}{z} \frac{\lambda_I P_R}{P_I}\right) x\right) I_0\left(2\sqrt{\frac{K(1+K)x}{\Omega_X}}\right) dx \\ &\overset{t=\sqrt{x}}{=} 2\Gamma_1 \int_0^\infty t \exp\left(-\left(\Gamma_2 + \frac{\Gamma_3}{z}\right) t^2\right) I_0\left(2\Gamma_4^{\frac{1}{2}} t\right) dt \\ &= \Gamma_1 \left(\Gamma_2 + \frac{\Gamma_3}{z}\right)^{-1} {}_1F_1\left(1; 1; \Gamma_4\left(\Gamma_2 + \frac{\Gamma_3}{z}\right)^{-1}\right), \end{aligned} \tag{4.22}$$

where $\Gamma_1 = \frac{1+K}{\Omega_X} \exp(-K)$, $\Gamma_2 = \frac{1+K}{\Omega_X}$, $\Gamma_3 = \frac{\lambda_I P_R}{P_I}$, and $\Gamma_4 = \frac{K(1+K)}{\Omega_X}$.

Then, the proof is completed. □

**Lemma 4.2** *When terminal* $D_i$ *is uniformly distributed in the circle with original O and radius L, and interfering node I is uniformly distributed in the circle with original O and radius T, the CDF of* $\gamma_{RD}$ *can be derived as*

$$F_{\gamma_{\mathrm{RD}}}(x) = \frac{\Gamma_1}{4\pi}\sum_{g}\sum_{j}\sum_{k}\Theta_{g,j,k1}F_1\left(1;1;\Gamma_4\Theta_{g,j,k}\right), \tag{4.23}$$

*where* $\sum_{g}\sum_{j}\sum_{k}$ *denotes* $\sum_{g=1}^{G}\vartheta_g\sum_{j=1}^{H}\zeta_j\sqrt{1-\nu_j^2}\left(\zeta_j+1\right)\sum_{k=1}^{J}\iota_k\sqrt{1-\kappa_k^2}\left(\kappa_k+1\right)$, $\varsigma_g=\cos\left(\frac{2g-1}{2G}\pi\right)$, $\vartheta_g=\frac{\pi}{G}$, $\eta_1=\frac{T}{2}$, $\zeta_j=\cos\left(\frac{2j-1}{2H}\pi\right)$, $\nu_j=\frac{\pi}{H}$, $\eta_2=\frac{L}{2}$, $\kappa_k=\cos\left(\frac{2k-1}{2H}\pi\right)$, $\Theta_{g,j,k}(x)=\left(\Gamma_2+\frac{\Gamma_3}{x}\frac{{\eta_2}^2(\kappa_k+1)^2+{\eta_1}^2(\zeta_j+1)^2-2\eta_1\eta_2\varsigma_g(\zeta_j+1)(\kappa_k+1)}{{\eta_2}^2(\kappa_k+1)^2+{H_1}^2}\right)^{-1}$, *and* $\iota_k=\frac{\pi}{J}$.

***Proof*** Please refer to Appendix I. □

**Theorem 4.1** *Using Lemma 4.2, the CP for interference scenarios can be presented as*

$$p_c = \Pr\left\{\gamma_{\mathrm{RD}} \geq \gamma_{\mathrm{th}}\right\} = 1 - F_{\gamma_{\mathrm{RD}}}\left(\gamma_{\mathrm{th}}\right). \tag{4.24}$$

***Proof*** It is easy to obtain (4.24) by using the relationship between CP and the CDF. □

### *4.1.3 Outage Analysis for S-R Link*

In this subsection, we will investigate the impact of the randomness of the satellite's position on the outage performance of the information transmissions over S-R link. It is assumed that S works in the circular orbit with the earth's center, E, which is a sphere with radius, $R_{\mathrm{E}} = 6371$ km.

As shown in Fig. 4.3, the satellite, S, is assumed to be uniformly distributed in the space, which is a part that the spherical cone with radius $U_1$ minus the one with radius $U_2$ ($U_1 > U_2$). The two spherical cones are respectively with radius $U_1$ and $U_2$, as well as with sphere central, E, and the same apex angle, $\Psi/2$.[6] The distance between R and E is $H_{\mathrm{R}}$, which can be written as $H_{\mathrm{R}} = H_1 + R_{\mathrm{E}}$. For example, as low earth orbit (LEO) satellites normally revolve at an altitude from the earth's surface between 160 and 2000 km, then one can have $U_1 = 8371$ km and $U_2 = 6531$ km.

In order to facilitate the following analysis, spherical coordinates are adopted, while E is set as the original. Then, the coordinate of S can be presented as $(r_S, \theta, \psi)$, where $U_2 \leq r_S \leq U_1$, $0 \leq \theta \leq \Psi/2$ and $0 \leq \psi \leq 2\pi$, and the coordinate of R can be written as $(H_{\mathrm{R}}, 0, 0)$, where $R_{\mathrm{E}} \leq H_{\mathrm{R}} \leq R_{\mathrm{E}} + H_{\mathrm{max,R}}$ and $H_{\mathrm{max,R}}$ is the maximum height of R.

[6] $\Psi$ is the apex angle determined by the distribution space of S in which S is able to successfully communicate with R. Hence, we can see that, for practical scenarios, $\Psi$ is decided some system parameters over S-R link, e.g., the transmit power at S, the antenna gains at S and R, the receiver sensitivity at R, the power levels of the noise and interference at R, etc.

Therefore, the distance between S and R, $d_0$, can be presented as

$$d_0 = \sqrt{r_S^2 + H_R^2 - 2r_S H_R \cos\theta}. \tag{4.25}$$

So it is obvious that $U_2 - H_R = d_{0,\min} \le d_0 \le d_{0,\max} = \sqrt{U_1^2 + H_R^2 - 2U_1 H_R \cos\frac{\Psi}{2}}$. In the following, we consider the case that path-loss factor is 2 to simplify the analysis.

**Theorem 4.2** *When the position of R is fixed and S is uniformly distributed in the space shown in Fig. 4.3, the PDF of $d_0^2$ can derived as*

$$f_{d_0^2}(x) = \frac{3}{4H_R}\frac{1}{1-\cos\frac{\Psi}{2}}\frac{1}{U_1^3 - U_2^3}\left[\omega^2(x) - \rho^2(x)\right] = \tau\left[\omega^2(x) - \rho^2(x)\right], \tag{4.26}$$

*where* $\omega(x) = \min\{U_1, H_R + \sqrt{x}\}$, $\rho(x) = \max\left\{U_2, H_R\cos\frac{\Psi}{2} + \sqrt{x - {H_R}^2\sin^2\frac{\Psi}{2}}\right\}$, $\tau = \frac{3}{4H_R}\frac{1}{1-\cos\frac{\Psi}{2}}\frac{1}{U_1^3-U_2^3}$, *and* $x - {H_R}^2\cos^2\frac{\Psi}{2} \ge 0$.

***Proof*** Please refer to Appendix II. □

Using (4.3) and (4.5), the outage probability (OP) for S-R link can be given as

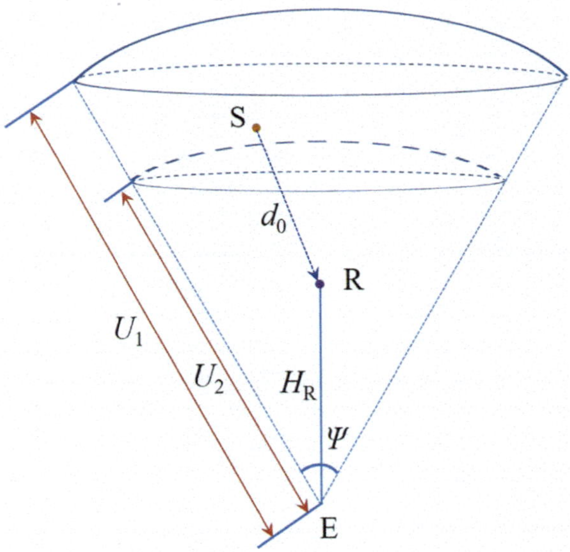


**Fig. 4.3** Satellite-relay link

$$
\begin{aligned}
P_{\text{out,SR}} &= \Pr\left\{\gamma_{\text{SR}} \leq \gamma_{\text{out}}\right\} \\
&= \Pr\left\{\frac{P_{\text{S}}}{\sigma^2}\frac{|h_{\text{SR}}|^2}{d_0^2} = \frac{\lambda_{\text{S}}}{d_0^2} \leq \gamma_{\text{out}}\right\} \\
&= 1 - \alpha \sum_{k=0}^{m-1} \frac{\varsigma(k)}{\bar{\lambda}^{k+1}} \sum_{p=0}^{k} \frac{k!}{p!}\left(\frac{\beta-\delta}{\bar{\lambda}}\right)^{-(k+1-p)} {\gamma_{\text{out}}}^p d_0^{2p} \exp\left(-\frac{\beta-\delta}{\bar{\lambda}}\gamma_{\text{out}} d_0^2\right),
\end{aligned}
\tag{4.27}
$$

where $\gamma_{\text{out}}$ is the threshold for the outage events.

Taking the randomness of $d_0^2$ in to account and using Theorem 2, the OP for S-R link can be presented as

$$
\begin{aligned}
P_{\text{out,SR}} =& 1 - \alpha \sum_{k=0}^{m-1} \frac{\varsigma(k)}{\bar{\lambda}^{k+1}} \sum_{p=0}^{k} \frac{k!}{p!}\left(\frac{\beta-\delta}{\bar{\lambda}}\right)^{-(k+1-p)} \\
&\times {\gamma_{\text{out}}}^p \int_{d_{0,\min}^2}^{d_{0,\max}^2} x^p \exp(-\Delta x) f_{d_0^2}(x)\, dx \\
=& 1 - \alpha\tau \sum_{k=0}^{m-1} \frac{\varsigma(k)}{\bar{\lambda}^{k+1}} \sum_{p=0}^{k} \frac{k!}{p!}\left(\frac{\beta-\delta}{\bar{\lambda}}\right)^{-(k+1-p)} \\
&\times {\gamma_{\text{out}}}^p \int_{d_{0,\min}^2}^{d_{0,\max}^2} x^p \left[\omega^2(x) - \rho^2(x)\right] \exp(-\Delta x)\, dx,
\end{aligned}
\tag{4.28}
$$

where $\Delta = \frac{\beta-\delta}{\bar{\lambda}}\gamma_{\text{out}}$.

Again, employing Chebyshev-Gauss quadrature in the first case, the OP for S-R link can be finally written as

$$
\begin{aligned}
P_{\text{out,SR}} =& 1 - \alpha\tau b_1 \sum_{k=0}^{m-1} \frac{\varsigma(k)}{\bar{\lambda}^{k+1}} \sum_{p=0}^{k} \frac{k!}{p!}\left(\frac{\beta-\delta}{\bar{\lambda}}\right)^{-(k+1-p)} \gamma_{\text{out}}^p \int_{-1}^{1} (b_1 t + b_2)^p \\
&\times \left[\omega^2(b_1 t + b_2) - \rho^2(b_1 t + b_2)\right] \exp\left(-\Delta\,(b_1 t + b_2)\right) dt \\
=& 1 - \alpha\tau b_1 \sum_{k=0}^{m-1} \frac{\varsigma(k)}{\bar{\lambda}^{k+1}} \sum_{p=0}^{k} \frac{k!}{p!}\left(\frac{\beta-\delta}{\bar{\lambda}}\right)^{-(k+1-p)} \gamma_{\text{out}}^p \\
&\times \sum_{j=1}^{Q} \varpi_j \sqrt{1-\chi_j^2}\,(b_1\chi_j + b_2)^p \\
&\times \left[\omega^2(b_1\chi_j + b_2) - \rho^2(b_1\chi_j + b_2)\right] \exp\left(-\Delta\left(b_1\chi_j + b_2\right)\right),
\end{aligned}
\tag{4.29}
$$

where $b_1 = \frac{d_{0,\max}^2 - d_{0,\min}^2}{2}$, $b_2 = \frac{d_{0,\max}^2 + d_{0,\min}^2}{2}$, $\chi_i = \cos\left(\frac{2j-1}{2Q}\pi\right)$ and $\varpi_i = \frac{\pi}{Q}$.

**Corollary 4.1** *When DF scheme is adopted at R, the e2e OP for S-R-*$D_i$ *link can be finally obtained as*

$$P_{e2e} = 1 - p_c(1 - P_{\text{out,SR}}), \tag{4.30}$$

*where* $p_c$ *is presented as (4.14) and (4.24) for non-interference and interference scenarios, respectively, and* $P_{\text{out,SR}}$ *is given as (4.29).*

### 4.1.4 The E2e Energy Efficiency Optimization

It is well-known that satellites and aerial relays are resource-constrained systems, especially suffering rigid power budget to realize the designed operating life. Though satellites are normally equipped with solar panels to charge the batteries, the energy renewing process is restricted by the satellites' orbiting movements. Therefore, it is vital for the considered CSATC systems to efficiently and effectively exploit the limited power resource at both satellites and aerial relays.

As presented in Fig. 4.1, intuitively, there exists an optimal transmit power and transmission time allocation over S-R and R-D hops to achieve optimal e2e energy efficiency for the considered CSATC system. Because the transmit power and transmission time over S-R and R-$D_i$ hops are directly related to the total energy and time consumption to realize the delivery of the same amount of information bits over both S-R and R-$D_i$ hops. Then, in this subsection, we will present an optimization problem in terms of the transmit power and transmission time over S-R and R-$D_i$ hops to realize the optimal e2e energy efficiency performance of S-R-$D_i$ link.

Furthermore, in this study, we only concern the energy consumed on transmitting information bits and ignore that spent on processing the signals (e.g., encoding and decoding), as radio frequency communications have been proved as the main parts of energy consumption at the terminals.

#### 4.1.4.1 Problem Formation

Similar to [8], in this work energy efficiency is adopted to evaluate the efficiency of the energy consumption on delivering the information bits, which is defined as the number of delivered bits over S-R-$D_i$ link with unit-joule consumption. Therefore, the following optimization problem is considered: How to optimally allocate the transmit power and transmission time over S-R and R-$D_i$ links so that the energy efficiency of the considered system is maximized.

$$\textbf{OPT}-\textbf{1}: \quad \max_{\mathcal{P},\mathcal{T}} \ \eta = \frac{D_{\text{SD}}}{P_{\text{S}}T_{\text{S}} + P_{\text{R}}T_{\text{R}}} \tag{4.31}$$

Subject to

$$
\begin{aligned}
C1 :& T_{\mathrm{S}} B_{\mathrm{S}} \log_2(1 + P_{\mathrm{S}} \gamma_{\mathrm{SR}}) = D_{\mathrm{SD}}, \\
C2 :& T_{\mathrm{R}} B_{\mathrm{R}} \log_2(1 + P_{\mathrm{R}} \gamma_{\mathrm{RD}}) = D_{\mathrm{SD}}, \\
C3 :& T_{\mathrm{S}} + T_{\mathrm{R}} \leq T, \\
C4 :& T_{\mathrm{S}} \geq \frac{D_{\mathrm{SD}}}{B_{\mathrm{S}} \log_2(1 + P_{\mathrm{S}}^{\max} \gamma_{\mathrm{SR}})}, \\
C5 :& T_{\mathrm{R}} \geq \frac{D_{\mathrm{SD}}}{B_{\mathrm{R}} \log_2(1 + P_{\mathrm{R}}^{\max} \gamma_{\mathrm{RD}})},
\end{aligned}
$$

where $\mathcal{P} = \{P_{\mathrm{S}}, P_{\mathrm{R}}\}$ is the power allocation policy, and $\mathcal{T} = \{T_{\mathrm{S}}, T_{\mathrm{R}}\}$ is the time allocation policy. $P_{\mathrm{S}}^{\max}$ and $P_{\mathrm{R}}^{\max}$ denotes the maximum transmit power at S and R, respectively. $\gamma_{\mathrm{SR}} = \frac{|h_{\mathrm{SR}}|^2}{\sigma^2 d_0^{n_1}}$, $\gamma_{\mathrm{RD}} = \frac{|h_{\mathrm{RD}}|^2}{\sigma^2 d_i^{n_2}}$ for non-interference case, and $\gamma_{\mathrm{RD}} = \frac{|h_{\mathrm{RD}}|^2}{P_{\mathrm{I}} |h_{\mathrm{ID}}|^2} \frac{u_i^{n_2}}{d_i^{n_2}}$ for interference case. C1 and C2 are the transmission rate constraints, which guarantee the amount of the transmitted data over the two hops is equal to the requirement, $D_{\mathrm{SD}}$. C3 restricts that the total transmission time for the two hops is bounded by the maximum time $T$ for the e2e transmission. C4 and C5 ensures the low bounds of the transmission time for the two hops, respectively. The coupling between $\mathcal{P}$ with $\mathcal{T}$ results in an obstacle to address the problem $\mathbf{OPT-1}$. However, by exploiting Lemma 4.3 in the following, we can derive the optimal solution by solving the convex optimization problem $\mathbf{OPT-2}$ as follows

$$
\mathbf{OPT-2} : \min_{\mathcal{T}} \eta_t = \frac{T_{\mathrm{S}}}{\gamma_{\mathrm{SR}}} g\left(\frac{D_{\mathrm{SD}}}{T_{\mathrm{S}}}, B_{\mathrm{S}}\right) + \frac{T_{\mathrm{R}}}{\gamma_{\mathrm{RD}}} g\left(\frac{D_{\mathrm{SD}}}{T_{\mathrm{R}}}, B_{\mathrm{R}}\right) \tag{4.32}
$$

Subject to C3, C4, and C5.

**Lemma 4.3** *We can derive the solution of problem* $\mathbf{OPT-1}$ *by solving the convex optimization problem* $\mathbf{OPT-2}$.

***Proof*** We first define a function $g(x, B) = 2^{\frac{x}{B}} - 1$. Then, $P_{\mathrm{S}} = \frac{1}{\gamma_{\mathrm{SR}}} g\left(\frac{D_{\mathrm{SD}}}{T_{\mathrm{S}}}, B_{\mathrm{S}}\right)$ and $P_{\mathrm{R}} = \frac{1}{\gamma_{\mathrm{RD}}} g\left(\frac{D_{\mathrm{SD}}}{T_{\mathrm{R}}}, B_{\mathrm{R}}\right)$.

Further, we can rewrite the objective function of $\mathbf{OPT-1}$ as

$$
\max_{\mathcal{P},\mathcal{T}} \quad \eta = \frac{D_{\mathrm{SD}}}{\frac{T_{\mathrm{S}}}{\gamma_{\mathrm{SR}}} g\left(\frac{D_{\mathrm{SD}}}{T_{\mathrm{S}}}, B_{\mathrm{S}}\right) + \frac{T_{\mathrm{R}}}{\gamma_{\mathrm{RD}}} g\left(\frac{D_{\mathrm{SD}}}{T_{\mathrm{R}}}, B_{\mathrm{R}}\right)}. \tag{4.33}
$$

It is obvious that the maximum energy efficiency $\eta^*$ is equal to $\frac{D_{\mathrm{SD}}}{\eta_t^*}$.

Since the second derivative of $g(x, B)$, i.e., $\frac{\mathrm{d}^2 g(x,B)}{\mathrm{d}x}$, is always non-negative, $g(x, B)$ is a convex function w.r.t $x$. The perspective function of $g(x, B)$, i.e., $xg\left(\frac{1}{x}, B\right)$, is also a convex function [9]. Thus, the objective function of $\mathbf{OPT-2}$ is the sum of two convex functions, and the constraints are linear. As a result, the transformed problem $\mathbf{OPT-2}$ is a convex optimization problem w.r.t $\mathcal{T}$. □

#### 4.1.4.2 Solution of the Problem OPT-2

The Lagrangian function of the problem **OPT − 2** is given by

$$L(\lambda, T_S, T_R) = \frac{T_S}{\gamma_{SR}} g\left(\frac{D_{SD}}{T_S}, B_S\right) + \frac{T_R}{\gamma_{RD}} g\left(\frac{D_{SD}}{T_R}, B_R\right) + \lambda(T_S + T_R - T), \tag{4.34}$$

where $\lambda$ is the Lagrangian multiplier corresponds to the total time constraint.

Then, the dual problem of (4.34) can be presented as

$$\max_{\lambda \geq 0} \min_{\mathcal{T}} L(\lambda, T_S, T_R). \tag{4.35}$$

By solving the dual problem (4.35), we can derive the solution of problem (4.32). To this end, we first solve the inner problem for fixed $\lambda$ and then update the Lagrangian multiplier $\lambda$ by subgradient method until the Lagrangian multiplier converges.

Using KKT conditions, the time allocation policy for the first phase transmission, i.e., from S to R, is given by

$$T_S^* = \left[\hat{T}_S\right]_{T_{S\min}}, \tag{4.36}$$

where $[x]_a$ is an operator which is defined as $\max(a, x)$, $T_{S\min} = \frac{D_{SD}}{B_S \log 2(1+P_S^{\max}\gamma_{SR})}$, and $\hat{T}_S$ denotes the solution of the following equation

$$1 - \frac{\ln 2 D_{SD}}{T_S B_S} = (1 - \gamma_{SR}\lambda)\left(\frac{1}{2}\right)^{\frac{D_{SD}}{B_S T_S}}. \tag{4.37}$$

Similarly, the time allocation policy for the second phase transmission, i.e., from S to $D_i$, is given by

$$T_R^* = \left[\hat{T}_R\right]_{T_{R\min}}, \tag{4.38}$$

where $T_{R\min} = \frac{D_{SD}}{B_R \log 2(1+P_R^{\max}\gamma_{RD})}$, and $\hat{T}_R$ denotes the solution of the following equation

$$1 - \frac{\ln 2 D_{SD}}{T_R B_R} = (1 - \gamma_{RD}\lambda)\left(\frac{1}{2}\right)^{\frac{D_{SD}}{B_R T_R}}. \tag{4.39}$$

Since (4.37) and (4.39) are transcendental equations, which do not have a closed-form solution generally, we can use some numerical methods (e.g., Newton method) to find the approximate solution. Moreover, there are some packets on various platforms to solve such a simple form transcendental equation, such as Mathematica.

As for the high-level problem of (4.35), we update the Lagrangian multiplier via the subgradient method

$$\lambda(j+1) = \left[\lambda(j) + \varphi(j)\left(T_{\rm S} + T_{\rm R} - T\right)\right]_0, \tag{4.40}$$

where $j \geq 0$ denotes the iteration index, and $\varphi(j)$ represents the positive diminishing step size.

The pseudocode of the proposed algorithm is shown in Algorithm 1. The proposed iterative algorithm consists of only one loop, it has a polynomial time complexity, i.e., $O(N)$.

**Algorithm 1** Iterative Algorithm for **OPT-2** Problem

```
Initialization:Initialize λ(0), and sets j = 0
repeat
    Calculate T*_S and T*_R based on (4.36) and (4.38)
    Update λ(j) by using (4.40) ;
    j = j + 1
until Sequence λ converges
Calculate the η*_i by (4.32)
return T*_S, T*_R and η* = D_SD / η*_i
```

## 4.1.5 Numerical Results

In this subsection, numerical results will be provided to study the performance of the considered system, as well as to verify the proposed analytical models and optimal designs. In the following, we use D instead of $D_i$ for simplification, and path-loss factors for S-R and R-D links are set as 2. Furthermore, we run $1 \times 10^6$ times of the realizations of the considered system and $1 \times 10^6$ trials of Monte-Carlo simulations, to model the randomness of the positions of the considered terminals and channel gains over each link.

### 4.1.5.1 Coverage Performance Over S-R Link

In this part, $\gamma_{\rm th} = 0$ dB, and two scenarios, non-interference and interference scenarios, will be studied for various the average channel power gain of R-D link ($\Omega_{\rm R}$), respectively.

*(1) Non-interference scenarios:* Main parameters adopted in this case are set as: $P_{\rm R} = 1$ dB, $K = -10$ dB, $L = 20$ km, and $H_1 = 5$ km.

In Figs. 4.4 and 4.5, the CP over R-D links are presented for various of $H_1$ and $L$, respectively. One can see that in both figures CP can be improved while $\Omega_{\rm R}$ increasing. Because a large $\Omega_{\rm R}$ presents a large average power of the received signal, which means that the received SNR at D increases.

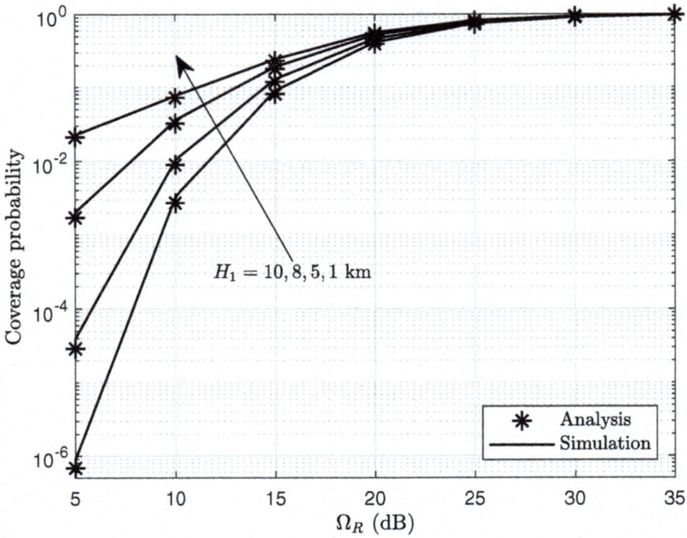


**Fig. 4.4** CP over R-D link for various $H_1$

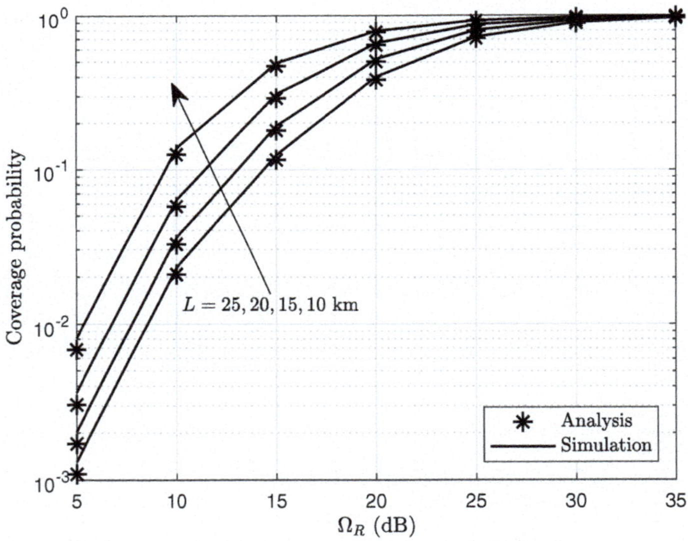


**Fig. 4.5** CP over R-D link for various $L$

Also, it is obvious that a small $H_1/L$ leads to a large CP. This observation comes from the fact that a small $H_1$ denotes a low path-loss for the signal transmitted over R-D link, and a small $L$ shows a small distributed area for D, leading to small path-loss over R-D link.

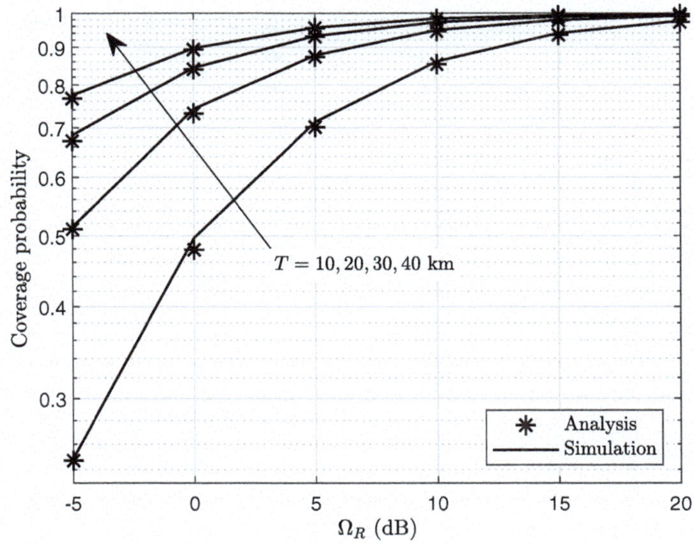


**Fig. 4.6** CP over R-D link for various $T$

*(2) Interference Scenarios:* Main parameters adopted in this case are set as: $G = H = J = 50$, $P_R = 1$ dB, $K = -10$ dB, $L = 5$ km, $H_I = 5$ km, and $T = 30$ km.

Figure 4.6 shows the impact of the size of the distribution area for the interfering node on the coverage performance of the considered system. We can observe that the CP with a large $T$ outperforms the one with a small $T$. It can be explained that a large $T$ represents a large distribution area for the interfering node, which exhibits a large probability of large path-loss over the interfering link.

In Fig. 4.7, the influence of $H_I$ on the coverage performance of the considered system is presented for the interference scenario. It is easy to find out that $H_I$ shows a negative effect on the coverage performance of the considered system, as the CP with a low $H_I$ outperforms the one with a large $H_I$. This finding can also be explained by the reason proposed for the observations in Fig. 4.4.

Furthermore, one can clearly find that simulation and numerical results agree with each other very well, which verifies the accuracy of the proposed analytical model presented in Sect. 4.1.2.

#### 4.1.5.2 Outage Performance

In this part, we set $\gamma_{out} = 0$ dB, $\Psi = \pi/3$, $m = 2$, $R_E = 6371$ km, $U_1 = 8371$ km and $U_2 = 6531$ km, and the outage performance over S-R link and the e2e outage performance of the considered system will be investigated for various the average

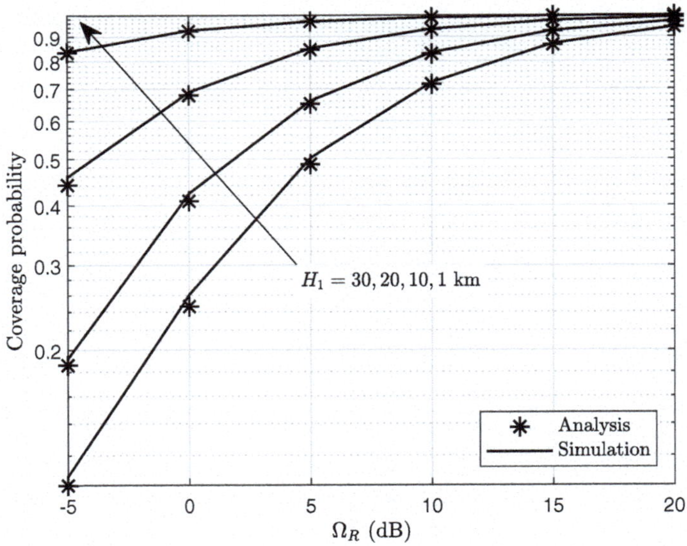


**Fig. 4.7** CP over R-D link for various $H_1$

power of the LOS components over S-R link ($\Omega$) and various the average channel power gain of R-D link ($\Omega_R$), respectively.

*(1) OP over S-R link:* Main parameters adopted here are set as: $P_S = 30$ dB, $b = 10$ dB, $P = 50$.

In Fig. 4.8, the outage performance over S-R link is studied for various $H_1$. It can be seen that the height of R, $H_1$, exhibits quite a weak impact on the outage performance over S-R link since the differences among the OP for various $H_1$ from 5 km to 20 km is very narrow (roughly on the orders of $10^{-3}$, as suggested by Fig. 4.8). Since the distance of S-R link is on the order of hundreds to thousands of kilometers, the variations of the height of R, $H_1$, from tens to thousands of meters cannot produce a significant effect on the outage performance over S-R link.

Figure 4.9 presents the outage performance over S-R link for various $\Psi$, while $\Omega$ increasing. Obviously, $\Psi$ shows a negative impact on the outage performance over S-R link, as a large $\Psi$ means a large distribution space for S, resulting in large path-loss for S-R link.

The influence of the transmit power at S, $P_S$, on the outage performance over S-R link is depicted in Fig. 4.10. As expected, $P_S$ exhibits a positive effect, since increasing $P_S$ can definitely improve the received SNR at R with no doubts.

Moreover, as presented in Figs. 4.8, 4.9 and 4.10, the OP over S-R link is improved, while $\Omega$ increasing. Because $\Omega$ represents the average power of the LOS components of the received signal at R.

Finally, we can also easily see that simulation and numerical results match well with each other, which confirms the correctness of the proposed analytical model presented in Sect. 4.1.3.

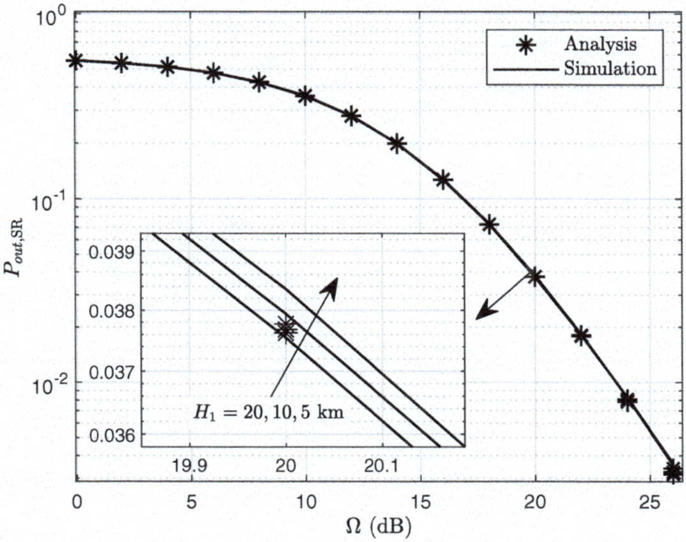


**Fig. 4.8** OP over S-R link for various $H_1$

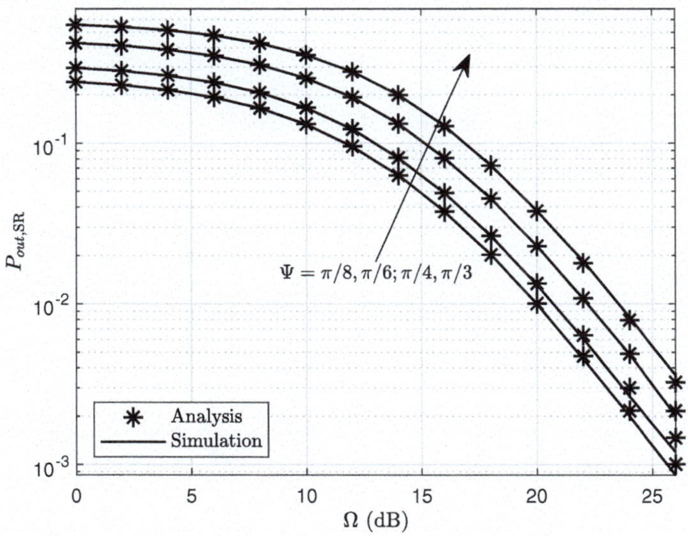


**Fig. 4.9** OP over S-R link for various $\Psi$

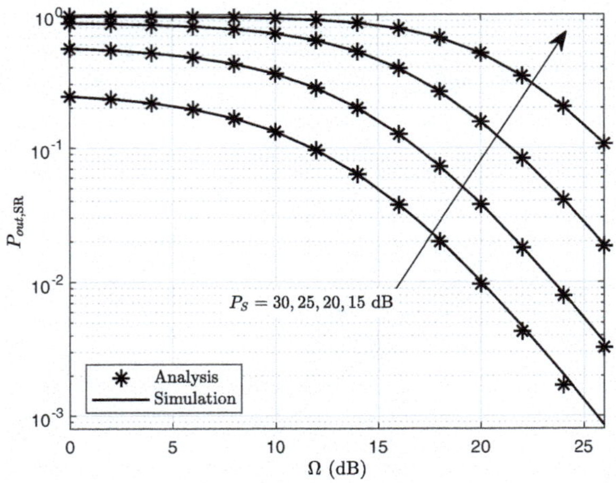


**Fig. 4.10** OP over S-R link for various $P_S$

*(2) The e2e OP:* In this part, we will present some simulation results of the e2e outage performance of the considered system shown in Fig. 4.1. Main parameters adopted in this part are set as: $H_1 = 10\,\text{km}$, $L = 5$ km, $T = 50\,\text{km}$, $P_I = P_R = 1$ dB, $P_S = 30$ dB, $K = -10$ dB.

As depicted in Figs. 4.11 and 4.12, the size of the distribution area of D, $L$, shows a negative influence on the e2e OP of the considered system in both interference

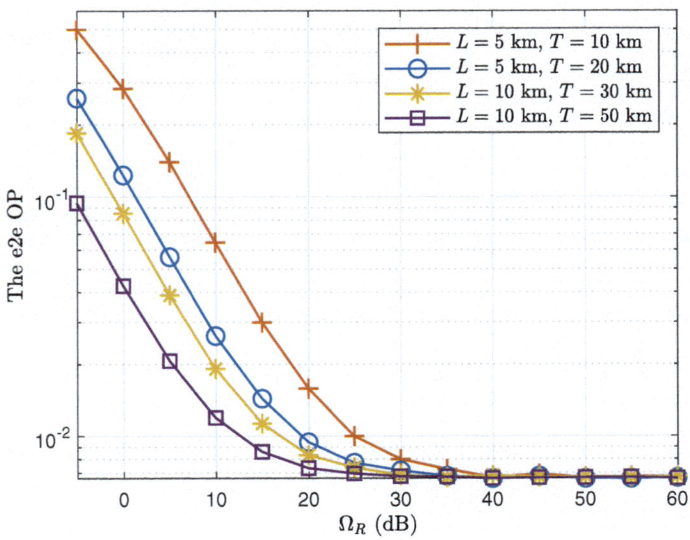


**Fig. 4.11** The e2e OP for interference scenarios with various combinations of $L$ and $T$

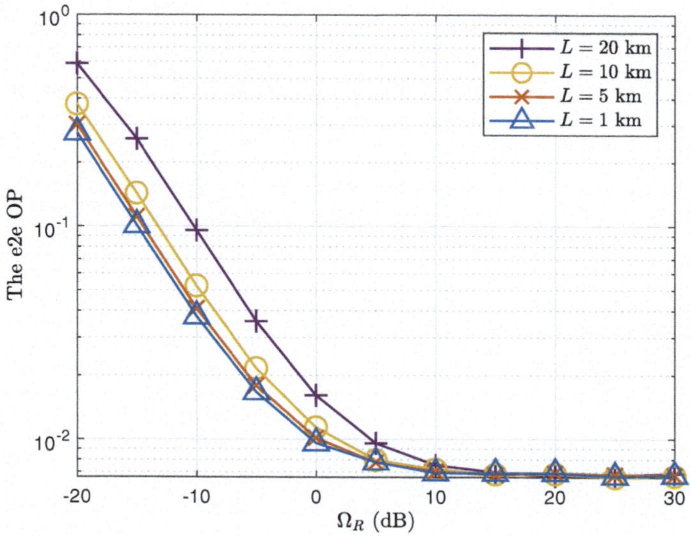


**Fig. 4.12** The e2e OP for non-interference scenarios with various $L$

and non-interference scenarios, since a small $L$ means a large probability of small path-loss over R-D link.

In Fig. 4.11, we can see that increasing the range of the distribution area of the interfering node can improve the e2e OP of the considered system, as the probability that the interfering distance from the interfering node to D increases will enlarge, leading to the decreased interfering power and further the improved SIR at D.

#### 4.1.5.3 Optimal Energy Efficiency Design

In this part, we will present the e2e energy efficiency for non-interference and interference scenarios by using the proposed optimization method, which is given in Sect. 4.1.4. The main parameters adopted here are set as $\sigma^2 = -64$ dB, $d_0 = 1500$ km, $L = 15$ km, $R_{\mathrm{E}} = 6371$ km, $U_1 = 8371$ km, $U_2 = 6531$ km, $P_{\mathrm{S}}^{\max} = 50$ W, and $P_{\mathrm{R}}^{\max} = 10$ W. In the following two figures, for simplification purpose, we adopt "NI" and "IS" to represent non-interference and interference scenarios, respectively.

We first evaluate the system performance for the proposed algorithm in terms of convergence. Figure 4.13 depicts the evolution of energy efficiency for different permutations of $L$ and $H$. We can observe that the proposed Algorithm 1 achieves convergence within five iterations.

As observed in Figs. 4.14 and 4.15, energy efficiency gets worse while the height of the aerial relay, $H_1$, increases in both non-interference and interference scenarios. Because a large $H_1$ will result in large path-loss over R-D link, which leads to increased energy consumption for the information transmissions over R-D link. Moreover, it is obvious that the energy efficiency in the non-interference scenario

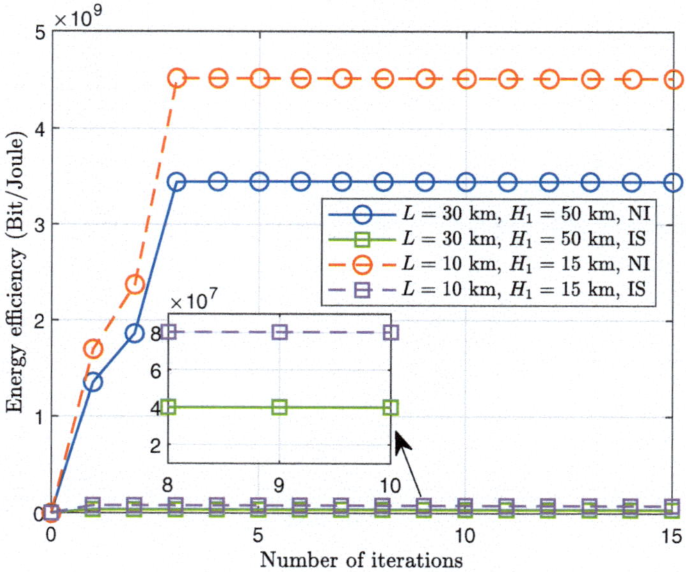


**Fig. 4.13** The optimal energy efficiency versus the number of the iterations for Algorithm 1

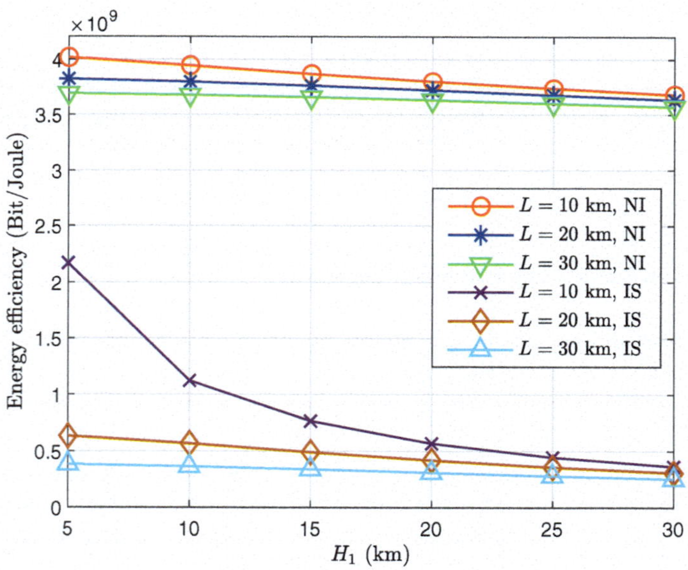


**Fig. 4.14** The optimal energy efficiency for various $L$

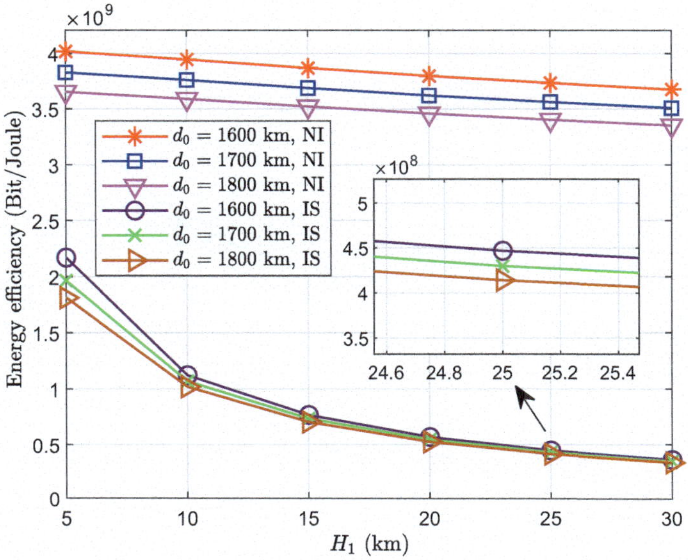


**Fig. 4.15** The optimal energy efficiency for various $d_0$

is much higher than that in the interference scenario. This finding can be easily understood by the fact that more energy consumption is needed in the interference scenario to overcome the negative impact of the interfering signal on the information transmission over R-D link.

In Fig. 4.14, one can see that the energy efficiency with a small $L$ outperforms that with a large $L$ in both non-interference and interference scenarios. This observation can also be explained by the reason provided in the previous paragraph. Namely, a large $L$ will also incur large path-loss over R-D link, and more energy is required to realize the information delivery over R-D link.

From Fig. 4.15, we can easily find that the energy efficiency with a small $d_0$ outperforms that with a large $d_0$ in both considered scenarios. Because a large $d_0$ means a large transmission distance over S-R link, resulting in large path-loss over S-R link. Furthermore, it is seen from Fig. 4.15 that the energy efficiency for various $d_0$ in the interference scenario is similar. Especially, when the height of R is too large (e.g., $H_1 > 20$ km), the differences among the energy efficiency for various $d_0$ are on the order of $10^7$ bits/Joule. This observation can be explained the fact that the path loss over S-R link will dominate the main power consumption over S-R-D link when $d_0$ is sufficiently large, e.g., the value of $d_0$ ranges from 1600 km to 1800 km considered in Fig. 4.15, which is quite larger than the one of $H_1$ ranging from 5 km to 30 km.

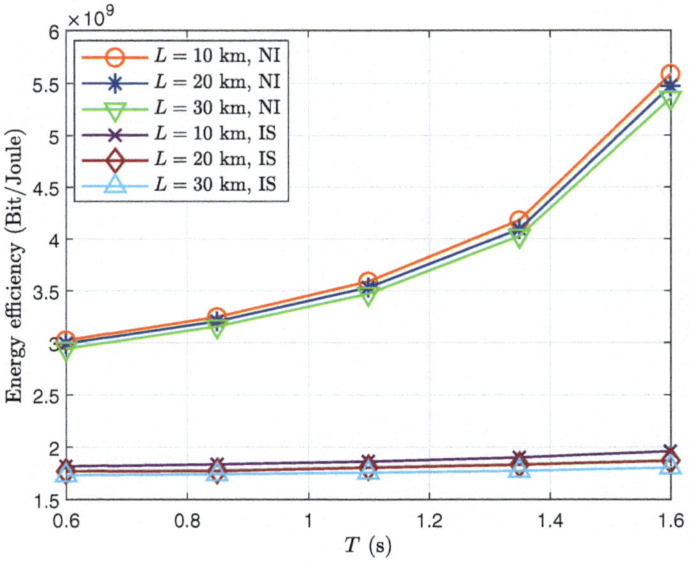


**Fig. 4.16** The optimal energy efficiency for various $T$

Figure 4.16 indicates that the energy efficiency of the system increases with $T$ increasing in every considered scenario. With the time $T$ increasing, both satellite and relay can adopt a lower transmit power to achieve a better energy efficiency performance. For fixed $T$, the energy efficiency for non-interference scenario is one order higher than that for interference scenario.

It can be observed from Fig. 4.17 that the maximum power allowance (i.e., $P_S^{\max}$) has a negligible impact on the energy efficiency of the target system. This observation can be explained as follows: the long-distance transmission over S-R link, which is normally on the order of 1000 km, leads to negligible improvements on the path-loss of S-R link, e.g., the average received SNR of S-R link ranges from $3.5 \times 10^{-11}$ to $6 \times 10^{-11}$ when the maximum power allowance at S increases from 35 W to 60 W and path-loss factor is 2, and further exhibits no obvious influence on the transmission capacity of S-R link and the energy efficiency of the whole system. With our practical simulation setup, we find that the solution to the optimization problem always works in a reasonable region of the power allowance.

### *4.1.6 Conclusion*

In this section, we first have studied the outage performance of a CSATC system with DF relay scheme and derived the approximated analytical expressions for the coverage probability over R-D link and the e2e OP in both non-interference and interference scenarios, while considering the randomness of the positions of the

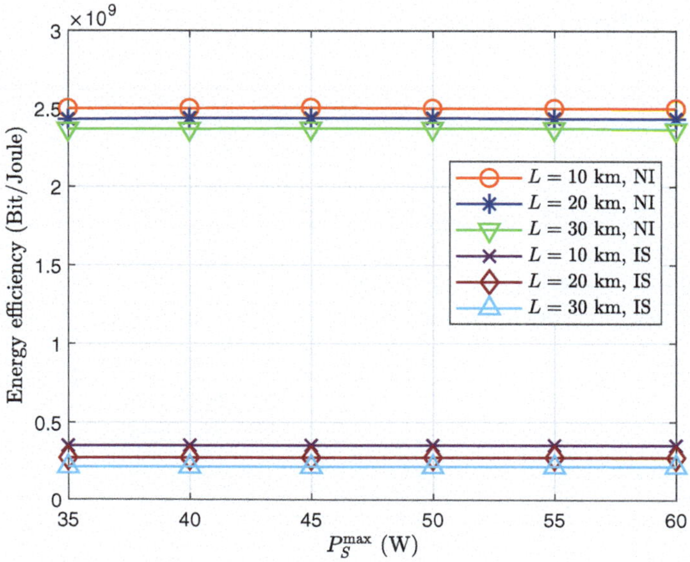


**Fig. 4.17** The optimal energy efficiency for various $P_s^{\max}$

satellite and the terrestrial receivers. Next, the e2e energy efficiency of the considered system has been optimized by employing the proposed optimization model.

Observing from the numerical results, some remarks can be achieved as follows:

(1) The radius of the coverage space of the aerial relay shows a negative impact on the coverage/outage/energy efficiency performance in both non-interference and interference scenarios.

(2) The radius of the distribution area of the interfering node exhibits a positive influence on the coverage/outage/energy efficiency performance.

(3) The height of the aerial relay shows a clear negative effect on the coverage performance over R-D link and a weak impact on the outage performance over S-R link.

(4) Adjusting the height of aerial relay cannot always improve the e2e outage/energy efficiency performance of the considered system, which has a performance limitation arisen from the transmission distance from the satellite to the aerial relay. Because the transmission distance over S-R link is quite larger than that over R-D link.

(5) The size of the distribution space of the satellite shows an obvious negative effect on the outage performance over S-R link.

(6) Increasing the transmit power at the satellite is not an efficient way to improve the performance of the whole system, due to the huge path-loss arisen from the long-distance transmission over S-R link normally on the order of thousands of kilometers.

## *Appendix I: Proof of Lemma 2*

Let $t = \cos\theta$ $(-1 \leq t \leq 1)$, we have $\theta = arc\cos y$. Then, the PDF of $\cos\theta$ can be written as

$$f_{\cos\theta}(t) = \frac{2}{2\pi}\left|-\frac{1}{\sqrt{1-t^2}}\right| = \frac{1}{\pi\sqrt{1-t^2}}. \tag{4.41}$$

Therefore, considering the randomness of the position of $\cos\theta$ and using Chebyshev-Gauss quadrature in the first case, which is presented as $\int_{-1}^{1}\frac{f(x)}{\sqrt{1-x^2}}dx \approx \sum_{i=1}^{W_T} w_i f(x_i)$ with $x_i = \cos\left(\frac{2i-1}{2T}\pi\right)$ and the weight $w_i = \frac{\pi}{T}$, the CDF of $\gamma_{\mathrm{RD}}$ can be rewritten as

$$\begin{aligned} F_{\gamma_{\mathrm{RD}}}(x) &= \Gamma_1 \int_{-1}^{1}\left(\Gamma_2 + \frac{\Gamma_3}{x}\frac{r_i^2 + r_I^2 - 2r_i r_I t}{r_i^2 + {H_1}^2}\right)^{-1} \\ &\quad \times {}_1F_1\left(1; 1; \Gamma_4\left(\Gamma_2 + \frac{\Gamma_3}{x}\frac{r_i^2 + r_I^2 - 2r_i r_I t}{r_i^2 + {H_1}^2}\right)^{-1}\right) f_{\cos\theta}(t)\,dt \\ &= \frac{\Gamma_1}{\pi}\sum_{g=1}^{G}\vartheta_g\left(\Gamma_2 + \frac{\Gamma_3}{x}\frac{r_i^2 + r_I^2 - 2r_i r_I \varsigma_g}{r_i^2 + {H_1}^2}\right)^{-1} \\ &\quad \times {}_1F_1\left(1; 1; \Gamma_4\left(\Gamma_2 + \frac{\Gamma_3}{x}\frac{r_i^2 + r_I^2 - 2r_i r_I \varsigma_g}{r_i^2 + {H_1}^2}\right)^{-1}\right), \end{aligned} \tag{4.42}$$

where $\varsigma_g = \cos\left(\frac{2g-1}{2G}\pi\right)$ and $\vartheta_g = \frac{\pi}{G}$.

As shown in Fig. 4.2, $\mathrm{D}_i$ is uniformly distributed in the circle with origin O and radius $L$, the PDF of $r_i$ can be given as

$$f_{r_i}(x) = \begin{cases} \frac{2x}{L^2}, & \text{if } 0 \leq x \leq L; \\ 0, & \text{else} \end{cases}. \tag{4.43}$$

Similarly, the PDF of $r_I$ can be written as

$$f_{r_I}(x) = \begin{cases} \frac{2x}{T^2}, & \text{if } 0 \leq x \leq T; \\ 0, & \text{else} \end{cases}. \tag{4.44}$$

Similarly, taking the randomness of the position of I into account and applying Chebyshev-Gauss quadrature in the first case, the CDF of $\gamma_{\mathrm{RD}}$ can be presented as

$$F_{\gamma_{\mathrm{RD}}}(x) = \frac{2\Gamma_1}{\pi T^2}\sum_{g=1}^{G}\vartheta_g\int_0^T z\left(\Gamma_2 + \frac{\Gamma_3}{x}\frac{r_i^2 + z^2 - 2r_i z\varsigma_g}{r_i^2 + H_1{}^2}\right)^{-1}$$
$$\times {}_1F_1\left(1;1;\Gamma_4\left(\Gamma_2 + \frac{\Gamma_3}{x}\frac{r_i^2 + z^2 - 2r_i z\varsigma_g}{r_i^2 + H_1{}^2}\right)^{-1}\right)dz$$
$$= \frac{\Gamma_1}{2\pi}\sum_{g=1}^{G}\vartheta_g\sum_{j=1}^{H}\varsigma_j\sqrt{1-v_j^2}\left(\zeta_j+1\right)$$
$$\times\left(\Gamma_2 + \frac{\Gamma_3}{x}\frac{r_i^2 + \eta_1{}^2(\zeta_j+1)^2 - 2\eta_1 r_i\left(\zeta_j+1\right)\varsigma_g}{r_i^2 + H_1{}^2}\right)^{-1}$$
$$\times {}_1F_1\left(1;1;\Gamma_4\left(\Gamma_2 + \frac{\Gamma_3}{x}\frac{r_i^2 + \eta_1{}^2(\zeta_j+1)^2 - 2\eta_1 r_i\left(\zeta_j+1\right)\varsigma_g}{r_i^2 + H_1{}^2}\right)^{-1}\right), \tag{4.45}$$

where $\eta_1 = \frac{T}{2}$, $\zeta_j = \cos\left(\frac{2j-1}{2H}\pi\right)$, and $v_j = \frac{\pi}{H}$.

Again, considering the randomness of the position of $D_i$ and using Chebyshev-Gauss quadrature in the first case, the CDF of $\gamma_{\mathrm{RD}}$ can be further derived as

$$F_{\gamma_{\mathrm{RD}}}(x) = \frac{\Gamma_1}{\pi L^2}\sum_{g=1}^{G}\vartheta_g\sum_{j=1}^{H}\varsigma_j\sqrt{1-v_j^2}\left(\zeta_j+1\right)$$
$$\times\int_0^L\left(\Gamma_2 + \frac{\Gamma_3}{x}\frac{z^2 + \eta_1{}^2(\zeta_j+1)^2 - 2\eta_1\varsigma_g\left(\zeta_j+1\right)z}{z^2 + H_1{}^2}\right)^{-1}$$
$$\times {}_1F_1\left(1;1;\Gamma_4\left(\Gamma_2 + \frac{\Gamma_3}{x}\frac{z^2 + \eta_1{}^2(\zeta_j+1)^2 - 2\eta_1\varsigma_g\left(\zeta_j+1\right)z}{z^2 + H_1{}^2}\right)^{-1}\right)f_{r_i}(z)\,dz$$
$$= \frac{\Gamma_1}{4\pi}\sum_{g=1}^{G}\vartheta_g\sum_{j=1}^{H}\varsigma_j\sqrt{1-v_j^2}\left(\zeta_j+1\right)\int_{-1}^{1}(y+1)$$
$$\times\left(\Gamma_2 + \frac{\Gamma_3}{x}\frac{\eta_2{}^2(y+1)^2 + \eta_1{}^2(\zeta_j+1)^2 - 2\eta_1\eta_2\varsigma_g\left(\zeta_j+1\right)(y+1)}{\eta_2{}^2(y+1)^2 + H_1{}^2}\right)^{-1}$$
$$\times {}_1F_1\,(1;1;$$
$$\Gamma_4\left(\Gamma_2 + \frac{\Gamma_3}{x}\frac{\eta_2{}^2(y+1)^2 + \eta_1{}^2(\zeta_j+1)^2 - 2\eta_1\eta_2\varsigma_g\left(\zeta_j+1\right)(y+1)}{\eta_2{}^2(y+1)^2 + H_1{}^2}\right)^{-1}\Bigg)dy$$
$$= \frac{\Gamma_1}{4\pi}\sum_{g=1}^{G}\vartheta_g\sum_{j=1}^{H}\zeta_j\sqrt{1-v_j^2}\left(\zeta_j+1\right)$$
$$\times\sum_{k=1}^{J}\iota_k\sqrt{1-\kappa_k^2}\left(\kappa_k+1\right)\Theta_{g,j,k}(x)\,{}_1F_1\left(1;1;\Gamma_4\Theta_{g,j,k}(x)\right), \tag{4.46}$$

where $\eta_2 = \frac{L}{2}$, $\kappa_i = \cos\left(\frac{2i-1}{2H}\pi\right)$, $\iota_i = \frac{\pi}{J}$, and
$\Theta_{g,j,k}(x) = \left(\Gamma_2 + \frac{\Gamma_3}{x}\frac{{\eta_2}^2(\kappa_k+1)^2+{\eta_1}^2(\zeta_j+1)^2-2\eta_1\eta_2\varsigma_g(\zeta_j+1)(\kappa_k+1)}{{\eta_2}^2(\kappa_k+1)^2+{H_1}^2}\right)^{-1}$.

To facilitate the presentation, we adopt $\sum\limits_g\sum\limits_j\sum\limits_k$ to denote $\sum\limits_{g=1}^{G}\vartheta_g\sum\limits_{j=1}^{H}\zeta_j\sqrt{1-\nu_j^2}\left(\zeta_j+1\right)\sum\limits_{k=1}^{J}\iota_k\sqrt{1-\kappa_k^2}\left(\kappa_k+1\right)$. Then, the CDF of $\gamma_{\mathrm{RD}}$ can be finally presented as (4.23).

So, the Proof of Lemma 2 is finished.

## *Appendix II: Proof of Theorem 2*

The volume of target space shown in Fig. 4.3, namely, the volume of the difference space between the spherical cones with radius $U_1$ and $U_2$ can be expressed as

$$V = \frac{2\pi}{3}\left(1-\cos\frac{\Psi}{2}\right)\left({U_1}^3-{U_2}^3\right). \tag{4.47}$$

Thus, employing the method adopted in Appendix A of [10], and considering that S is uniformly distributed, the CDF of $\theta$ can be given as

$$\begin{aligned} F_\theta(x) &= \int\limits_{U_2}^{U_1}\int\limits_{0}^{2\pi}\int\limits_{0}^{x}\frac{1}{V}d^2\sin\theta d\theta d\varphi d(d) \\ &= \frac{1}{V}\frac{2\pi}{3}\left({U_1}^3-{U_2}^3\right)(1-\cos x) \\ &= \begin{cases} \frac{1-\cos x}{1-\cos\frac{\Psi}{2}}, & \text{if } 0 \le x \le \frac{\Psi}{2}; \\ 0, & \text{else} \end{cases}. \end{aligned} \tag{4.48}$$

So, it is easy to have the PDF of $\theta$ as

$$f_\theta(x) = \frac{\sin x}{1-\cos\frac{\Psi}{2}}. \tag{4.49}$$

Similarly, employing the method adopted in Appendix A of [10], and considering that S is uniformly distributed, the PDF of $r_{\mathrm{S}}$ can be written as

$$f_{r_{\mathrm{S}}}(x) = \begin{cases} \frac{3x^2}{U_1^3-U_2^3}, & \text{if } U_2 \le x \le U_1; \\ 0, & \text{else} \end{cases}. \tag{4.50}$$

Thus, the joint PDF of $r_S$ and $\theta$ can be presented as

$$f_{r_S,\theta}(x, y) = \frac{3x^2}{U_1^3 - U_2^3} \frac{\sin y}{1 - \cos \frac{\Psi}{2}}, \tag{4.51}$$

where $0 \le \theta \le \Psi/2$ and $U_2 \le x \le U_1$.

Since $d_0 = \sqrt{r_S^2 + H_R^2 - 2r_S H_R \cos\theta}$, one can have

$$\left|\frac{\partial\left(d_0^2, r_S\right)}{\partial\left(r_S, \theta\right)}\right| = \begin{vmatrix} 2r_S - 2H_R\cos\theta & 2r_S H_R \sin\theta \\ 1 & 0 \end{vmatrix} = 2r_S H_R \sin\theta. \tag{4.52}$$

Then, the joint PDF of $d_0^2$ and $r_S$ can be written as

$$f_{d_0^2,r_S}(x, y) = \frac{f_{r_S,\theta}(x, y)}{\left|\frac{\partial(d_0^2, r_S)}{\partial(r_S,\theta)}\right|} = \frac{1}{2H_R} \frac{1}{1 - \cos\frac{\Psi}{2}} \frac{3y}{U_1^3 - U_2^3}, \tag{4.53}$$

where $U_2 \le y \le U_1$ and $\cos\frac{\Psi}{2} \le \frac{y^2+H_R^2-x}{2yH_R} \le 1$.

So, we can calculate the PDF of $d_0^2$ as

$$\begin{aligned} f_{d_0^2}(x) &= \frac{3}{4H_R} \frac{1}{1 - \cos\frac{\Psi}{2}} \frac{1}{U_1^3 - U_2^3} \left[\omega^2(x) - \rho^2(x)\right] \\ &= \tau\left[\omega^2(x) - \rho^2(x)\right], \end{aligned} \tag{4.54}$$

where $\omega(x) = \min\left\{U_1, H_R + \sqrt{x}\right\}$, $\rho(x) = \max\left\{U_2, H_R\cos\frac{\psi}{2} + \sqrt{x - H_R{}^2\sin^2\frac{\psi}{2}}\right\}$, $\tau = \frac{3}{4H_R}\frac{1}{1-\cos\frac{\Psi}{2}}\frac{1}{U_1^3-U_2^3}$ and $x - H_R{}^2\cos^2\frac{\Psi}{2} \ge 0$.

With this result, the proof of Theorem 2 is completed.

## 4.2 Outage Analysis of Cooperative Satellite-Aerial-Terrestrial Systems in Presence of Spatially Random Interference

### *4.2.1 System Model*

In this paper, we consider a cooperative satellite-aerial-terrestrial network (CSATN), which consists of a satellite (S), an aerial relay (R), and a terrestrial terminal (D), as shown in Fig. 4.1. All nodes are equipped with a single antenna and operate in half-duplex mode. Specifically, S delivers its signal to both R and D over S-R and

R-D links, respectively.[7] Assuming that R adopts the DF scheme, in the second hop, R will forward the decoded signal to D if D is within its serving area. We adopt the DF scheme at the aerial relay to make the analysis more concise and further facilitate the readers to understand our proposed analysis method. Furthermore, though the cochannel interference (CI) may exist among satellites, satellite constellations, cochannel aerial relays, or ground transmitters [11, 12], in this work, the impacts of CI will be ignored to make the following analysis feasible and concise since this work mainly focuses on the influences of the location of the aerial relay and the ground transmitter over the system.

According to the fading types experienced by the transmitted signal over each channel, in this work, we classify S-R, S-D, and R-D links into two categories.

#### 4.2.1.1 S-R and S-D Links

SR fading model presented in [3] is adopted to describe the statistical distribution of the link from S to R and S to D, which has been proved to be accurate and practical in various frequencies bands. The PDF of the corresponding channel power gain $\left|h_{Sg}\right|^2$ ($g \in \{R, D\}$) is given by [13]

$$f_{\left|h_{Sg}\right|^2}(x) = \alpha_{Sg} \exp\left(-\beta_{Sg} x\right) {}_1F_1\left(m_{Sg}; 1; \delta_{Sg} x\right), x \geq 0, \tag{4.55}$$

where $\alpha_{Sg} = \left(\frac{2b_{Sg} m_{Sg}}{2b_{Sg} m_{Sg} + \Omega_{Sg}}\right)^{m_{Sg}} / \left(2b_{Sg}\right)$, $\beta_{Sg} = \frac{1}{2b_{Sg}}$, $\delta_{Sg} = \frac{\Omega_{Sg}}{2b_{Sg}\left(2b_{Sg} m_{Sg} + \Omega_{Sg}\right)}$, $\Omega_{Sg}$ and $2b_{Sg}$ are the average power of the LoS and multi-path components, respectively, $m_{Sg}$ is the fading severity parameter, and ${}_1F_1(;1;)$ is the confluent hyper-geometric function of the first kind [14, Eq(9.14.1)].

Then it is easy to obtain the received SNR at R and D as

$$\gamma_{SR} = \frac{P_S}{\sigma^2} \frac{\left|h_{SR}\right|^2}{d_{SR}^{n_1}} = \frac{\lambda_{SR}}{d_{SR}^{n_1}} \tag{4.56}$$

and

$$\gamma_{SD} = \frac{P_S}{\sigma^2} \frac{\left|h_{SD}\right|^2}{d_{SD}^{n_2}} = \frac{\lambda_{SD}}{d_{SD}^{n_2}}, \tag{4.57}$$

respectively, where $\lambda_{SR} = P_S|h_{SR}|^2/\sigma^2$, $\lambda_{SD} = P_S|h_{SD}|^2/\sigma^2$, $P_S$ and $\sigma^2$ are the transmit power at S and the average power of the additive white Gaussian noise (AWGN), which is assumed to follow the same distribution at both R and D; $d_{SR}$ is the distance between S and R, $n_1$ is the path-loss factor over S-R link, which is assumed to be a constant ranging from 2 to 4 depending on the radio propagation

[7] Here, the effects of Doppler shifts at the receiver have been ignored for simplification purposes, which have been well studied and are out of the main focus of our work.

environment and terrain, $d_{SD}$ is the distance between S and D, $n_2$ is the path-loss factor over S-D link.

For simplicity, we suppose that $m_{Sg}$ takes integer values and rewrite the hypergeometric function ${}_1F_1\left(m_{Sg}; 1; \delta_{Sg}x\right)$ through [15, (41)] as

$$ {}_1F_1\left(m_{Sg}; 1; \delta_{Sg}x\right) = \exp\left(\delta_{Sg}x\right) \sum_{k=0}^{m_{Sg}-1} \frac{(-1)^k\left(1-m_{Sg}\right)_k\left(\delta_{Sg}x\right)^k}{(k!)^2}, \tag{4.58} $$

where $(t)_k = t\,(t+1)\,s\,(t+k-1)$ is the Pochhammer symbol [14].

By substituting (4.58) into (4.55), the PDF and CDF of $\lambda_{Sg}$ can be calculated as

$$ f_{\lambda_{Sg}}(x) = \alpha_{Sg} \sum_{k=0}^{m_{Sg}-1} \frac{\zeta(k)}{\overline{\lambda}_{Sg}^{k+1}} x^k \exp\left(-\frac{\beta_{Sg}-\delta_{Sg}}{\overline{\lambda}_{Sg}}x\right), \tag{4.59} $$

and

$$ F_{\lambda_{Sg}}(x) = 1 - \alpha_{Sg} \sum_{k=0}^{m_{Sg}-1} \frac{\zeta(k)}{\overline{\lambda}_{Sg}^{k+1}} \sum_{p=0}^{k} \frac{k!}{p!} \left(\frac{\beta_{Sg}-\delta_{Sg}}{\overline{\lambda}_{Sg}}\right)^{-(k+1-p)} x^p \exp\left(-\frac{\beta_{Sg}-\delta_{Sg}}{\overline{\lambda}_{Sg}}x\right), \tag{4.60} $$

respectively, where $\zeta(k) = \frac{(-1)^k\left(1-m_{Sg}\right)_k\left(\delta_{Sg}x\right)^k}{(k!)^2}$ and $\overline{\lambda}_{Sg} = P_S/\sigma^2$.

#### 4.2.1.2 R-D Link

Rician distribution is adopted to model the channel gain over R-D link due to the line-of-sight propagation of aerial-terrestrial transmissions. Then, the PDF and CDF of the channel power gain $|h_{RD}|^2$ can be given by

$$ f_{|h_{RD}|^2}(x) = \frac{(1+K)}{\Omega_{RD}} \exp\left(-K - \frac{(1+K)\,x}{\Omega_{RD}}\right) I_0\left(2\sqrt{\frac{K\,(1+K)\,x}{\Omega_{RD}}}\right) \tag{4.61} $$

and

$$ F_{|h_{RD}|^2}(x) = 1 - Q_1\left(\sqrt{2K}, \sqrt{\frac{2\,(1+K)\,x}{\Omega_{RD}}}\right), \tag{4.62} $$

respectively, where $\Omega_{RD}$ is the variance of the signal, $K$ is the Rician factor which corresponds to the ratio of the power of the LoS (specular) component to the average power of the scattered component, $I_0()$ is the modified Bessel function of order 0, $Q_1(a,b) = \int_b^{\infty} x \exp\left(-\frac{x^2+a^2}{2}\right) I_0(ax)dx$ is Marcum-$Q$ function.

Similarly, for R-D link, the instantaneous received SNR at D can be written as

$$\gamma_{RD} = \frac{P_R}{\sigma^2}\frac{|h_{RD}|^2}{d_{RD}^{n_3}} = \frac{\lambda_{RD}}{d_{RD}^{n_3}}, \tag{4.63}$$

where $\lambda_{RD} = P_R|h_{RD}|^2/\sigma^2$, $P_R$ is the transmit power at R, $d_{RD}$ is the distance between R and D, $n_3$ is the path-loss factor over R-D link.

Therefore, the PDF and CDF of $\lambda_{RD}$ can be calculated as

$$f_{\lambda_{RD}}(x) = \frac{(1+K)}{\Omega_R}\exp\left(-K - \frac{(1+K)x}{\Omega_R}\right) I_0\left(2\sqrt{\frac{K(1+K)x}{\Omega_R}}\right) \tag{4.64}$$

and

$$F_{\lambda_{RD}}(x) = 1 - Q_1\left(\sqrt{2K}, \sqrt{\frac{2(1+K)x}{\Omega_R}}\right), \tag{4.65}$$

respectively, where $\Omega_R = \frac{\Omega_{RD}P_R}{\sigma^2}$.

### 4.2.2 *Outage Analysis*

In this subsection, we study the outage performance of the considered system. Considering the beam shape of the transmission signal from S, in this work we model the operation space of the aerial relay as a circular truncated cone, the height of which is decided by the upper and lower limitations of the operation height of R. As presented in Fig. 4.1, signals are delivered from S to D through two links: (1) S-D link, through which D receives the signal directly from S, (2) S-R-D link, through which S delivers signals to D via the help of R if D locates in R's coverage area. For the considered cooperative system, one can easily obtain the following principle: D will experience an outage and cannot successfully decode the signal transmitted by S, when an outage occurs over both links. Thus, we can further investigate the outage performance of the target system in the following two cases:

(1) In the first case, R can receive the transmitted signal from S, outage event occurs over both R-D and S-D links, which means that the received SNR at R is larger than the outage threshold $\gamma_{\text{out},R}$ at R while the SNR at D is smaller than the outage threshold $\gamma_{\text{out},D}$ at D.

(2) In the second case, the received SNR at both R and D is smaller than the outage thresholds.

Further, we define the OP for S-R, S-D, and R-D links as $P_{\text{out},SR}$, $P_{\text{out},SD}$ and $P_{\text{out},RD}$, respectively. Then we can derive the two cases as

$$P_{\text{out},1} = \left(1 - P_{\text{out},SR}\right) P_{\text{out},RD} P_{\text{out},SD}, \tag{4.66a}$$

$$P_{\text{out},2} = P_{\text{out},SR} P_{\text{out},SD}. \tag{4.66b}$$

Therefore, the end-to-end OP of the considered system can be given as $P_{\text{out}} = P_{\text{out},1} + P_{\text{out},2}$.

In the following subsections, we will study the outage performance of S-R, S-D, and R-D links, respectively.

#### 4.2.2.1 Outage Over S-R Link

As shown in Fig. 4.18, R is assumed to be uniformly distributed in the coverage space of S, which is a circular truncated cone with lower radius $R_{SL} = H_{SL} \tan \Phi$ and upper radius $R_{SH} = H_{SH} \tan \Phi$, where $H_{SL}$ is the altitude from S to the minimum flight altitude, $H_{SH}$ is the altitude from S to the maximum flight altitude, and $\Phi$ is the

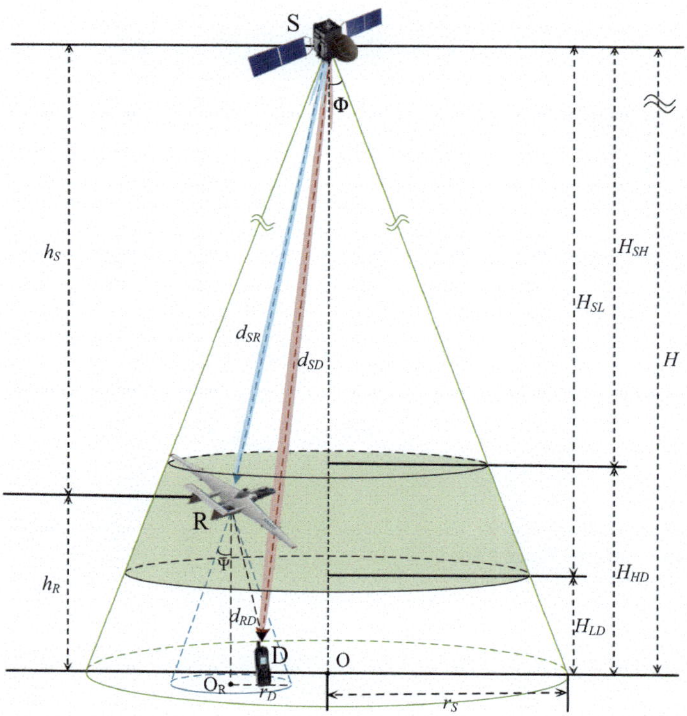


**Fig. 4.18** System model

beam angle of S. In the following, we name the circular truncated cone as $S_1$ and its volume as $V_S = \int_0^{2\pi} \int_{H_{SH}}^{H_{SL}} \int_0^{z \tan \Phi} r dr dz d\theta = \frac{1}{3}\pi \left( H_{SL}^3 - H_{SH}^3 \right)$.

**Theorem 4.3** *When R is uniformly distributed in $S_1$, the CDF and PDF of $d_{SR}$ can be derived as follow. Considering the preference of $H_{SH}$, $H_{SL}$ and $\Phi$, this scenario can be discussed in two cases:*

***Case 1**: $\frac{H_{SH}}{\cos \Phi} < H_{SL}$. Then, the CDF and PDF of $d_{SR}$ can be calculated as*

$$F_{d_{SR}}(x) = \begin{cases} \frac{\pi}{3V_S} \left( 2x^3 - 3H_{SH}x^2 + H_{SH}^3 \right), & \text{if } H_{SH} < x < \frac{H_{SH}}{\cos \Phi}; \\ \frac{\pi}{3V_S} \left( 2x^3 - 2x^3 \cos \Phi - \frac{H_{SH}^3}{\cos^2 \Phi} + H_{SH}^3 \right), & \text{if } \frac{H_{SH}}{\cos \Phi} < x < H_{SL}; \\ \frac{\pi}{3V_S} \left( 3H_{SL}x^2 - 2x^3 \cos \Phi - \frac{H_{SH}^3}{\cos^2 \Phi} - H_{SL}^3 + H_{SH}^3 \right), & \text{if } H_{SL} < x < \frac{H_{SL}}{\cos \Phi} \end{cases} \tag{4.67}$$

*and*

$$f_{d_{SR}}(x) = \begin{cases} \frac{2\pi}{V_S} \left( x^2 - H_{SH}x \right), & \text{if } H_{SH} < x < \frac{H_{SH}}{\cos \Phi}; \\ \frac{2\pi}{V_S} \left( x^2 - x^2 \cos \Phi \right), & \text{if } \frac{H_{SH}}{\cos \Phi} < x < H_{SL}; \\ \frac{2\pi}{V_S} \left( H_{SL}x - x^2 \cos \Phi \right), & \text{if } H_{SL} < x < \frac{H_{SL}}{\cos \Phi} \end{cases}, \tag{4.68}$$

*respectively.*

***Case 2**: $\frac{H_{SH}}{\cos \Phi} > H_{SL}$. Thus, we can derive the CDF and PDF of $d_{SR}$ as*

$$F_{d_{SR}}(x) = \begin{cases} \frac{\pi}{3V_S} \left( 2x^3 - 3H_{SH}x^2 + H_{SH}^3 \right), & \text{if } H_{SH} < x < H_{SL}; \\ \frac{\pi}{3V_S} \left( 3H_{SL}x^2 - 3H_{SH}x^2 - H_{SL}^3 + H_{SH}^3 \right), & \text{if } H_{SL} < x < \frac{H_{SH}}{\cos \Phi}; \\ \frac{\pi}{3V_S} \left( 3H_{SL}x^2 - 2x^3 \cos \Phi - \frac{H_{SH}^3}{\cos^2 \Phi} - H_{SL}^3 + H_{SH}^3 \right), & \text{if } \frac{H_{SH}}{\cos \Phi} < x < \frac{H_{SL}}{\cos \Phi} \end{cases} \tag{4.69}$$

*and*

$$f_{d_{SR}}(x) = \begin{cases} \frac{2\pi}{V_S} \left( x^2 - H_{SH}x \right), & \text{if } H_{SH} < x < H_{SL}; \\ \frac{2\pi}{V_S} \left( H_{SL}x - H_{SH}x \right), & \text{if } H_{SL} < x < \frac{H_{SH}}{\cos \Phi}; \\ \frac{2\pi}{V_S} \left( H_{SL}x - x^2 \cos \Phi \right), & \text{if } \frac{H_{SH}}{\cos \Phi} < x < \frac{H_{SL}}{\cos \Phi} \end{cases}, \tag{4.70}$$

*respectively.*

***Proof*** Please refer to Appendix I. □

Using (4.56) and (4.60), $P_{\text{out},SR}$ can be calculated as

$$\begin{aligned}P_{\text{out},SR} &= \Pr\left\{\gamma_{SR} \le \gamma_{\text{out},R}\right\} = \Pr\left\{\lambda_S \le \gamma_{\text{out},R} d_{SR}^{n_1}\right\} \\ &= 1 - \alpha_{SR} \sum_{k=0}^{m_{SR}-1} \frac{\zeta(k)}{\overline{\lambda}_{SR}^{k+1}} \sum_{p=0}^{k} \frac{k!}{p!} \left(\frac{\beta_{SR} - \delta_{SR}}{\overline{\lambda}_{SR}}\right)^{-(k+1-p)} \gamma_{\text{out},R}{}^{p} \\ &\quad \times d_{SR}^{n_1 p} \exp\left(-\frac{\beta_{SR} - \delta_{SR}}{\overline{\lambda}_{SR}} \gamma_{\text{out},R} d_{SR}^{n_1}\right). \end{aligned} \tag{4.71}$$

Taking the randomness of $d_{SR}^{n_1}$ into account, OP can be further calculated in the following two cases:

**Case 1**: $\frac{H_{SH}}{\cos\Phi} < H_{SL}$. Using (4.69), the OP over S-R link can be presented as

$$\begin{aligned}P_{\text{out},SR} &= 1 - \frac{2\pi\alpha_{SR}}{V_S} \sum_{k=0}^{m_{SR}-1} \frac{\zeta(k)}{\overline{\lambda}_{SR}^{k+1}} \sum_{p=0}^{k} \frac{k!}{p!} \left(\frac{\beta_{SR} - \delta_{SR}}{\overline{\lambda}_{SR}}\right)^{-(k+1-p)} \gamma_{\text{out},R}{}^{p} \\ &\quad \times \left[\int_{H_{SH}}^{\frac{H_{SH}}{\cos\Phi}} \left(x^2 - H_{SH}x\right) x^{\xi} \exp\left(\Lambda x^{n_1}\right) dx \right. \\ &\quad + \int_{\frac{H_{SH}}{\cos\Phi}}^{H_{SL}} \left(x^2 - x^2\cos\Phi\right) x^{\xi} \exp\left(\Lambda x^{n_1}\right) dx \\ &\quad \left. + \int_{H_{SL}}^{\frac{H_{SL}}{\cos\Phi}} \left(H_{SL}x - x^2\cos\Phi\right) x^{\xi} \exp\left(\Lambda x^{n_1}\right) dx \right] \\ &= 1 - \frac{2\pi\alpha_{SR}}{V_S} \sum_{k=0}^{m_{SR}-1} \frac{\zeta(k)}{\overline{\lambda}_{SR}^{k+1}} \sum_{p=0}^{k} \frac{k!}{p!} \left(\frac{\beta_{SR} - \delta_{SR}}{\overline{\lambda}_{SR}}\right)^{-(k+1-p)} \gamma_{\text{out},R}{}^{p} \\ &\quad \times \left\{ \frac{1}{n_1 \Lambda^{\frac{\xi+3}{n_1}}} \left[\Gamma\left(\frac{\xi+3}{n_1}, \Xi_1\right) - \Gamma\left(\frac{\xi+3}{n_1}, \Xi_3\right)\right] \right. \\ &\quad - \frac{H_{SH}}{n_1 \Lambda^{\frac{\xi+2}{n_1}}} \left[\Gamma\left(\frac{\xi+2}{n_1}, \Xi_1\right) - \Gamma\left(\frac{\xi+2}{n_1}, \Xi_2\right)\right] \\ &\quad + \frac{H_{SL}}{n_1 \Lambda^{\frac{\xi+2}{n_1}}} \left[\Gamma\left(\frac{\xi+2}{n_1}, \Xi_3\right) - \Gamma\left(\frac{\xi+2}{n_1}, \Xi_4\right)\right] \\ &\quad \left. - \frac{\cos\Phi}{n_1 \Lambda^{\frac{\xi+3}{n_1}}} \left[\Gamma\left(\frac{\xi+3}{n_1}, \Xi_2\right) - \Gamma\left(\frac{\xi+3}{n_1}, \Xi_4\right)\right] \right\} \end{aligned} \tag{4.72}$$

where $\Lambda = \frac{\beta_{SR}-\delta_{SR}}{\overline{\lambda}_{SR}} \gamma_{\text{out},R}$, $\xi = n_1 p$ and in the third line, we define $y = \Lambda x^{n_1}$, $\Xi_1 = \Lambda H_{SH}^{n_1}$, $\Xi_2 = \Lambda\left(\frac{H_{SH}}{\cos\Phi}\right)^{n_1}$, $\Xi_3 = \Lambda H_{SL}^{n_1}$, $\Xi_4 = \Lambda\left(\frac{H_{SL}}{\cos\Phi}\right)^{n_1}$, and $\Gamma(a, x) = \int_x^{\infty} t^{a-1} \exp(-t)\, dt$ is the upper incomplete gamma function.

**Case 2**: $\frac{H_{SH}}{\cos\Phi} > H_{SL}$. Using (4.70), the OP for S-R link in the second case can be easily presented as

$$\begin{aligned}P_{\text{out},SR} =& 1-\frac{2\pi\alpha_{SR}}{V_S}\sum_{k=0}^{m_{SR}-1}\frac{\zeta(k)}{\overline{\lambda}_{SR}^{k+1}}\sum_{p=0}^{k}\frac{k!}{p!}\left(\frac{\beta_{SR}-\delta_{SR}}{\overline{\lambda}_{SR}}\right)^{-(k+1-p)}\gamma_{\text{out},R}{}^{p}\\&\times\left\{\frac{1}{n_1\Lambda^{\frac{\xi+3}{n_1}}}\left[\Gamma\left(\frac{\xi+3}{n_1},\Xi_1\right)-\Gamma\left(\frac{\xi+3}{n_1},\Xi_3\right)\right]\right.\\&-\frac{H_{SH}}{n_1\Lambda^{\frac{\xi+2}{n_1}}}\left[\Gamma\left(\frac{\xi+2}{n_1},\Xi_1\right)-\Gamma\left(\frac{\xi+2}{n_1},\Xi_2\right)\right]\\&-\frac{\cos\Phi}{n_1\Lambda^{\frac{\xi+3}{n_1}}}\left[\Gamma\left(\frac{\xi+3}{n_1},\Xi_2\right)-\Gamma\left(\frac{\xi+3}{n_1},\Xi_4\right)\right]\\&\left.+\frac{H_{SL}}{n_1\Lambda^{\frac{\xi+2}{n_1}}}\left[\Gamma\left(\frac{\xi+2}{n_1},\Xi_3\right)-\Gamma\left(\frac{\xi+2}{n_1},\Xi_4\right)\right]\right\}.\end{aligned}\tag{4.73}$$

#### 4.2.2.2 Outage Over S-D Link

In this work, we treat the coverage area of S on the ground as a circle, $S_S$, with radius, $R_S$, where the projection of S, O, is located at the center of the circle. As D is uniformly distributed in the coverage area of S, the CDF of $r_S$ can be written as

$$F_{r_S}(x)=\frac{\int_0^{2\pi}\int_0^{x} r dr d\theta}{\pi H^2\tan^2\Phi}=\frac{x^2}{H^2\tan^2\Phi}.\tag{4.74}$$

Therefore, the PDF of $r_S$ can be obtained as $f_{r_S}(x)=\frac{2x}{H^2\tan^2\Phi}, 0\le x\le H\tan\Phi$. Further, we can achieve the PDF of $d_{SD}$ as $f_{d_{SD}}(x)=\frac{2x}{H^2\tan^2\Phi}, H\le x\le\frac{H}{\cos\Phi}$.

Similar to S-R link, we can obtain $P_{\text{out},SD}$ as

$$\begin{aligned}P_{\text{out},SD} &= \Pr\left\{\frac{2\gamma_{SD}}{H^2\tan^2\Phi}\le\gamma_{\text{out},D}\right\}=\Pr\left\{\lambda_{SD}\le\gamma_{\text{out},D}d_{SD}^{n_2}\right\}\\&=1-\frac{2\alpha_{SD}}{H^2\tan^2\Phi}\sum_{k=0}^{m_{SD}-1}\frac{\zeta(k)}{\overline{\lambda}_{SD}^{k+1}}\sum_{p=0}^{k}\frac{k!}{p!}\left(\frac{\beta_{SD}-\delta_{SD}}{\overline{\lambda}_{SD}}\right)^{-(k+1-p)}\\&\times\frac{1}{n_2\left(\frac{\beta_{SD}-\delta_{SD}}{\overline{\lambda}_{SD}}\gamma_{\text{out},D}\right)^{\frac{n_2p+2}{n_2}}}\gamma_{\text{out},D}^{p}\left[\Gamma\left(\frac{n_2p+2}{n_2},z_{\min}\right)-\Gamma\left(\frac{n_2p+2}{n_2},z_{\max}\right)\right],\end{aligned}\tag{4.75}$$

where $z = \frac{\beta_{SD}-\delta_{SD}}{\bar{\lambda}_{SD}} \gamma_{\text{out},D} x^{n_2}$, $z_{\min} = \frac{\beta_{SD}-\delta_{SD}}{\bar{\lambda}_{SD}} \gamma_{\text{out},D} H^{n_2}$, and $z_{\max} = \frac{\beta_{SD}-\delta_{SD}}{\bar{\lambda}_{SD}} \gamma_{\text{out},D} \left(\frac{H}{\cos\Phi}\right)^{n_2}$.

#### 4.2.2.3 Outage Over R-D Link

As depicted in Fig. 4.1, the coverage area of R on the ground is also a circle with variable radius $R_D$, which is determined by the height of R, and center $O_R$, which is the projection of R. We denote the height of R as $h_R$, the distance between the maximum flight altitude of R and the ground terminal as $H_{HD}$, the distance between the minimum flight altitude of R and D as $H_{LD}$, and the height between S and R as $h_S = H - h_R$. Then it is easy to obtain the radius $R_D$ and as $R_D = h_R \tan\Psi$, where $\Psi$ is the beam angle of R.

Then, one can derive the CDF of $h_S$ as $F_{h_S}(x) = \frac{\int_0^{2\pi}\int_{H_{SH}}^{x}\int_0^{z\tan\Phi} r dr dz d\theta}{V_S} = \frac{\frac{1}{3}\pi(x^3-H_{SH}^3)\tan^2\Phi}{V_S}$.

Further, the CDF and PDF of $h_R$ can be written as

$$\begin{aligned} F_{h_R}(x) &= \Pr\{h_R \le x\} = \Pr\{H - h_S \le x\} \\ &= \Pr\{h_S \ge H - x\} = 1 - F_{h_S}(H-x), \end{aligned} \tag{4.76}$$

and

$$f_{h_R}(x) = \frac{\pi(H-x)^2\tan^2\Phi}{V_S}, \; H_{LD} \le x \le H_{HD}, \tag{4.77}$$

respectively.

When D is uniformly distributed within R's serving space, conditioned on $h_R$, the conditional PDF of the length of OD, can be presented as

$$f_{r_D|h_R}(x) = \frac{\int_0^{2\pi}\int_0^{x} r dr d\theta}{\pi h_R^2 \tan^2\Psi} = \frac{x^2}{h_R^2\tan^2\Psi}, \tag{4.78}$$

where $0 \le x \le h_R \tan\Psi$.

**Theorem 4.4** *Observing Fig. 4.1, the distance between R and D can be written as $d_{RD} = \sqrt{r_D^2 + h_R^2}$. Further, the PDF of $d_{RD}$ can be derived in the following two cases as*

***Case 1:*** *$\frac{H_{LD}}{\cos\Psi} < H_{HD}$. Then, $d_{RD}$ can be calculated as*

$$f_{d_{RD}}(d_{RD}) = \begin{cases} \frac{2\pi}{V_S}\frac{\tan^2\Phi}{\tan^2\Psi} d_{RD}\left[\mathcal{F}(d_{RD},0) - \mathcal{F}(H_{LD},0)\right], & \text{if } H_{LD} < d_{RD} < \frac{H_{LD}}{\cos\Psi}; \\ \frac{2\pi}{V_S}\frac{\tan^2\Phi}{\tan^2\Psi} d_{RD}\left[\mathcal{F}(d_{RD},0) - \mathcal{F}(d_{RD},\Psi)\right], & \text{if } \frac{H_{LD}}{\cos\Psi} < d_{RD} < H_{HD}; \\ \frac{2\pi}{V_S}\frac{\tan^2\Phi}{\tan^2\Psi} d_{RD}\left[\mathcal{F}(H_{HD},0) - \mathcal{F}(d_{RD},\Psi)\right], & \text{if } H_{HD} < d_{RD} < \frac{H_{HD}}{\cos\Psi} \end{cases} \tag{4.79}$$

*where* $\mathcal{F}(M,\Psi) = -\frac{H^2}{M\cos\Psi} - 2H\ln(M\cos\Psi) + M\cos\Psi$, $M \in \{H_{LD}, H_{HD}, d_{RD}\}$.

***Case 2:*** $\frac{H_{LD}}{\cos\Psi} > H_{HD}$. *Thus, we can derive* $d_{RD}$ *as*

$$f_{d_{RD}}(d_{RD}) = \begin{cases} \frac{2\pi}{V_S}\frac{\tan^2\Phi}{\tan^2\Psi} d_{RD}\left[\mathcal{F}(d_{RD},0) - \mathcal{F}(H_{LD},0)\right], & \text{if } H_{LD} < d_{RD} < H_{HD}; \\ \frac{2\pi}{V_S}\frac{\tan^2\Phi}{\tan^2\Psi} d_{RD}\left[\mathcal{F}(H_{HD},0) - \mathcal{F}(H_{LD},0)\right], & \text{if } H_{HD} < d_{RD} < \frac{H_{LD}}{\cos\Psi}; \\ \frac{2\pi}{V_S}\frac{\tan^2\Phi}{\tan^2\Psi} d_{RD}\left[\mathcal{F}(H_{HD},0) - \mathcal{F}(d_{RD},\Psi)\right], & \text{if } \frac{H_{LD}}{\cos\Psi} < d_{RD} < \frac{H_{HD}}{\cos\Psi} \end{cases}. \tag{4.80}$$

***Proof*** Please refer to Appendix II. □

**Corollary 4.2** *Assuming that D is within R's serving area, and when* $\frac{H_{LD}}{\cos\Psi} < H_{HD}$, *then the OP over R-D link can be calculated as*

$$\begin{aligned} \mathcal{P}_{RD} = 1 - \frac{2\pi}{V_S}\frac{\tan^2\Phi}{\tan^2\Psi}\Bigg\{ & \frac{a_1\pi}{P_1}\sum_{i=1}^{P_1}\sqrt{1-\chi_i{}^2}\,(a_1\chi_i + a_2) \\ & \times \mathcal{H}(a_1\chi_i + a_2)\,\mathcal{F}(a_1\chi_i + a_2, 0) \\ & - \frac{2}{n_2\mu} e^{-\frac{4v}{n_3\mu}}\Theta^{-\frac{2}{n_3}}\mathcal{F}(H_{LD},0)\left[\Gamma\left(\frac{4}{n_3\mu},\Delta_1\right) - \Gamma\left(\frac{4}{n_3\mu},\Delta_2\right)\right] \\ & + \frac{a_3\pi}{P_2}\sum_{j=1}^{P_2}\sqrt{1-\delta_j{}^2}\,(a_3\delta_j + a_4)\,\mathcal{H}(a_3\delta_j + a_4) \\ & \times \left[\mathcal{F}(a_3\delta_j + a_4, 0) - \mathcal{F}(a_3\delta_j + a_4, \Psi)\right] \\ & + \frac{2}{n_2\mu} e^{-\frac{4v}{n_3\mu}}\Theta^{-\frac{2\mu}{n_3\mu}+\frac{\mu}{2}}\mathcal{F}(H_{HD},0)\left[\Gamma\left(\frac{4}{n_3\mu},\Delta_3\right) - \Gamma\left(\frac{4}{n_3\mu},\Delta_4\right)\right] \\ & - \frac{a_5\pi}{P_3}\sum_{k=1}^{P_3}\sqrt{1-\varepsilon_k{}^2}\,(a_5\varepsilon_k + a_6)\,\mathcal{H}(a_5\varepsilon_k + a_6)\,\mathcal{F}(a_5\varepsilon_k + a_6, \Psi)\Bigg\} \end{aligned} \tag{4.81}$$

*where* $w = e^{v}\Theta^{\frac{\mu}{2}}x^{\frac{n_3\mu}{2}}$, $\mathcal{H}(x) = \exp(w)$, $\Delta_1 = e^{v}\Theta^{\frac{\mu}{2}}H_{LD}^{\frac{n_3\mu}{2}}$, $\Delta_2 = e^{v}\Theta^{\frac{\mu}{2}}\left(\frac{H_{LD}}{\cos\Psi}\right)^{\frac{n_3\mu}{2}}$, $\Delta_3 = e^{v}\Theta^{\frac{\mu}{2}}H_{HD}^{\frac{n_3\mu}{2}}$, $\Delta_4 = e^{v}\Theta^{\frac{\mu}{2}}\left(\frac{H_{HD}}{\cos\Psi}\right)^{\frac{n_3\mu}{2}}$, $\chi_i = \cos\left(\frac{2i-1}{2P_1}\pi\right)$, $\delta_j = \cos\left(\frac{2j-1}{2P_2}\pi\right)$, $\varepsilon_k = \cos\left(\frac{2k-1}{2P_3}\pi\right)$, $a_1 = \frac{H_{LD}/\cos\Psi - H_{LD}}{2}$, $a_2 = \frac{H_{LD}/\cos\Psi + H_{LD}}{2}$, $a_3 = \frac{H_{HD} - H_{LD}/\cos\Psi}{2}$, $a_4 = \frac{H_{HD} + H_{LD}/\cos\Psi}{2}$, $a_5 = \frac{H_{HD}/\cos\Psi - H_{HD}}{2}$, *and* $a_6 = \frac{H_{HD}/\cos\Psi + H_{HD}}{2}$.

***Proof*** Please refer to Appendix III. □

Similarly, when $\frac{H_{LD}}{\cos\Psi} > H_{HD}$, we can obtain $\mathcal{P}_{RD}$ as

$$\begin{aligned}\mathcal{P}_{RD} = 1 - \frac{2\pi}{V_S}\frac{\tan^2\Phi}{\tan^2\Psi}\Bigg\{&\frac{b_1\pi}{Q_1}\sum_{l=1}^{Q_1}\sqrt{1-{\eta_l}^2}\,(b_1\eta_l + b_2)\\&\times\mathcal{H}(b_1\eta_l + b_2)\mathcal{F}(b_1\eta_l + b_2, 0)\\&-\frac{2}{n_2\mu}e^{-\frac{4v}{n_3\mu}}\Theta^{-\frac{2\mu}{n_3\mu}+\frac{\mu}{2}}\mathcal{F}(H_{LD}, 0)\left[\Gamma\left(\frac{4}{n_2\mu}, \Delta_1\right) - \Gamma\left(\frac{4}{n_2\mu}, \Delta_2\right)\right]\\&+\frac{2}{n_2\mu}e^{-\frac{4v}{n_3\mu}}\Theta^{-\frac{2}{n_3}}\mathcal{F}(H_{HD}, 0)\left[\Gamma\left(\frac{4}{n_3\mu}, \Delta_3\right) - \Gamma\left(\frac{4}{n_3\mu}, \Delta_4\right)\right]\\&-\frac{b_3\pi}{Q_2}\sum_{m=1}^{Q_2}\sqrt{1-{\iota_m}^2}\,(b_3\iota_m + b_4)\mathcal{H}(b_3\iota_m + b_4)\mathcal{F}(b_3\iota_m + b_4, \Psi)\Bigg\}\end{aligned} \tag{4.82}$$

where $\iota_m = \cos\left(\frac{2m-1}{2Q_2}\pi\right)$, $\eta_l = \cos\left(\frac{2l-1}{2Q_1}\pi\right)$, $b_1 = \frac{H_{HD}-H_{LD}}{2}$, $b_2 = \frac{H_{HD}+H_{LD}}{2}$, $b_3 = \frac{H_{HD}/\cos\Psi - H_{LD}/\cos\Psi}{2}$, and $b_4 = \frac{H_{HD}/\cos\Psi + H_{LD}/\cos\Psi}{2}$.

Then, we consider the scenario that D is uniformly distributed within S's serving space. In this scenario, the probability that D is uniformly distributed within R's serving space can be calculated as

$$P_{DR} = \frac{S_R}{S_S} = \frac{\pi(h_R\tan\Psi)^2}{\pi(H\tan\Phi)^2} = \frac{{h_R}^2\tan^2\Psi}{H^2\tan^2\Phi}, \tag{4.83}$$

where $S_R$ and $S_S$ are the coverage areas of R and S, respectively.

As shown in Fig. 4.18, $h_R = d_{RD}\cos\Psi$. Then, we can further get $P_{DR} = \frac{{h_R}^2\tan^2\Psi}{H^2\tan^2\Phi} = \frac{d_{RD}^2\cos^2\Psi\tan^2\Psi}{H^2\tan^2\Phi}$.

Therefore, when D is uniformly distributed within S's serving area, the OP over R-D link can be further calculated as $P_{\text{out},RD} = 1 - P_{DR}\mathcal{P}_{RD} = 1 - P_{DR}\Pr\left\{\lambda_{RD} > \gamma_{\text{out},D}d_{RD}^{n_3}\right\}$.

Similar to the derivation of $\mathcal{P}_{RD}$, $P_{\text{out},RD}$ can be further derived as (4.84), when $\frac{H_{LD}}{\cos\Psi} < H_{HD}$, and (4.85), when $\frac{H_{LD}}{\cos\Psi} > H_{HD}$,

$$\begin{aligned}P_{\text{out},RD} &= 1 - \int_{d_{RD,\min}}^{d_{RD,\min}}\frac{x^2\cos^2\Psi\tan^2\Psi}{H^2\tan^2\Phi}\mathcal{H}(x)f_{d_{RD}}(x)dx\\&= 1 - \frac{2\pi}{V_S}\frac{\cos^2\Psi}{H^2}\frac{\tan^2\Psi}{\tan^2\Phi}\Bigg\{\frac{a_1\pi}{R_1}\sum_{n=1}^{R_1}\sqrt{1-{\kappa_n}^2}\,(a_1\kappa_n + a_2)^3\\&\quad\times\mathcal{H}(a_1\kappa_n + a_2)\mathcal{F}(a_1\kappa_n + a_2, 0)\\&\quad-\frac{2}{n_2\mu}e^{-\frac{8v}{n_3\mu}}\Theta^{-\frac{4}{n_3}}\mathcal{F}(H_{LD}, 0)\left[\Gamma\left(\frac{8}{n_3\mu}, \Delta_1\right) - \Gamma\left(\frac{8}{n_3\mu}, \Delta_2\right)\right]\end{aligned}$$

$$+\frac{a_3\pi}{R_2}\sum_{p=1}^{R_2}\sqrt{1-{\nu_p}^2}\left(a_3\nu_p+a_4\right)^3\mathcal{H}\left(a_3\nu_p+a_4\right)$$
$$\times\left[\mathcal{F}\left(a_3\nu_p+a_4,0\right)-\mathcal{F}\left(a_3\nu_p+a_4,\Psi\right)\right]$$
$$-\frac{a_5\pi}{R_3}\sum_{q=1}^{R_3}\sqrt{1-{\rho_q}^2}\left(a_5\rho_q+a_6\right)^3\mathcal{H}\left(a_5\rho_q+a_6\right)\mathcal{F}\left(a_5\rho_q+a_6,\Psi\right)$$
$$+\frac{2}{n_2\mu}e^{-\frac{8v}{n_3\mu}}\Theta^{-\frac{4}{n_3}}\mathcal{F}\left(H_{HD},0\right)\left[\Gamma\left(\frac{8}{n_3\mu},\Delta_3\right)-\Gamma\left(\frac{8}{n_3\mu},\Delta_4\right)\right]\Bigg\} \tag{4.84}$$

$$P_{\text{out},RD}=1-\frac{2\pi}{V_S}\frac{\cos^2\Psi}{H^2}\frac{\tan^2\Psi}{\tan^2\Phi}\Bigg\{\frac{b_1\pi}{S_1}\sum_{r=1}^{S_1}\sqrt{1-{\tau_r}^2}(b_1\tau_r+b_2)^3$$
$$\times\mathcal{H}\left(b_1\tau_r+b_2\right)\mathcal{F}\left(b_1\tau_r+b_2,0\right)$$
$$-\frac{2}{n_2\mu}e^{-\frac{8v}{n_3\mu}}\Theta^{-\frac{4}{n_3}}\mathcal{F}\left(H_{LD},0\right)\left[\Gamma\left(\frac{8}{n_2\mu},\Delta_1\right)-\Gamma\left(\frac{8}{n_2\mu},\Delta_2\right)\right]$$
$$+\frac{2}{n_2\mu}e^{-\frac{8v}{n_3\mu}}\Theta^{-\frac{4}{n_3}}\mathcal{F}\left(H_{HD},0\right)\left[\Gamma\left(\frac{8}{n_3\mu},\Delta_3\right)-\Gamma\left(\frac{8}{n_3\mu},\Delta_4\right)\right]$$
$$-\frac{b_3\pi}{S_2}\sum_{t=1}^{S_2}\sqrt{1-{\upsilon_t}^2}(b_3\upsilon_t+b_4)^3\mathcal{H}\left(b_3\upsilon_t+b_4\right)\mathcal{F}\left(b_3\upsilon_t+b_4,\Psi\right)\Bigg\} \tag{4.85}$$

respectively, where $\kappa_n=\cos\left(\frac{2n-1}{2R_1}\pi\right)$, $\nu_p=\cos\left(\frac{2p-1}{2R_2}\pi\right)$, $\rho_q=\cos\left(\frac{2q-1}{2R_3}\pi\right)$, $\tau_r=\cos\left(\frac{2r-1}{2S_1}\pi\right)$, and $\upsilon_t=\cos\left(\frac{2t-1}{2S_2}\pi\right)$.

#### 4.2.2.4 Joint OP Over S-R and R-D Links

It should be noted that we cannot obtain $\left(1-P_{\text{out,SR}}\right)P_{\text{out,RD}}$ just by multiplying $1-P_{\text{out,SR}}$ and $P_{\text{out,RD}}$ together. The reason is that $1-P_{\text{out,SR}}$ and $P_{\text{out,RD}}$ share the same random variables $h_S$ when considering the randomness of the aerial platform. In other words, the PDF of $d_{SR}$ and $d_{RD}$ are correlated. By adopting the expressions of $1-P_{\text{out,SR}}$ conditioned on $d_{SR}$ and $P_{\text{out,RD}}$ conditioned on $d_{RD}$, we can derive the expression of $\left(1-P_{\text{out,SR}}\right)P_{\text{out,RD}}$ as

$$\left(1-P_{\text{out,SR}|\text{d}_{\text{SR}}}\right)P_{\text{out,RD}|\text{d}_{\text{RD}}}=\left[1-\alpha_{Sg}\sum_{k=0}^{m_{Sg}-1}\frac{\zeta(k)}{\overline{\lambda}_{Sg}^{k+1}}\sum_{p=0}^{k}\frac{k!}{p!}\left(\frac{\beta_{Sg}-\delta_{Sg}}{\overline{\lambda}_{Sg}}\right)^{-(k+1-p)}\right.$$
$$\left.\times\,d_{SR}^{p}\exp\left(-\frac{\beta_{Sg}-\delta_{Sg}}{\overline{\lambda}_{Sg}}d_{SR}\right)\right]\exp\left(-e^{v}\Theta^{\frac{\mu}{2}}d_{RD}^{\frac{n_3\mu}{2}}\right) \tag{4.86}$$

Since $d_{SR}$ and $d_{RD}$ can be re-expressed as

$$d_{SR} = \sqrt{h_S^2 + (h_S \psi_S)^2} \tag{4.87}$$

and

$$d_{DR} = \sqrt{(1 - h_S)^2 + [(1 - h_S)\, \psi_R]^2}, \tag{4.88}$$

where $\psi_S$ and $\psi_R$ are the polar angles of S and R, respectively.

Therefore, the expression of $\left(1 - P_{\text{out,SR}}\right) P_{\text{out,RD}}$ can be calculated by removing the randomness of $h_S$ as

$$\begin{aligned} &\left(1 - P_{\text{out,SR}|\text{d}_{\text{SR}}}\right) P_{\text{out,RD}|\text{d}_{\text{RD}}} \\ &= \int_{H_{SL}}^{H_{SH}} \left[ 1 - \alpha_{Sg} \sum_{k=0}^{m_{Sg}-1} \frac{\zeta(k)}{\overline{\lambda}_{Sg}^{k+1}} \sum_{p=0}^{k} \frac{k!}{p!} \left( \frac{\beta_{Sg} - \delta_{Sg}}{\overline{\lambda}_{Sg}} \right)^{-(k+1-p)} \right. \\ &\quad \left. \times \left[h_S^2 + (h_S \psi_S)^2\right]^{\frac{p}{2}} \exp\left( -\frac{\beta_{Sg} - \delta_{Sg}}{\overline{\lambda}_{Sg}} \sqrt{h_S^2 + (h_S \psi_S)^2} \right) \right] \\ &\quad \times \exp\left( -e^{v} \Theta^{\frac{\mu}{2}} \left\{ (1 - h_S)^2 + [(1 - h_S)\, \psi_R]^2 \right\}^{\frac{n_3 \mu}{4}} \right) dh_S. \end{aligned} \tag{4.89}$$

Further, we use Chebyshev-Gauss quadrature to obtain the following approximation as

$$\begin{aligned} &\left(1 - P_{\text{out,SR}|\text{d}_{\text{SR}}}\right) P_{\text{out,RD}|\text{d}_{\text{RD}}} \\ &= \sum_{j=1}^{W} \frac{c_1 \pi}{W} \sqrt{1 - \vartheta_j} \times \left[ 1 - \alpha_{Sg} \sum_{k=0}^{m_{Sg}-1} \frac{\zeta(k)}{\overline{\lambda}_{Sg}^{k+1}} \sum_{p=0}^{k} \frac{k!}{p!} \left( \frac{\beta_{Sg} - \delta_{Sg}}{\overline{\lambda}_{Sg}} \right)^{-(k+1-p)} \right. \\ &\quad \left. \times \exp\left( -\frac{\beta_{Sg} - \delta_{Sg}}{\overline{\lambda}_{Sg}} \sqrt{\left(c_1 \vartheta_j + c_2\right)^2 + \left[\left(c_1 \vartheta_j + c_2\right) \psi_S\right]^2} \right) \right] \\ &\quad \times \exp\left( -e^{v} \Theta^{\frac{\mu}{2}} \left\{ \left[1 - \left(c_1 \vartheta_j + c_2\right)\right]^2 \right. \right. \\ &\quad \left. \left. + \left\{ \left[1 - \left(c_1 \vartheta_j + c_2\right)\right] \psi_R \right\}^2 \right\}^{\frac{n_3 \mu}{4}} \right) \left[h_S^2 + (h_S \psi_S)^2\right]^{\frac{p}{2}}, \end{aligned} \tag{4.90}$$

where $c_1 = \frac{H_{SH} - H_{LD}}{2}$, $c_2 = \frac{H_{SH} + H_{LD}}{2}$, $\vartheta_j = \cos\left(\frac{2j-1}{2W}\pi\right)$.

Substituting $P_{\text{out},SR}$, $P_{\text{out},SD}$, $P_{\text{out},RD}$, and $\left(1 - P_{\text{out,SR}}\right) P_{\text{out,RD}}$ into Eqs. (4.66a) and (4.66b), we can finally approximate the OP of the considered system in closed form.

#### 4.2.2.5 SNR Received at D

When selection combining (SC) scheme is adopted at D to process the multiple copies of the received signal, the received SNR can be expressed as $\gamma_D = \max\left\{\min\left\{\gamma_{SR}, \gamma_{RD}\right\}, \gamma_{SD}\right\}$.

While maximum ratio combining (MRC) scheme is adopted, the SNR received at D can be expressed as $\gamma_{MRC} = \sum_{l=1}^{L} \gamma_l = \min\left\{\gamma_{SR}, \gamma_{RD}\right\} + \gamma_{SD}$.

Moreover, in this paper we only analyze the system performance considering one terrestrial terminal, however, the analysis can be expanded to multiple terminals. In that case, the CDF/PDF of the e2e SNR under multi-user scenarios with $N$ users can be derived as $\gamma_{out,\text{non-coop}} = \max\{\gamma_1, \gamma_2, \cdots, \gamma_N\}$ while these users work in non-cooperative mode, and $\gamma_{out,\text{coop}} = \gamma_1 + \gamma_2 + \cdots + \gamma_N$ while these users work in collaborating mode, which are essentially the cases of adopting SC and MRC schemes, respectively.

#### 4.2.2.6 Asymptotic Analysis

To verify our theoretical results and to provide applicable expressions of OP with low computational complexity and acceptable accuracy that can be applied in engineering applications, in this subsection we will derive the asymptotic expression in the case of high SNR. Namely, in the following asymptotic expressions will be derived by assuming $\overline{\lambda}_{SR}, \overline{\lambda}_{SD}$, and $\overline{\lambda}_{RD} \to \infty$.

(1) $P^{\infty}_{\text{out},SR}$: The PDF of $\lambda_{SR} = \overline{\lambda}_{SR}|h_{SR}|^2$ can be expressed as

$$f_{\overline{\lambda}_{SR}|h_{SR}|^2}(x) = \frac{\alpha_{SR}}{\overline{\lambda}_{SR}} \exp\left(-\beta_{SR}\frac{x}{\overline{\lambda}_{SR}}\right) {}_1F_1\left(m_{SR}; 1; \delta_{SR}\frac{x}{\overline{\lambda}_{SR}}\right). \tag{4.91}$$

Adopting the series representations of the exponential function, $\exp\left(-\beta_{SR}\frac{x}{\overline{\lambda}_{SR}}\right) = \sum_{n=0}^{\infty} \frac{\left(-\beta_{SR}\frac{x}{\overline{\lambda}_{SR}}\right)^n}{n!}$, and keeping the first two terms while ignoring the higher order term, the asymptotic CDF and PDF of $\lambda_{SR}$ can be written as

$$\begin{aligned} F^{\infty}_{\overline{\lambda}_{SR}|h_{SR}|^2}(x) =& 1 - \alpha_{SR} \sum_{k=0}^{m_{SR}-1} \frac{\zeta(k)}{\overline{\lambda}_{SR}^{k+1}} \sum_{p=0}^{k} \frac{k!}{p!} \\ & \times \left(\frac{\beta_{SR}-\delta_{SR}}{\overline{\lambda}_{SR}}\right)^{-(k+1-p)} x^p \left(1 - \frac{\beta_{SR}-\delta_{SR}}{\overline{\lambda}_{SR}}x\right) \end{aligned} \tag{4.92}$$

and

$$f^{\infty}_{\overline{\lambda}_{SR}|h_{SR}|^2}(x) = \alpha_{SR} \sum_{k=0}^{m_{SR}-1} \frac{\zeta(k)}{\overline{\lambda}_{SR}^{k+1}} x^k \left(1 - \frac{\beta_{SR} - \delta_{SR}}{\overline{\lambda}_{SR}} x\right), \tag{4.93}$$

respectively.

Accordingly, the asymptotic OP over S-R link, $P^{\infty}_{\text{out},SR}$, of two cases can be calculated as (4.94), when $\frac{H_{SH}}{\cos\Phi} < H_{HL}$, and (4.95), when $\frac{H_{SH}}{\cos\Phi} > H_{HL}$, respectively,

$$\begin{aligned}
P^{\infty}_{\text{out},SR} &= \int_0^{\infty} F^{\infty}_{\overline{\lambda}_{SR}|h_{SR}|^2}\left(\gamma_{\text{out},R} x^{n_1}\right) f_{d_{SR}}(x)\, dx \\
&= 1 - \alpha_{SR} \sum_{k=0}^{m_{SR}-1} \frac{\zeta(k)}{\overline{\lambda}_{SR}^{k+1}} \sum_{p=0}^{k} \frac{k!}{p!} \left(\frac{\beta_{SR} - \delta_{SR}}{\overline{\lambda}_{SR}}\right)^{-(k+1-p)} \gamma_{\text{out},R}{}^p \int_{d_{SR,\min}}^{d_{SR,\max}} x^{\xi} \\
&\quad \times \left(1 - \frac{\beta_{SR} - \delta_{SR}}{\overline{\lambda}_{SR}} \gamma_{\text{out},R} x^{n_1}\right) f_{d_{SR}}(x)\, dx \\
&= 1 - \alpha_{SR} \sum_{k=0}^{m_{SR}-1} \frac{\zeta(k)}{\overline{\lambda}_{SR}^{k+1}} \sum_{p=0}^{k} \frac{k!}{p!} \left(\frac{\beta_{SR} - \delta_{SR}}{\overline{\lambda}_{SR}}\right)^{-(k+1-p)} \gamma_{\text{out},R}{}^p \\
&\quad \times \frac{2\pi}{V_S} \left[\frac{1}{\xi+3}\left({t_2}^{\xi+3} - {t_1}^{\xi+3}\right) - \frac{\Lambda H_{SH}}{\xi+2}\left({t_2}^{\xi+2} - {t_1}^{\xi+2}\right)\right. \\
&\quad - \frac{\Lambda}{\xi+n_1+3}\left({t_2}^{\xi+n_1+3} - {t_1}^{\xi+n_1+3}\right) + \frac{H_{SH}}{\xi+n_1+2}\left({t_2}^{\xi+n_1+2} - {t_1}^{\xi+n_1+2}\right) \\
&\quad - \frac{\cos\Phi}{\xi+3}\left({t_3}^{\xi+3} - {t_2}^{\xi+3}\right) + \frac{1}{\xi+3}\left({t_3}^{\xi+3} - {t_2}^{\xi+3}\right) \\
&\quad - \frac{\Lambda}{\xi+n_1+3}\left({t_3}^{\xi+n_1+3} - {t_2}^{\xi+n_1+3}\right) - \frac{\cos\Phi}{\xi+3}\left({t_4}^{\xi+3} - {t_3}^{\xi+3}\right) \\
&\quad + \frac{\Lambda\cos\Phi}{\xi+n_1+3}\left({t_3}^{\xi+n_1+3} - {t_2}^{\xi+n_1+3}\right) + \frac{H_{SL}}{\xi+2}\left({t_4}^{\xi+2} - {t_3}^{\xi+2}\right) \\
&\quad \left. + \frac{\Lambda\cos\Phi}{\xi+n_1+3}\left({t_4}^{\xi+n_1+3} - {t_3}^{\xi+n_1+3}\right) - \frac{\Lambda H_{SL}}{\xi+n_1+2}\left({t_4}^{\xi+n_1+2} - {t_3}^{\xi+n_1+2}\right)\right],
\end{aligned} \tag{4.94}$$

when $\frac{H_{SH}}{\cos\Phi} < H_{HL}$, and

$$
\begin{aligned}
P_{\text{out},SR}^{\infty} = 1 - \alpha_{SR} \sum_{k=0}^{m_{SR}-1} \frac{\zeta(k)}{\overline{\lambda}_{SR}^{k+1}} \sum_{p=0}^{k} \frac{k!}{p} \left( \frac{\beta_{SR} - \delta_{SR}}{\overline{\lambda}_{SR}} \right)^{-(k+1-p)} {\gamma_{\text{out},R}}^{p} \\
\times \frac{2\pi}{V_S} \left[ \frac{1}{\xi+3} \left( {t_3}^{\xi+3} - {t_1}^{\xi+3} \right) - \frac{H_{SH}}{\xi+2} \left( {t_3}^{\xi+2} - {t_1}^{\xi+2} \right) \right. \\
- \frac{\Lambda}{\xi+n_1+3} \left( {t_3}^{\xi+n_1+3} - {t_1}^{\xi+n_1+3} \right) + \frac{H_{SH}\Lambda}{\xi+n_1+2} \left( {t_3}^{\xi+n_1+2} - {t_1}^{\xi+n_1+2} \right) \\
+ \frac{H_{SL} - H_{SH}}{\xi+2} \left( {t_2}^{\xi+2} - {t_3}^{\xi+2} \right) - \frac{(H_{SL} - H_{SH})\Lambda}{\xi+n_1+2} \left( {t_2}^{\xi+n_1+2} - {t_3}^{\xi+n_1+2} \right) \\
- \frac{\cos\Phi}{\xi+3} \left( {t_4}^{\xi+3} - {t_2}^{\xi+3} \right) + \frac{H_{SL}}{\xi+2} \left( {t_4}^{\xi+2} - {t_2}^{\xi+2} \right) \\
\left. + \frac{\Lambda\cos\Phi}{\xi+n_1+3} \left( {t_4}^{\xi+n_1+3} - {t_2}^{\xi+n_1+3} \right) - \frac{H_{SL}\Lambda}{\xi+n_1+2} \left( {t_4}^{\xi+n_1+2} - {t_2}^{\xi+n_1+2} \right) \right],
\end{aligned}
\tag{4.95}
$$

when $\frac{H_{SH}}{\cos\Phi} > H_{HL}$, respectively, where $t_1 = H_{SH}$, $t_2 = \frac{H_{SH}}{\cos\Phi}$, $t_3 = H_{SL}$, $t_4 = \frac{H_{SL}}{\cos\Phi}$.

(2) $P_{\text{out},SD}^{\infty}$: Similarly, $P_{\text{out},SD}^{\infty}$ can be obtained directly as (4.96):

$$
\begin{aligned}
P_{\text{out},SD}^{\infty} &= \int_0^{\infty} F_{\overline{\lambda}_{SD}|h_{SD}|^2}^{\infty} \left( \gamma_{\text{out},D} x^{n_2} \right) f_{d_{SD}}(x)\, dx \\
&= 1 - \alpha_{SD} \sum_{k=0}^{m_{SD}-1} \frac{\zeta(k)}{\overline{\lambda}_{SD}^{k+1}} \sum_{p=0}^{k} \frac{k!}{p!} \left( \frac{\beta_{SD} - \delta_{SD}}{\overline{\lambda}_{SD}} \right)^{-(k+1-p)} {\gamma_{\text{out},D}}^{p} \\
&\quad \times \int_{d_{SD,\min}}^{d_{SD,\max}} x^{n_2 p} \left( 1 - \frac{\beta_{SD} - \delta_{SD}}{\overline{\lambda}_{SD}} \gamma_{\text{out},D} x^{n_2} \right) f_{d_{SD}}(x)\, dx \\
&= 1 - \alpha_{SD} \sum_{k=0}^{m_{SD}-1} \frac{\zeta(k)}{\overline{\lambda}_{SD}^{k+1}} \sum_{p=0}^{k} \frac{k!}{p!} \left( \frac{\beta_{SD} - \delta_{SD}}{\overline{\lambda}_{SD}} \right)^{-(k+1-p)} {\gamma_{\text{out},D}}^{p} \\
&\quad \times \frac{2}{\tan^2\Phi} \left\{ \frac{1}{n_2 p + 2} \left[ \left( \frac{H}{\cos\Phi} \right)^{n_2 p + 2} - H^{n_2 p + 2} \right] \right. \\
&\quad \left. - \frac{\Lambda}{n_2 p + n_2 + 2} \left[ \left( \frac{H}{\cos\Phi} \right)^{n_2 p + n_2 + 2} - H^{n_2 p + n_2 + 2} \right] \right\}
\end{aligned}
\tag{4.96}
$$

(3) $P_{\text{out},RD}^{\infty}$: The PDF of $\lambda_{RD} = \overline{\lambda}_{RD}|h_{RD}|^2$ can be expressed as (4.64). Adopting the series representations of the exponential function and the modified Bessel function of the first kind with order 0, we can have $\exp\left(-\frac{1+K}{\Omega}\frac{x}{\overline{\lambda}_{RD}}\right) \approx 1 - \frac{1+K}{\Omega}\frac{x}{\overline{\lambda}_{RD}}$ and $I_0\left(2\sqrt{\frac{K(1+K)}{\Omega\overline{\lambda}_{RD}}}\right) \approx 1 + \frac{K(1+K)}{\Omega\overline{\lambda}_{RD}}$ when $\overline{\lambda}_{RD} \to \infty$. Thus, the approximated CDF and PDF of $\lambda_{RD} = \overline{\lambda}_{RD}|h_{RD}|^2$ can be further simplified as

$$
f_{\lambda_{RD}}(x) = \frac{(1+K)}{\Omega_R} \exp(-K) \times \left( 1 - \frac{(1+K)x}{\Omega_R} \right) \left( 1 + \frac{K(1+K)x}{\Omega_R} \right) \tag{4.97}
$$

and

$$F_{\lambda_{RD}}(x) = \frac{(1+K)}{\Omega_R}\exp(-K)\left(x - \frac{K(1+K)^2 x^3}{3\Omega_R^2} + \frac{(1-K)(1+K)x^2}{2\Omega_R}\right), \tag{4.98}$$

respectively.

Then, $P^{\infty}_{\text{out},RD}$ is determined by (4.99) as follows:

$$\begin{aligned}
P^{\infty}_{\text{out},RD} &= \Pr\left\{\lambda_{RD} < d_{RD}^{n_3}\gamma_{\text{out},D}\right\} \\
&= \int_{d_{RD,\min}}^{d_{RD,\min}} D \times \left(\gamma_{\text{out},D}x^{n_3} - Ex^{3n_3} + Fx^{2n_3}\right) f_{d_{RD}}(x)dx \\
&= D \times \frac{2\pi}{V_S}\frac{\tan\Phi}{\tan\Psi} \times \left\{\int_{H_{LD}}^{\frac{H_{LD}}{\cos\Psi}} C\left[\mathcal{F}(d_{RD},0) - \mathcal{F}(H_{LD},0)\right]dx\right. \\
&+ \int_{\frac{H_{LD}}{\cos\Psi}}^{H_{HD}} C\left[\mathcal{F}(d_{RD},0) - \mathcal{F}(d_{RD},\Psi)\right]dx \\
&+ \left.\int_{H_{HD}}^{\frac{H_{HD}}{\cos\Psi}} C\left[\mathcal{F}(H_{HD},0) - \mathcal{F}(d_{RD},\Psi)\right]dx\right\} \\
&= \frac{2\pi\tan\Phi}{V_S\tan\Psi}D\left\{\left[\mathcal{G}(H_{HD}) - \mathcal{G}(H_{LD})\right] - \mathcal{F}(H_{LD},0)\left[\mathcal{I}\left(\frac{H_{LD}}{\cos\Psi}\right) - \mathcal{I}(H_{LD})\right]\right. \\
&- \left.\mathcal{J}\left(\frac{H_{HD}}{\cos\Psi}\right) + \mathcal{J}\left(\frac{H_{LD}}{\cos\Psi}\right) + \mathcal{F}(H_{HD},0)\left[\mathcal{I}\left(\frac{H_{HD}}{\cos\Psi}\right) - \mathcal{I}(H_{HD})\right]\right\}
\end{aligned} \tag{4.99}$$

where $\frac{H_{LD}}{\cos\Psi} < H_{HD}$, $D = \frac{(1+K)}{\Omega_R}\exp(-K)$, $E = \frac{K(1+K)^2\gamma_{\text{out},D}{}^3}{3\Omega_R^2}$, $F = \frac{(1-K)(1+K)\gamma_{\text{out},D}{}^2}{2\Omega_R}$, $C = \gamma_{\text{out},D}x^{n_3+1} - Ex^{3n_3+1} + Fx^{2n_3+1}$, $\mathcal{I}(x) = \frac{1}{n_3+2}\gamma_{\text{out},D}x^{n_3+2} + \frac{1}{3n_3+2}Cx^{3n_3+2} + \frac{1}{2n_3+2}Fx^{2n_3+2}$, $\mathcal{G}(x) = \gamma_{\text{out},D}\mathcal{K}(1,x) + E\mathcal{K}(3,x) + F\mathcal{K}(2,x)$, $\mathcal{K}(c,x) = \frac{1}{cn_3+3}x^{cn_3+3} - \frac{1}{cn_3+1}H^2x^{cn_3+1} - \frac{2H\left[(cn_3+2)x^{cn_3+2}\log x - x^{cn_3+2}\right]}{(cn_3+2)^2}$, $c \in \{1,2,3\}$, $\mathcal{J}(x) = \gamma_{\text{out},D}\mathcal{L}(1,x) + E\mathcal{L}(3,x) + F\mathcal{L}(2,x)$, $\mathcal{L}(d,x) = \frac{1}{dn_3+3}x^{dn_3+3}\cos\Psi - \frac{1}{(dn_3+1)\cos\Psi}H^2x^{dn_3+1} - \frac{2H\left[(dn_3+2)x^{dn_3+2}(\log x + \log(\cos\Psi)) - x^{dn_3+2}\right]}{(dn_3+2)^2}$, $d \in \{1,2,3\}$.

When $\frac{H_{LD}}{\cos\Psi} < H_{HD}$, $P^{\infty}_{\text{out},RD}$ can be presented as

$$\begin{aligned}
P^{\infty}_{\text{out},RD} = \frac{2\pi}{V_S}\frac{\tan\Phi}{\tan\Psi}D \times \{&\mathcal{G}(H_{HD}) - \mathcal{G}(H_{LD}) \\
&+ \mathcal{F}(H_{LD},0)\left[\mathcal{I}(H_{LD}) - \mathcal{I}\left(\frac{H_{LD}}{\cos\Psi}\right)\right] \\
&+ \mathcal{F}(H_{HD},0)\left[\mathcal{I}\left(\frac{H_{HD}}{\cos\Psi}\right) - \mathcal{I}(H_{HD})\right] \\
&- \left[\mathcal{J}\left(\frac{H_{HD}}{\cos\Psi}\right) - \mathcal{J}\left(\frac{H_{LD}}{\cos\Psi}\right)\right]\Bigg\}.
\end{aligned} \tag{4.100}$$

### 4.2.3 Numerical Results

In this section, we will provide numerical results to verify the proposed analytical models.

#### 4.2.3.1 Outage Performance Over S-R Link

Following the parameters adopted in [3, 16, 17], in this subsection, the main parameters are set as $m_{SR} = 10$, $b_{SR} = 0.126$ , $\Omega_{SR} = 0.835$ (for average shadowing), $\gamma_{\text{out},R} = 0$ dB, $H = 2000$ km, $H_{SH} = 1980$ km, $H_{SL} = 1990$ km, $H_{HD} = 280$ km, $H_{LD} = 10$ km, $\Phi = \pi/30$ rad, and $n_1 = 2$.

In Fig. 4.19, we plot the OP of the SR link versus $P_S$ with various heights of the satellite and a fixed height of the aerial relay. It is easy to observe that OP can be improved when $P_S$ increases. We can also see that the OP decreases as $H$ increases. The reason is that the increasing $H$ leads to the enlarged transmission distance between S and R, which can make the received SNR at R worse without a doubt.

The influence of the height of R, $H_{SH}$ and $H_{SL}$, on the outage performance of R-D link is presented in Fig. 4.20. Compared with Fig. 4.19, it is obvious that $H_{SH}$ and $H_{SL}$ exhibit quite a weak impact on the outage performance over the S-R link. Because the distance between S and R is on the order of hundreds to thousands of kilometers, while the variation range of the height of R is only on the order of tens/hundreds/thousands of meters. Thus, $H_{SH}$ and $H_{SL}$ cannot produce a significant

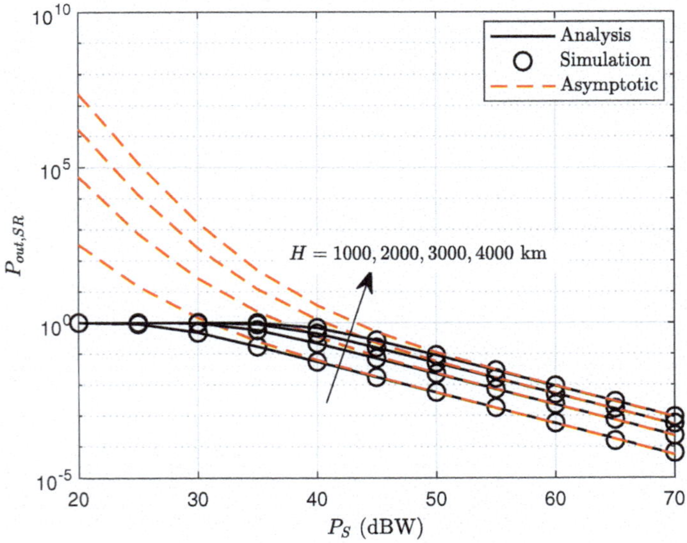


**Fig. 4.19** OP over S-R link for various $H$

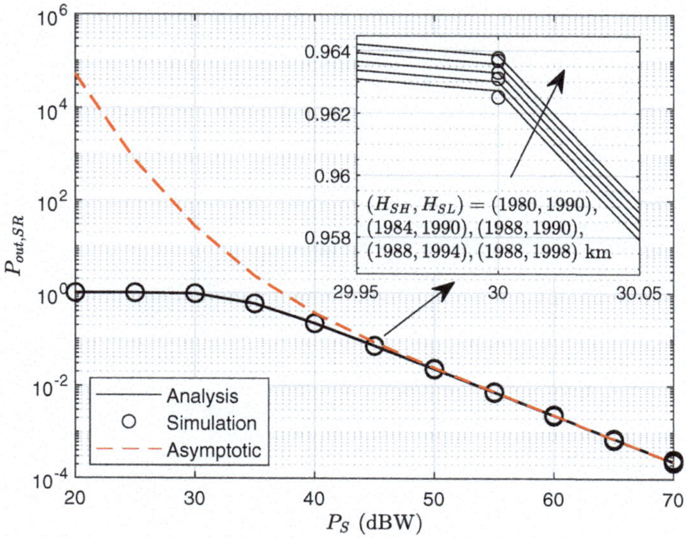


**Fig. 4.20** OP over S-R link for various $H_{SH}$ and $H_{SL}$

effect on the outage performance over the S-R link. However, it still can be observed that $H_{HD}$ and $H_{LD}$ show a negative influence on the OP of the R-D link.

As depicted in Figs. 4.19 and 4.20, the asymptotic results for the OP obtained can accurately match the simulation and analysis in the high SNR regime.

#### 4.2.3.2 Outage Performance Over S-D Link

In this subsection, the main parameters are set the same as the ones for S-R link, where we set $\gamma_{\text{out},D} = 0$ dB, $H = 2000$ km, $n_2 = 2$, and $\Phi = \pi/30$ rad. However, the fading channel of S-D link is a heavy shadowing condition, thus, the SR fading parameters are set to be different from those for the S-R link, where we defined $m_{SD} = 1$, $b_{SD} = 0.0063$, and $\Omega_{SR} = 0.0005$.

In Fig. 4.21, the OP over S-D links is presented for various of $H$. It is obvious to see that a small $H$ leads to a small OP. This observation comes from the fact that a small $H$ denotes a low path loss for the signal transmitted over the S-D link. Also, compared with Fig. 4.19, the outage performance of the S-D link is worse than that of the S-R link, due to the heavy shadowing between the satellite and the terrestrial terminal.

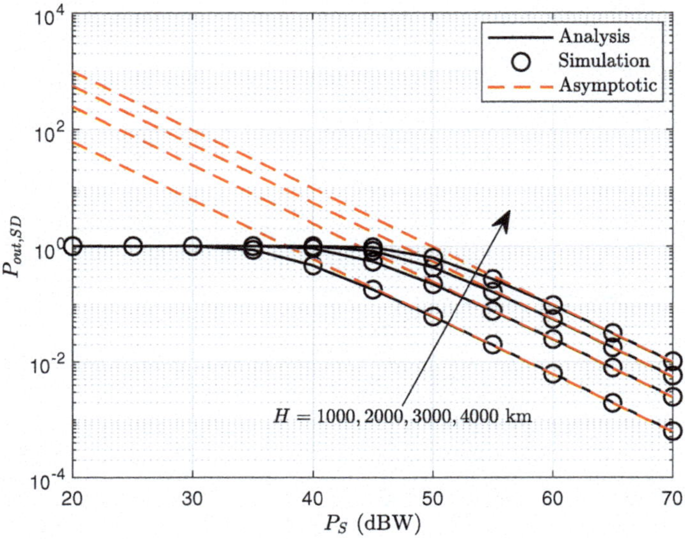


**Fig. 4.21** OP over S-D link for various $H$

#### 4.2.3.3 Outage Performance Over R-D Link

In this subsection, we will present the OP of R-D link, the main parameters adopted here are set as $\gamma_{\text{out},D} = 0$ dB, $H = 2000$ km, $K = 0$ dB, $\Phi = \pi/30$ rad, $\Psi = \pi/30$ rad, $n_3 = 2$, and $\Omega_{RD} = 0.278$.

In Fig. 4.22, the influence of the height of R, $H_{HD}$ and $H_{LD}$, on the outage performance of R-D link, is presented. It is easy to find out that $H_{HD}$ and $H_{LD}$ show a negative influence on the OP of the R-D link. This can also be explained by the reason proposed for the observations in Fig. 4.20. It also shows the impact of the transmit power at S on the outage performance of the R-D link. One can see that increasing $P_S$ can improve the OP of the R-D link.

The asymptotic results can also fit well with the simulation and analysis in the high SNR region shown in Fig. 4.22.

#### 4.2.3.4 The E2e OP

In this subsection, we will present some simulation results of the e2e outage performance of the considered system. The main parameters adopted here are set as $m_{SR} = 10, b_{SR} = 0.126, \Omega_{SR} = 0.835, \gamma_{\text{out},R} = 0$ dB, $H = 500$ km, $P_S = 50$ dBW, $P_R = 30$ dBW, $K = -10$ dB, $\Phi = \pi/30$ rad, $\Psi = \pi/30$ rad, $\Omega_{RD} = 0.278$, and $n_1 = n_3 = 2$.

As depicted in Fig. 4.23, the impacts of the transmit power at S and R are demonstrated. This figure illustrates the curves of the e2e OP for various $P_S$ and $P_R$. It

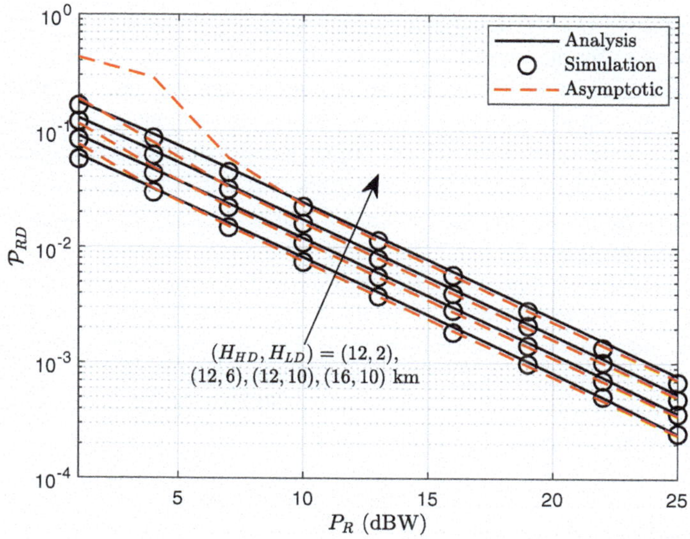


**Fig. 4.22** OP over R-D link for various $H_{HD}$ and $H_{LD}$

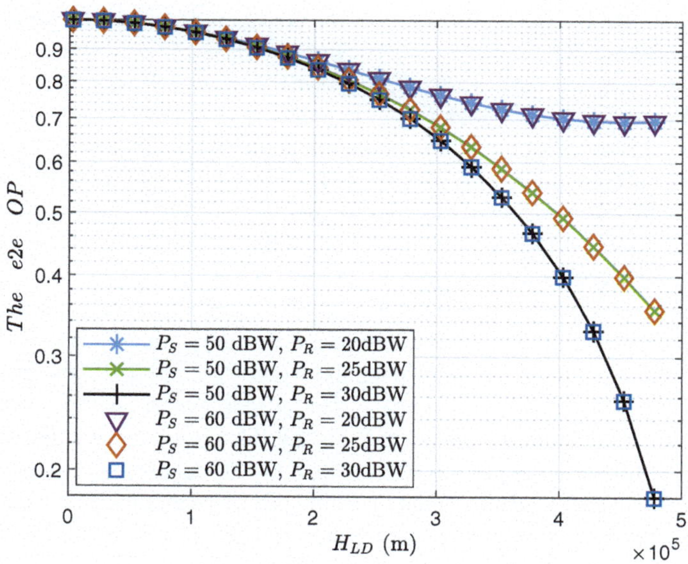


**Fig. 4.23** The e2e OP for various $P_S$ and $P_R$

is easy to observe that OP can be improved when $P_R$ increases. This observation can be explained by the reason provided in the previous paragraph. However, the outage performance of the system with different $P_S$ is similar when $P_R$ is fixed. This is because when $H_{LD}$ is small, the main factor that dominates the e2e OP is the coverage area of the aerial relay, which is much smaller than the coverage area of the satellite, and thus, the influence of the OP of the S-R link is rather small; when $H_{LD}$ is large, the distance between the satellite and the aerial relay is not large which can perform well with small $P_S$.

Also, it can be seen that, when $P_S$ and $P_R$ are fixed, the e2e OP decreases with a large $H_{LD}$. However, when $P_S = 50$ dBW and $P_R = 20$ dBW, the trend of the e2e OP is to decrease first and then increase. This finding can be explained by the following fact: When the distance $H_{LD}$ becomes larger and $P_R$ is not large enough to ensure the reliability of the communication, the OP of R-D link will degrade, which further leads to the deterioration of the e2e OP.

#### 4.2.3.5 OP of the CSATN System

In this subsection, some simulation results of the whole CSATN system will be presented to show the outage performance of the considered system. The main parameters adopted here are set as $m_{SR} = 10$, $b_{SR} = 0.126$ , $\Omega_{SR} = 0.835$, $\gamma_{\mathrm{out},R} = 0$ dB, $H = 500$ km, $P_S = 50$ dBW, $P_R = 30$ dBW, $K = -10$ dB, $\Phi = \pi/30$ rad, $\Psi = \pi/30$ rad, $\Omega_{RD} = 0.278$, and $n_1 = n_3 = 2$.

In Fig. 4.24, the OP of the whole system degrades while $P_S$ decreases. This can also be explained by the reason given for the observations before. Also, as shown in Fig. 4.24, the OP of the whole system outperforms the OP of the S-D link, which shows that an aerial relay can be a promising solution to improve the reliability of the satellite-terrestrial transmission.

The analytical result of the OP of the CSATN system versus $H_{LD}$ for various $P_R$ has been presented in Fig. 4.25. It can be seen that $H_{LD}$ and $P_R$ show a negative impact on the OP of the CSATN system, and when $P_R = 20$ dBW, the trend of the OP is to decrease first and then increase. These findings can also be explained by the discussions for the observations in Fig. 4.23. Also, without any doubt, we can find that the MRC scheme performs better than the SC scheme, as all copies of the received signal are utilized during the decoding process under the MRC scheme and then a larger diversity gain is achieved. However, when $H_{LD}$ is small, MRC and SC schemes exhibit a similar influence on the overall OP.

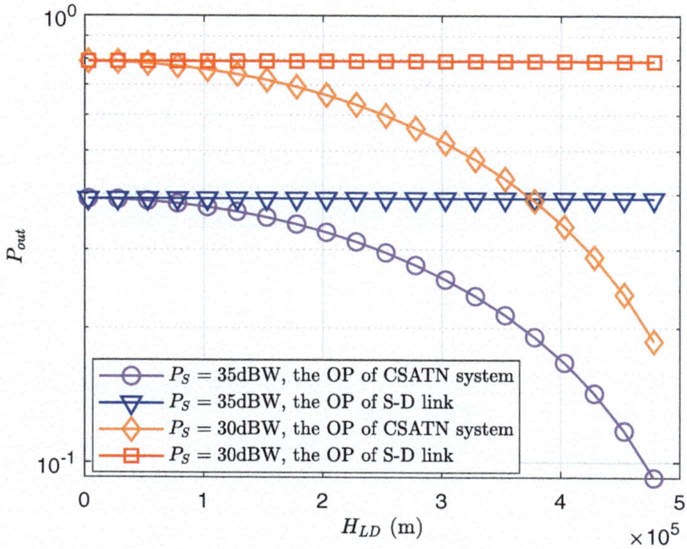


**Fig. 4.24** OP of the CSATN system for various $P_S$

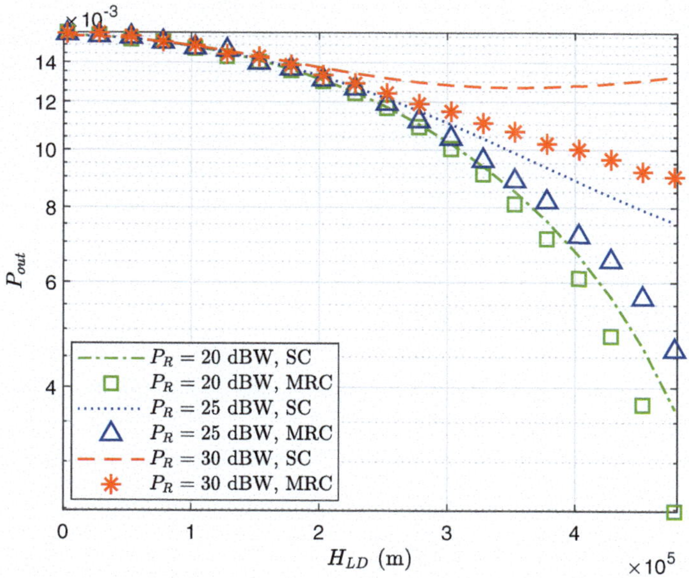


**Fig. 4.25** OP of the CSATN system for various $P_R$

### 4.2.4 Conclusion

This work investigates the outage performance of a CSATN system, where R assists the transmission from S to D. We first studied the approximated analytical expressions for the OP over S-R, R-D, and S-D links while considering the randomness of R and D. Then we analyzed the two outage scenarios over the three links. Finally, the asymptotic closed-form expressions for the OP of each link have been derived. The correctness of the outage performance analysis has been verified through Monte Carlo simulations.

Observing the numerical results, some meaningful remarks can be obtained as follows:

(1) The height of the satellite shows a strong negative impact on the outage performance over both S-R and S-D links.

(2) The height of R shows an obvious negative effect on the outage performance over the R-D link and a weak impact on the outage performance over the S-R link.

(3) The transmit power at S and R exhibits a positive influence on the outage performance over the S-R/S-D link and R-D link, respectively. However, when $P_R$ is of the same value, the influence of $P_S$ on the e2e OP is weak.

(4) The aerial relay can improve the reliability of the satellite-terrestrial transmission, and the impact of $P_S$ is more obvious than that of $P_R$.

(5) When $H_{LD}$ is large, MRC scheme outperforms SC scheme. However, MRC and SC schemes show a similar influence on the overall OP when $H_{LD}$ is small.

In this work, the outage performance of the considered system with a single terrestrial user was studied. However, the proposed analysis method and the adopted technical roadmap can serve as a meaningful reference to investigate the bit error/capacity performance of CSATN, or the OP in multi-terrestrial-user scenarios.

## Appendix I: Proof of Theorem 1

(1) For the first case ($\frac{H_{SH}}{\cos\Phi} < H_{SL}$), we can discuss the scenario in the following three kinds of situations.

When $H_{SH} < x < H_{SH}/\cos\Phi$, the CDF of the distance between S and R can be derived as

$$\begin{aligned} F_{d_{SR}}(x) &= \frac{1}{V_S}\int_0^{\arccos\frac{H_{SH}}{x}}\int_{\frac{H_{SH}}{\cos\phi}}^{x}\int_0^{2\pi}\delta^2\sin\phi d\theta d\delta d\phi \\ &= \frac{2\pi}{3V_S}\left(x^3 - \frac{3}{2}H_{SH}x^2 + \frac{1}{2}H_{SH}^3\right). \end{aligned} \tag{4.101}$$

For $\frac{H_{SH}}{\cos\Phi} < x < H_{SL}$, we can have

$$F_{d_{SR}}(x) = \frac{1}{V_S}\int_0^{\Phi}\int_{\frac{H_{SH}}{\cos\phi}}^{x}\int_0^{2\pi}\delta^2\sin\phi d\theta d\delta d\phi$$
$$= \frac{2\pi}{3V_S}\left(x^3 - x^3\cos\Phi - \frac{1}{2}\frac{H_{SH}^3}{\cos^2\Phi} + \frac{1}{2}H_{SH}^3\right). \tag{4.102}$$

While $H_{SL} < x < \frac{H_{SL}}{\cos\Phi}$, $d_{SR}$ is obtained as

$$F_{d_{SR}}(x) = \frac{1}{V_S}\left(\int_{\arccos\frac{H_{SL}}{x}}^{\Phi}\int_{\frac{H_{SH}}{\cos\phi}}^{x}\int_0^{2\pi}\delta^2\sin\phi d\theta d\delta d\phi\right.$$
$$\left.+\int_0^{\arccos\frac{H_{SL}}{x}}\int_{\frac{H_{SH}}{\cos\phi}}^{\frac{H_{SL}}{\cos\phi}}\int_0^{2\pi}\delta^2\sin\phi d\theta d\delta d\phi\right)$$
$$= \frac{2\pi}{3V_S}\left(\frac{3}{2}H_{SL}x^2 - x^3\cos\Phi - \frac{1}{2}\frac{H_{SH}^3}{\cos^2\Phi} - \frac{1}{2}H_{SL}^3 + \frac{1}{2}H_{SH}^3\right). \tag{4.103}$$

(2) As for $H_{SH} < x < H_{SL}$ and $\frac{H_{SH}}{\cos\Phi} < x < \frac{H_{SL}}{\cos\Phi}$, we can easily obtain $F_{d_{SR}}$ as same as Eqs. (4.101) and (4.103), respectively.

When $H_{SL} < x < \frac{H_{SH}}{\cos\Phi}$, the derivation is different from (4.102) and can be presented as

$$F_{d_{SR}}(x) = \frac{1}{V_S}\left(\int_{\arccos\frac{H_{SL}}{x}}^{\arccos\frac{H_{SH}}{x}}\int_{\frac{H_{SH}}{\cos\phi}}^{x}\int_0^{2\pi}\delta^2\sin\phi d\theta d\delta d\phi\right.$$
$$\left.+\int_0^{\arccos\frac{H_{SL}}{x}}\int_{\frac{H_{SH}}{\cos\phi}}^{\frac{H_{SL}}{\cos\phi}}\int_0^{2\pi}\delta^2\sin\phi d\theta d\delta d\phi\right)$$
$$= \frac{2\pi}{3V_S}\left(\frac{3}{2}H_{SL}x^2 - \frac{3}{2}H_{SH}x^2 - \frac{1}{2}H_{SL}^3 + \frac{1}{2}H_{SH}^3\right). \tag{4.104}$$

Then, the proof of Theorem 1 is finished.

## *Appendix II: Proof of Theorem 2*

Firstly, using Eqs. (4.77) and (4.78), we can obtain the joint PDF of $(r_D, h_R)$ as

$$f_{h_R,r_D}(h_R, r_D) = \frac{2\pi r_D(H-h_R)^2\tan^2\Phi\mathbb{I}(\text{con})}{V_S h_R^2\tan^2\Psi}, \tag{4.105}$$

where con= $(H_{LD} \le h_R \le H_{HD}, 0 \le r_D \le h_R\tan\Psi)$, $\mathbb{I}()$ is the indicator function, i.e., $\mathbb{I}(\text{con}) = 1$ when con is true, or $\mathbb{I}(\text{con}) = 0$ when con is false.

Then, the joint PDF of $(d_{RD}, h_R)$ is given as

$$f_{h_R,d_{RD}}\left(h_R,d_{RD}\right)=f_{h_R,r_D}\left(h_R,g\left(h_R,d_{RD}\right)\right)|J|, \tag{4.106}$$

where $g\left(h_R,d_{RD}\right)=\sqrt{d_{RD}^2+h_R^2}$ and the matrix $J$ is the Jacobian matrix defined as

$$J=\begin{bmatrix} \frac{\partial h_R}{\partial h_R} & \frac{\partial h_R}{\partial d_{RD}} \\ \frac{\partial g(h_R,d_{RD})}{\partial h_R} & \frac{\partial g(h_R,d_{RD})}{\partial d_{RD}} \end{bmatrix}. \tag{4.107}$$

Substituting (4.107) into (4.106), one can get $f_{h_R,d_{RD}}\left(h_R,d_{RD}\right)$ as (4.108).

$$f_{h_R,d_{RD}}\left(h_R,d_{RD}\right) = \frac{2\pi d_{RD}(H-h_R)^2\tan^2\Phi\mathbb{I}\left(H_{LD}\le h_R\le H_{HD},0\le g\left(h_R,d_{RD}\right)\le h_R\tan\Psi\right)}{V_S h_R^2\tan^2\Psi} \tag{4.108}$$

Further, the PDF of $d_{RD}$ can be derived as

$$f_{d_{RD}}\left(d_{RD}\right)=\int f_{h_R,d_{RD}}\left(h_R,d_{RD}\right)dh_R. \tag{4.109}$$

To deal with the condition in the indicator function of (4.108) shown on the top of this page, we discuss the following four cases:

**Case 1**: $H_{LD}\le d_{RD}<\min\left\{\frac{H_{LD}}{\cos\Psi},H_{HD}\right\}$. Thus, the indicator function can be derived as $H_{LD}\le h_R\le d_{RD}$. Plugging into (42), we obtain that

$$\begin{aligned} f_{d_{RD}}\left(d_{RD}\right) &= \int_{H_{LD}}^{d_{RD}} f_{h_R,d_{RD}}\left(h_R,d_{RD}\right)dh_R \\ &= \frac{2\pi}{V_S}\frac{\tan^2\Phi}{\tan^2\Psi}d_{RD}\left(\frac{H^2}{H_{LD}}+2H\ln H_{LD}-H_{LD}\right. \\ &\quad \left.-\frac{H^2}{d_{RD}}-2H\ln d_{RD}+d_{RD}\right). \end{aligned} \tag{4.110}$$

**Case 2**: $\frac{H_{LD}}{\cos\Psi}\le d_{RD}<H_{HD}$. Then, we can rewrite the indicator function of (4.108) as $d_{RD}\cos\psi<h_R<d_{RD}$. Similarly, we obtain that

$$\begin{aligned} f_{d_{RD}}\left(d_{RD}\right) &= \int_{d_{RD}\cos\Psi}^{d_{RD}} f_{h_R,d_{RD}}\left(h_R,d_{RD}\right)dh_R \\ &= \frac{2\pi}{V_S}\frac{\tan^2\Phi}{\tan^2\Psi}d_{RD}\left(\frac{H^2}{d_{RD}\cos\Psi}+2H\ln\left(d_{RD}\cos\Psi\right)\right. \\ &\quad \left.-d_{RD}\cos\Psi-\frac{H^2}{d_{RD}}-2H\ln d_{RD}+d_{RD}\right). \end{aligned} \tag{4.111}$$

**Case 3**: $H_{HD} \leq d_{RD} < \frac{H_{LD}}{\cos\Psi}$. Thus, we can derive the indicator function $\mathbb{I}\left(H_{LD} \leq h_R \leq H_{HD},\ 0 \leq g\left(h_R, d_{RD}\right) \leq h_R \tan\Psi\right)$ as $H_{LD} < h_R < H_{HD}$. Then, we can obtain that

$$\begin{aligned} f_{d_{RD}}\left(d_{RD}\right) &= \int_{H_{LD}}^{H_{HD}} f_{h_R,d_{RD}}\left(h_R, d_{RD}\right) dh_R \\ &= \frac{2\pi}{V_S} \frac{\tan^2\Phi}{\tan^2\Psi} d_{RD} \left(\frac{H^2}{H_{LD}} + 2H\ln H_{LD} - H_{LD}\right. \\ &\quad \left. - \frac{H^2}{H_{HD}} - 2H\ln H_{HD} + H_{HD}\right). \end{aligned} \tag{4.112}$$

**Case 4**: $\max\left\{\frac{H_{LD}}{\cos\Psi}, H_{HD}\right\} < d_{RD} < H_{HD}/\cos\Psi$. The indicator function of (4.108) can be written as $d_{RD}\cos\Psi < h_R < H_{HD}$. Similarly, it gives

$$\begin{aligned} f_{d_{RD}}\left(d_{RD}\right) &= \int_{d_{RD}\cos\Psi}^{H_{HD}} f_{h_R,d_{RD}}\left(h_R, d_{RD}\right) dh_R \\ &= \frac{2\pi}{V_S} \frac{\tan^2\Phi}{\tan^2\Psi} d_{RD} \left(\frac{H^2}{d_{RD}\cos\Psi} + 2H\ln\left(d_{RD}\cos\Psi\right)\right. \\ &\quad \left. - d_{RD}\cos\Psi - \frac{H^2}{H_{HD}} - 2H\ln H_{HD} + H_{HD}\right). \end{aligned} \tag{4.113}$$

Thus, combining Eqs. (4.110), (4.111), (4.112), and (4.113) completes the proof.

## Appendix III: Proof of Corollary 1

Considering the randomness of the distance between R and D, OP can be derived as

$$\begin{aligned} \mathcal{P}_{RD} &= \Pr\left\{\gamma_{RD} < \gamma_{\text{out},D}\right\} = \Pr\left\{\lambda_{RD} < \gamma_{\text{out},D} d_{RD}^{n_3}\right\} \\ &= 1 - \int_{d_{RD,\min}}^{d_{RD,\min}} Q_1\left(\sqrt{2K}, \sqrt{\frac{2(1+K)\gamma_{\text{out},D} x^{n_3}}{\Omega_R}}\right) f_{d_{RD}}(x)\, dx. \end{aligned} \tag{4.114}$$

As suggested in [6], Marcum-$Q$ function can be approximated as $Q_1(a,b) \approx \exp\left(-e^{v(a)} b^{\mu(a)}\right)$, where $\mu(a) = 2.174 - 0.592a + 0.593a^2 - 0.092a^3 + 0.005a^4$, and $v(a) = -0.840 + 0.327a - 0.740a^2 + 0.083a^3 - 0.004a^4$. Then, we can obtain

$$Q_1\left(\sqrt{2K}, \sqrt{\frac{2(1+K)\gamma_{\text{out},D} x^{n_3}}{\Omega_R}}\right) \approx \exp\left(-e^{v} \Theta^{\frac{\mu}{2}} x^{\frac{n_3\mu}{2}}\right), \tag{4.115}$$

where $\Theta = \frac{2(1+K)\gamma_{\text{out},D}}{\Omega_R}$, $\mu = 2.174 - 0.937\sqrt{K} + 1.186K - 0.260K^{\frac{3}{2}} + 0.010K^2$, and $\upsilon = -0.840 + 0.809\sqrt{K} - 1.480K + 0.235K^{\frac{3}{2}} - 0.008K^2$.

We define $\mathcal{H}(x) = \exp\left(-e^{\upsilon}\Theta^{\frac{\mu}{2}} x^{\frac{n_3\mu}{2}}\right)$, then the OP of R-D link can be further derived as

$$
\begin{aligned}
\mathcal{P}_{RD} &\approx 1 - \int_{d_{RD,\min}}^{d_{RD,\min}} \mathcal{H}(x)\, f_{d_{RD}}(x)\, dx \\
&= 1 - \frac{2\pi}{V_S} \frac{\tan^2\Phi}{\tan^2\Psi} \left\{ \int_{H_{LD}}^{\frac{H_{LD}}{\cos\Psi}} x\mathcal{H}(x)\left[\mathcal{F}(d_{RD}, 0) - \mathcal{F}(H_{LD}, 0)\right] dx \right. \\
&\quad - \int_{\frac{H_{LD}}{\cos\Psi}}^{H_{HD}} x\mathcal{H}(x)\left[\mathcal{F}(d_{RD}, 0) - \mathcal{F}(d_{RD}, \Psi)\right] dx \\
&\quad \left. - \int_{H_{HD}}^{\frac{H_{HD}}{\cos\Psi}} x\mathcal{H}(x)\left[\mathcal{F}(H_{HD}, 0) - \mathcal{F}(d_{RD}, \Psi)\right] dx \right\}. \qquad (4.116)
\end{aligned}
$$

To facilitate the following derivation, we denote the first, second, and third integrals in (4.116) as $I_1$, $I_2$, and $I_3$, respectively, for simplicity. Then, $I_1$ can be derived as

$$
I_1 = \underbrace{\int_{H_{LD}}^{\frac{H_{LD}}{\cos\Psi}} x\mathcal{H}(x)\mathcal{F}(x, 0)dx}_{\Upsilon_1} - \underbrace{\int_{H_{LD}}^{\frac{H_{LD}}{\cos\Psi}} x\mathcal{H}(x)\mathcal{F}(H_{LD}, 0)dx}_{\Upsilon_2}, \qquad (4.117)
$$

where $\Upsilon_1 = \int_{H_{LD}}^{\frac{H_{LD}}{\cos\Psi}} x\mathcal{H}(x)\mathcal{F}(x, 0)dx$ and $\Upsilon_2 = \int_{H_{LD}}^{\frac{H_{LD}}{\cos\Psi}} x\mathcal{H}(x)\mathcal{F}(H_{LD}, 0)dx$.

Due to the complexity of the integral, we use Chebyshev-Gauss quadrate to facilitate further analysis with an adequate accuracy, which is presented as $\int_{-1}^{1} \frac{f(x)}{\sqrt{1-x^2}}dx \approx \sum_{i=1}^{T} \omega_i f(x_i)$ with $x_i = \cos\left(\frac{2i-1}{2T}\pi\right)$ and the weight $\omega_i = \frac{\pi}{T}$. Thus, $\Upsilon_1$ can be rewritten as

$$
\begin{aligned}
\Upsilon_1 &= \int_{H_{LD}}^{\frac{H_{LD}}{\cos\Psi}} x\mathcal{H}(x)\mathcal{F}(d_{RD}, 0)dx \\
&= \frac{a_1\pi}{P_1} \sum_{i=1}^{P_1} \sqrt{1 - {\chi_i}^2}\,(a_1\chi_i + a_2)\mathcal{H}(a_1\chi_i + a_2)\mathcal{F}(a_1\chi_i + a_2, 0), \qquad (4.118)
\end{aligned}
$$

where $a_1 = \frac{H_{LD}/\cos\Psi - H_{LD}}{2}$, $a_2 = \frac{H_{LD}/\cos\Psi + H_{LD}}{2}$, and $\chi_i = \cos\left(\frac{2i-1}{2P_1}\pi\right)$.

Then, $\Upsilon_2$ can be derived as

$$\begin{aligned}\Upsilon_2 &= \int_{H_{LD}}^{\frac{H_{LD}}{\cos\Psi}} x\mathcal{H}(x)\mathcal{F}(H_{LD},0)dx\\ &= \int_{\Delta_1}^{\Delta_2} \frac{2\pi}{V_S}\frac{2}{n_2\mu}\frac{\tan^2\Phi}{\tan^2\Psi}e^{-\frac{4v}{n_3\mu}}\Theta^{-\frac{2}{n_3}}\exp(-w)\,w^{\frac{2}{n_3\mu}(2-\frac{n_3\mu}{2})}\mathcal{F}(H_{LD},0)\,dw\\ &= \frac{2}{n_2\mu}e^{-\frac{4v}{n_3\mu}}\Theta^{-\frac{2}{n_3}}\mathcal{F}(H_{LD},0)\left[\Gamma\left(\frac{4}{n_3\mu},\Delta_1\right)-\Gamma\left(\frac{4}{n_3\mu},\Delta_2\right)\right],\end{aligned} \tag{4.119}$$

where in the second line we let $w = e^v\Theta^{\frac{\mu}{2}}x^{\frac{n_3\mu}{2}}$, $\Delta_1 = e^v\Theta^{\frac{\mu}{2}}H_{LD}^{\frac{n_3\mu}{2}}$, and $\Delta_2 = e^v\Theta^{\frac{\mu}{2}}\left(\frac{H_{LD}}{\cos\Psi}\right)^{\frac{n_3\mu}{2}}$.

Similarly, using Chebyshev-Gauss quadrate, $I_2$ can be rewritten as

$$\begin{aligned}I_2 &= \frac{a_3\pi}{P_2}\sum_{j=1}^{P_2}\sqrt{1-\delta_j{}^2}\left(a_3\delta_j+a_4\right)\mathcal{H}(x)\left(a_3\delta_j+a_4\right)\\ &\quad\times\left[\mathcal{F}\left(a_3\delta_j+a_4,0\right)-\mathcal{F}\left(a_3\delta_j+a_4,\Psi\right)\right],\end{aligned} \tag{4.120}$$

where $a_3 = \frac{H_{HD}-H_{LD}/\cos\Psi}{2}$, $a_4 = \frac{H_{HD}+H_{LD}/\cos\Psi}{2}$, and $\delta_j = \cos\left(\frac{2j-1}{2P_2}\pi\right)$.

Then, $I_3$ can be derived as

$$I_3 = \underbrace{\int_{H_{HD}}^{\frac{H_{HD}}{\cos\Psi}} x\mathcal{H}(x)\mathcal{F}(H_{HD},0)dx}_{\Upsilon_3} - \underbrace{\int_{H_{HD}}^{\frac{H_{HD}}{\cos\Psi}} x\mathcal{H}(x)\mathcal{F}(x,\Psi)dx}_{\Upsilon_4}, \tag{4.121}$$

where $\Upsilon_3 = \int_{H_{HD}}^{\frac{H_{HD}}{\cos\Psi}} x\mathcal{H}(x)\mathcal{F}(H_{HD},0)dx$ and $\Upsilon_4 = \int_{H_{HD}}^{\frac{H_{HD}}{\cos\Psi}} x\mathcal{H}(x)\mathcal{F}(x,\Psi)dx$.

The derivations of $\Upsilon_3$ and $\Upsilon_4$ are highly similar to those of $\Upsilon_1$ and $\Upsilon_2$, which can be directly obtained as

$$\Upsilon_3 = \frac{2}{n_2\mu}e^{-\frac{4v}{n_3\mu}}\Theta^{-\frac{2\mu}{n_3\mu}+\frac{\mu}{2}}\mathcal{F}(H_{HD},0)\left[\Gamma\left(\frac{4}{n_3\mu},\Delta_3\right)-\Gamma\left(\frac{4}{n_3\mu},\Delta_4\right)\right], \tag{4.122}$$

$$\begin{aligned}\Upsilon_4 &= \int_{H_{HD}}^{H_{HD}/\cos\Psi} \mathcal{H}(x)\,x\mathcal{F}(x,\Psi)dx\\ &= \frac{a_5\pi}{P_3}\sum_{k=1}^{P_3}\sqrt{1-\varepsilon_k{}^2}\,(a_5\varepsilon_k+a_6)\,\mathcal{H}(a_5\varepsilon_k+a_6)\,s\mathcal{F}(a_5\varepsilon_k+a_6,\Psi),\end{aligned} \tag{4.123}$$

respectively, where $\Delta_3 = e^{\upsilon}\Theta^{\frac{\mu}{2}}(H_{HD})^{\frac{n_3\mu}{2}}$, $\Delta_4 = e^{\upsilon}\Theta^{\frac{\mu}{2}}\left(\frac{H_{HD}}{\cos\Psi}\right)^{\frac{n_3\mu}{2}}$, $a_5 = \frac{H_{HD}/\cos\Psi - H_{HD}}{2}$, $a_6 = \frac{H_{HD}/\cos\Psi + H_{HD}}{2}$, and $\varepsilon_k = \cos\left(\frac{2k-1}{2P_3}\pi\right)$.

Finally, substituting Eqs. (4.117)–(4.123) into (4.116), we can finally approximate $P_{\text{out},RD,0}$ in closed form and complete the proof.

## References


1. G. Pan, J. Ye and Z. Ding, "On secure VLC systems with spatially random terminals," *IEEE Commun. Lett.*, vol. 21, no. 3, pp. 492–495, 2017.
2. J. Ye, C. Zhang, H. Lei, G. Pan and Z. Ding, "Secure UAV-to-UAV systems with spatially random UAVs," *IEEE Wireless Commun. Lett.*, vol. 8, no. 2, pp. 564–567, 2019.
3. A. Abdi, W. C. Lau, M. -S. Alouini and M. Kaveh, "A new simple model for land mobile satellite channels: First- and second-order statistics," *IEEE Trans. Wireless Commun.*, vol. 2, no. 3, pp. 519–528, 2003.
4. I.S. Gradshteyn and I.M. Ryzhik, *Table of Integrals, Series and Products,* 7th ed. San Diego, CA, USA: Academic Press, 2007.
5. M. K. Simon and M. -S. Alouini, *Digital Communication Over Fading Channels,* 2nd ed. New York, NY, USA: Wiley, 2005.
6. M. Z. Bocus, C. P. Dettmann and J. P. Coon, "An approximation of the first order Marcum Q-function with application to network connectivity analysis," *IEEE Commun. Lett.*, vol. 17, no. 3, pp. 499–502, 2013.
7. A. P. Prudnikov, Y. A. Brychkov, and O. I. Marichev, *Integrals and Series: Special Functions,* vol. 2, 3rd ed. New York, NY, USA: Gordon & Breach Science Publishers, 1992.
8. G. Pan, H. Lei, Y. Yuan and Z. Ding, "Performance Analysis and Optimization for SWIPT Wireless Sensor Networks," *IEEE Trans. Commun.*, vol. 65, no. 5, pp. 2291–2302, 2017.
9. S. Boyd and L. Vandenberghe, *Convex Optimization*. New York, NY, USA: Cambridge University Press, 2004.
10. G. Pan, H. Lei, Z. Ding and Q. Ni, "3-D Hybrid VLC-RF Indoor IoT Systems With Light Energy Harvesting," *IEEE Trans. Green Commun. Networking*, vol. 3, no. 3, pp. 853–865, 2019.
11. H. Jia, Z. Ni, C. Jiang, L. Kuang and J. Lu, "Uplink Interference and Performance Analysis for Megasatellite Constellation," *IEEE Internet Things J.*, vol. 9, no. 6, pp. 4318–4329, March, 2022.
12. Y. Ruan, Y. Li, C. Wang, R. Zhang and H. Zhang, "Outage performance of integrated satellite-terrestrial networks with hybrid CCI," *IEEE Commun. Lett.*, vol. 21, no. 7, pp. 1545–1548, 2017.
13. C. Loo, "A statistical model for a land mobile satellite link," *IEEE Trans. veh. tech.*, vol. 34, no. 3, pp. 122–127, 1985.
14. I.S. Gradshteyn and I. M. Ryzhik, *Table of Integrals, Products.* New York, NY, USA: Academic, 2014.
15. K. An, M. Lin, I. Ouyang, and W.-P. Zhu, "Secure transmission in cognitive satellite terrestrial networks," *IEEE J. Sel. Areas Commun.*, vol.34, no.11, pp. 3025–3037, 2016.
16. X. Yan, H. Xiao, K. An, G. Zheng, and W. Tao, "Hybrid satellite terrestrial relay networks with cooperative non-orthogonal multiple access," *IEEE Commun. Lett.*, vol. 22, no. 5, pp. 978–981, 2018.
17. K. Guo, M. Lin, B. Zhang, W. Zhu, J. Wang, and T. A. Tsiftsis, "On the performance of LMS communication with hardware impairments and interference," *IEEE Trans. Commun.*, vol. 67, no. 2, pp. 1490–1505, Feb. 2019.

# Chapter 5
# End-To-End Terrestrial-Satellite-Terrestrial Communications

**Zhe Song, Gaofeng Pan, Shuai Wang, and Mohamed-Slim Alouini**

## 5.1 Latency Versus Reliability in Multi-hop LEO Satellite Systems

### *5.1.1 System Model*

In this work, as presented in Fig. 5.1, a low Earth orbit (LEO) mega-constellation communication scenario is considered, which consists of a terrestrial source-destination (S-D) pair, $K$ ($K > 2$) LEO satellites, $1 \leq p < q \leq K$, a ground relay (G), an aerial relay (A), and a relay satellite ($S_0$) with higher orbit compared to the LEO satellites. More specifically, S and D can communicate with each other via the inter-LEO-satellite link according to the original design of the LEO satellite communication system.[1]

Moreover, there exist some candidate cooperative links via G, A, or $S_0$, which can offer some more choices for the data transmission between S and D to achieve better latency or reliability performance. Thus, in the following, we will analyze

[1] Normally, in practical LEO satellite systems LEO satellites are designed to be able to communicate with the neighboring visible LEO satellites. However, due to the main inherent properties of the LEO mega-constellation systems, e.g., rapidly-varying network topology, the distances of these visible inter-satellite links are time-varying, and then it is hard to get the accurate values of these distances along with the numerous hops of the long-distance transmissions for analysis and comparison purposes. In this work, we assume that the $K$ ($K > 2$) LEO satellites belong to the same plane to simplify the following analysis. The case that the $K$ ($K > 2$) LEO satellites come from different planes can also be analyzed by using the technical roadmap proposed in this study.

Z. Song · G. Pan (✉) · S. Wang
Beijing Institute of Technology, Beijing, China
e-mail: gfpan@bit.edu.cn

M.-S. Alouini
King Abdullah University of Science and Technology (KAUST), Thuwal, Saudi Arabia


S. Wang et al. (eds.), *Understanding Satellite Communications: The Stochastic Geometry Perspective*, https://doi.org/10.1007/978-981-97-7102-8_5

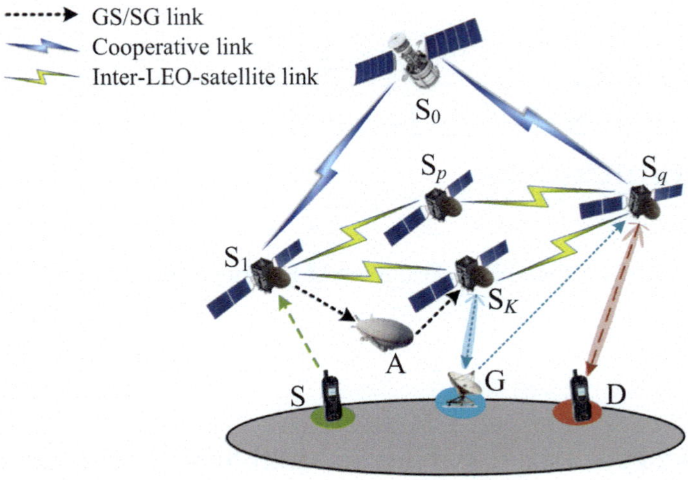


**Fig. 5.1** System model

and compare the delay and outage performance of the transmissions over inter-LEO-satellite links and each cooperative link, respectively, to clearly show when it is worthwhile to resort to cooperative links compared to inter-LEO-satellite links and which kind of cooperative links is the best candidate for the end-to-end (e2e) transmission between S and D.

Then, it is easy to obtain the received SNR over the link between nodes $i$ and $j$ (where $i, j \in \mathbb{W} = \{\mathrm{S, D, G, A, S}_k \,|0 \le k \le K\}$) in Fig. 5.1 as

$$\gamma_{ij} = \frac{P_i}{\sigma^2}\frac{|h_{ij}|^2}{d_{ij}^n} = \frac{\lambda_{ij}}{d_{ij}^n}, \tag{5.1}$$

where $\lambda_{ij} = P_i |h_{ij}|^2/\sigma^2 = \bar{\lambda}_i |h_{ij}|^2$, $|h_{ij}|^2$ is the power gain attributed to the small-scale fading,[2] $\bar{\lambda}_i = P_i/\sigma^2$ is the transmit SNR at node $i$, $P_i$ and $\sigma^2$ are the transmit power at $i$ and the average power of the additive white Gaussian noise at all nodes, respectively, $d_{ij}$ is the distance over the link between $i$ and $j$, $n$ is the path-loss factor, which is assumed to be the same for all links in Fig. 5.1 as all these links belong to line-of-sight (LoS) propagation scenarios.

[2] Differing from terrestrial propagation scenarios, there exist no multi-path components over the interlinks among the satellites and the links between satellites and HAP. Thus, we set $|h_{ij}|^2 = 1$ for these two kinds of aforementioned links in the following analysis.

In the following, we will adopt the e2e time delay and outage probability (OP) to quantitatively evaluate the latency and reliability quality of the e2e transmission between S and D, which are defined as[3]

$$\tau_{\text{int}}^{e2e} = \tau_{\text{S},\text{S}_1} + \tau_{\text{S}_K,\text{D}} + \sum_{p\in\{1\le p<K\}} \left(\tau_{p,p+1} + \epsilon_p\right) \tag{5.2}$$

$$\tau_r^{e2e} = \tau_{\text{S},\text{S}_1} + \tau_{\text{S}_K,\text{D}} + \sum_{k\in\{1\le k<p\}\cup\{q\le k<K\}} \left(\tau_{k,k+1} + \epsilon_k\right) + \epsilon_r + \tau_{\text{S}_p,r} + \tau_{r,q} \tag{5.3}$$

$$P_{\text{out,int}}^{e2e} = 1 - \left(1 - P_{\text{out}}^{\text{S},\text{S}_1}\right)\left(1 - P_{\text{out}}^{\text{S}_K,\text{D}}\right) \prod_{k\in\{1\le k<K\}} \left(1 - P_{\text{out}}^{k,k+1}\right) \tag{5.4}$$

$$\begin{aligned} P_{\text{out},r}^{e2e} = 1 &- \left(1 - P_{\text{out}}^{\text{S},\text{S}_1}\right)\left(1 - P_{\text{out}}^{\text{S}_K,\text{D}}\right)\left(1 - P_{\text{out}}^{\text{S}_p,r}\right) \\ &\times \left(1 - P_{\text{out}}^{r,\text{S}_q}\right) \prod_{k\in\{1\le k<p\}\cup\{q\le k<K\}} \left(1 - P_{\text{out}}^{k,k+1}\right), \end{aligned} \tag{5.5}$$

where $r \in \mathbb{R} = \{\text{A}, \text{G}, \text{S}_0\}$, $\tau_{i,j}$ (where $i, j \in \mathbb{W}$) denotes the transmission delay over the link between nodes $i$ and $j$, $\epsilon_i$ (where $i \in \mathbb{W}$) denotes the forwarding delay (like modulation/demodulation, coding/decoding, queuing, etc.) at node $i$, $P_{\text{out}}^{i,j}$ (where $i, j \in \mathbb{W}$) is the OP over the link between nodes $i$ and $j$.

### 5.1.2 *Inter LEO Satellite Transmissions*

In this section, we investigate the delay and outage performance for inter-LEO satellite transmissions without any kind of relays, as shown in Fig. 5.2, where $H$ is the height of LEO satellites from the ground, E denotes the center of the earth, $R_\text{E}$ is the radius of the earth, and $\theta$ is the included angle between the lines connecting E and any two neighboring LEO satellites.

Clearly, it is observed from Fig. 5.2 that two neighbors of any LEO satellite, namely, the next one ahead on the same orbital plane, and the one behind on that

[3] In this section, a regenerative forwarding scheme, namely, decode-and-forward (DF) relay scheme, is considered at each relay node, $i \in \mathbb{W}$, by exploiting the enhanced the enhanced onboard processing capabilities. Though the DF relay scheme exhibits worse latency performance compared to transparent forwarding relay schemes, like the amplify-and-forward (AF) relay scheme, the DF relay scheme can avoid noise/interference amplification problems along with the AF relay scheme to offer better e2e bit/packet error performance for long-distance inter-LEO satellite links transmission with numerous hops, as well as to provide a better comparable case to better show the purpose of this section: latency versus reliability transmission performance.

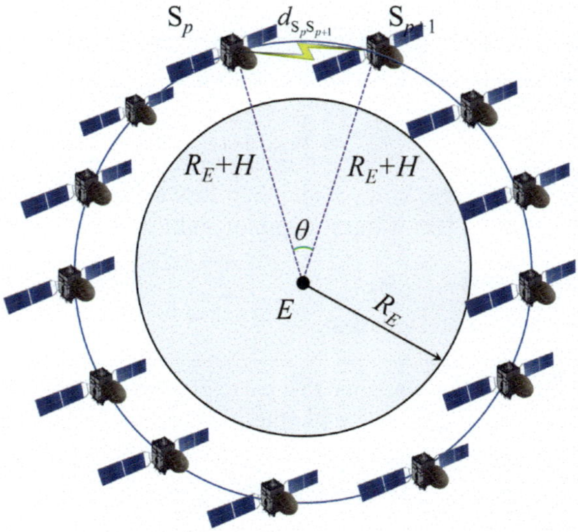


**Fig. 5.2** Inter LEO satellite transmissions

orbital plane, always remain in the same locations. In other words, the transmission distance between any two neighboring LEO satellites is fixed when the LEO satellite constellation is designed.

Here, we assume that there are $N$ $(0 < K < N)$ LEO satellites in each orbital plane. Then, we can have $\theta = 2\pi/N$.

By applying the knowledge of the elementary geometry, one can easily achieve the distance between two neighboring LEO satellites in the same plane as

$$\begin{aligned} d_{S_pS_{p+1}} &= 2(R_E + H)\sin\frac{\theta}{2} \\ &= 2(R_E + H)\sin\frac{\pi}{N}, \end{aligned} \tag{5.6}$$

where $1 \leq p < N$. Also, it is easy to expand our proposed analysis method/model to more general case that all these LEO satellites orbit in different planes by changing the value of $d_{S_pS_{p+1}}$ to the one under such cases.

Similarly, the distance from the centre of the terrestrial coverage area of two neighboring LEO satellites can be calculated as

$$d^{gs}_{S_pS_{p+1}} = \frac{2\pi R_E}{N}. \tag{5.7}$$

During the inter-LEO satellite transmissions, there are no multi-path components over intersatellite channels in the open outer space. Because there is no reflection, shadowing, diffraction, and scattering yield multi-path propagation. Thus, this obstacle-free channel of free space only includes the effect of the transmission distance which follows the free-space path-loss model. In other words, the received SNR over intersatellite links can be written as

$$\gamma_{ij} = \frac{P_i}{\sigma^2}\frac{1}{d_{ij}^n}, \tag{5.8}$$

where $i, j \in \mathbb{S} = \{S_k \mid 0 \leq k \leq K\}$.

#### 5.1.2.1 Transmission Delay

It is easy to obtain the transmission delay between nodes $i$ and $j$ (where $i, j \in \mathbb{W}$) as

$$\tau_{ij} = \frac{d_{ij}}{c}, \tag{5.9}$$

where $c$ is the transmission speed of the radio wave.

Therefore, the total transmission delay for inter-LEO satellite transmissions from $S_1$ to $S_K$ (shown in Fig. 5.1) can be given as

$$\begin{aligned}\sum_{k\in\{1\leq k<K\}} \tau_{k,k+1} &= \sum_{k\in\{1\leq k<K\}} \frac{d_{S_kS_{k+1}}}{c} \\ &= \frac{2(K-1)(R_E+H)}{c}\sin\frac{\theta}{2}.\end{aligned} \tag{5.10}$$

#### 5.1.2.2 Outage Performance

In this section, OP is defined as the probability that the received SNR is lower than a predefined threshold of SNR. Then, OP over the link between $i$ and $j$ ($i, j \in \mathbb{W}$) can be written as

$$P_{\text{out}}^{i,j} = \Pr\left\{\gamma_{ij} \leq \gamma_{\text{th}}\right\}, \tag{5.11}$$

where $\gamma_{\text{th}}$ is the predefined threshold.[4]

Therefore, the e2e OP for inter-LEO satellite transmissions from $S_1$ to $S_K$ (shown in Fig. 5.1) can be derived as

[4] In this section, we assume that the outage threshold at the terrestrial terminal and the LEO satellite is a same one for simplification.

$$\begin{aligned} P^{e2e}_{\text{int,LEO}} &= 1 - \prod_{k\in\{1\le k<K\}} \left(1 - P^{k,k+1}_{\text{out}}\right) \\ &= 1 - \prod_{k\in\{1\le k<K\}} \left(1 - \Pr\left\{\frac{P_k}{\sigma^2}\frac{1}{d^n_{\text{S}_k\text{S}_{k+1}}} \le \gamma_{\text{th}}\right\}\right) \\ &= 1 - \left(1 - \Pr\left\{\frac{P_{\text{LEO}}}{\sigma^2}\kappa \sin^{-n}\frac{\theta}{2} \le \gamma_{\text{th}}\right\}\right)^{K-1}, \end{aligned} \tag{5.12}$$

where we assume that all LEO satellites share a same transmit power, $P_{\text{LEO}}$, as a same transmission distance is observed from Fig. 5.1 for inter neighboring LEO satellite communications, and $\kappa = \frac{1}{2^n(K-1)^n(R_{\text{E}}+H)^n}$.

Observing (5.12), one can see that, unlike the traditional terrestrial wireless communication systems, the e2e OP for inter-LEO satellite transmissions has nothing to do with the small-scale fading of the wireless channel and the transmission distance between any two neighboring LEO satellites. This observation can be traced to the following two facts: (1) The propagation characteristic of RF signal in outer space, namely, no multi-path prorogation; (2) The transmission distance between any two neighboring LEO satellites is fixed once the number of satellites per plane is designed.

Thus, the e2e OP for inter-LEO satellite transmissions for a designed LEO mega-constellation system mainly depends on the available transmit power at each LEO satellite.

### *5.1.3 Satellite-Terrestrial Transmissions*

In this subsection, the analysis model for the transmissions between the LEO satellite ($\text{S}_1$/$\text{S}_K$) and the terrestrial terminals (S/D) will be set up, and then the transmission delay and outage performance will be studied while considering the randomness of the positions of the terrestrial terminals.

For tractability purposes, as depicted in Fig. 5.3, in this section we treat the coverage area of the LEO satellite ($\text{S}_1$/$\text{S}_K$) as a circle with radius $L$, where O is located at the centre of the circle. Terrestrial terminal (S/D) is uniformly distributed in the coverage area of $\text{S}_1$/$\text{S}_K$. Therefore, the probability density function (PDF) of the distance between the centre of the coverage area of the satellite and S/D, $r_{\text{S}}/r_{\text{D}}$, can be obtained as $f(x) = \frac{2x}{L^2}, 0 \le x \le L$.

As shown in Fig. 5.3, $\omega_{\text{S}}$/$\omega_{\text{D}}$ is the angle between line $\text{OS}_1$/$\text{OS}_K$ and the horizon, which is decided by the beam direction of the satellite and the range of which is $(0, \pi/2]$. Then, one can easily have $d_{\text{OS}_1} = H/\sin\omega_{\text{S}}$, $d_{\text{OS}_K} = H/\sin\omega_{\text{D}}$, and

$$d_{ij} = \sqrt{r_l^2 + \left(\frac{H}{\sin\omega_l}\right)^2 - 2r_l H \cot\omega_l}, \tag{5.13}$$

Fig. 5.3 Satellite-terrestrial transmissions

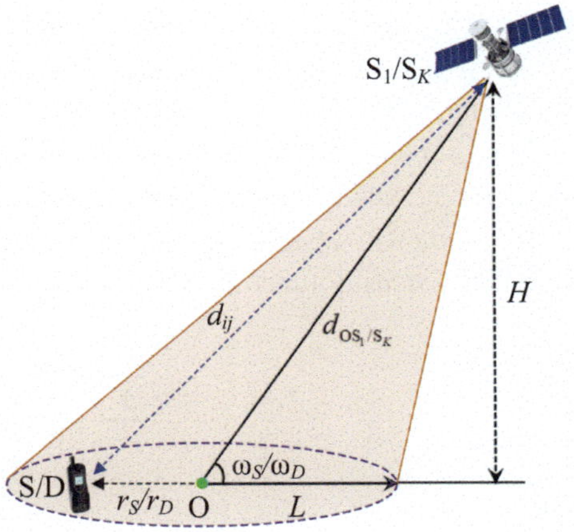


where $(i, j, l) \in \{(\mathrm{S,S_1,S}), (\mathrm{S}_K, \mathrm{D}, \mathrm{D})\}$, and $H$ is the orbit height of the LEO satellite.

Therefore, one can obtain the PDF of $d_{ij}$ as

$$f_{d_{ij}}(x) = \frac{2x}{L^2} \cdot \frac{H \cot \omega_l + \sqrt{x^2 + H^2 \cot^2 \omega_l - \left(\frac{H}{\sin \omega_l}\right)^2}}{\sqrt{x^2 + H^2 \cot^2 \omega_l - \left(\frac{H}{\sin \omega_l}\right)^2}}, \tag{5.14}$$

where $(i, j, l) \in \{(\mathrm{S,S_1,S}), (\mathrm{S}_K, \mathrm{D}, \mathrm{D})\}$, and $H \le x \le \sqrt{H^2 + L^2}$.

Observing the complexity of (5.14), for tractability purpose, in the reminder of this section we only focus on the cases that the projection of the LEO satellites is the the centre of their coverage areas. Under this case, $\omega_{\mathrm{S}} = \pi/2$ and $\omega_{\mathrm{D}} = \pi/2$, and then we can further achieve the PDF of $d_{ij} = \sqrt{H^2 + r_l^2}$ (where $(i, j, l) \in \{(\mathrm{S}, \mathrm{S}_1, \mathrm{S}), (\mathrm{S}_K, \mathrm{D}, \mathrm{D})\}$) as $f_{d_{ij}}(x) = \frac{2}{L^2}x$, $H \le x \le \sqrt{H^2 + L^2}$.

Furthermore, in this section, the shadowed-Rician (SR) model is adopted to describe the statistical distribution of the satellite link between the LEO satellite and the terrestrial terminal, which has been proved to be an accurate, practical, and applicable tool to evaluate the performance of the satellite propagation environments in various frequency bands, e.g., the UHF-band, L-band, S-band, Ku-band, and Ka-band [1].

Without loss of generality, the PDF of the power gain, $|h_{ij}|^2$ (where $(i, j) \in \{(\mathrm{S,S_1}), (\mathrm{S}_K, \mathrm{D})\}$),[5] for satellite-terrestrial links in SR fading is given as [1]

$$f_{|h_{ij}|^2}(x) = \alpha \exp(-\beta x)\, {}_1F_1(m; 1; \delta x), \quad x \ge 0, \tag{5.15}$$

[5] With channel reciprocal properties, in this section S-S$_0$ channel is also assumed to follow SR fading.

where $\alpha = \left(\frac{2bm}{2bm+\Omega}\right)^m \Big/ (2b)$, $\beta = \frac{1}{2b}$, and $\delta = \frac{\Omega}{2b(2bm+\Omega)}$, $\Omega$ and $2b$ are the average power of the LOS and multi-path components, respectively. $m$ is the fading severity parameter,[6] and ${}_1F_1(\cdot;\cdot;\cdot)$ is the confluent hypergeometric function of the first kind.

Then, it is easy to obtain the received SNR over satellite-terrestrial links in Fig. 5.3 as $\gamma_{ij} = \frac{\lambda_{ij}}{d_{ij}^n}$, where $(i, j) \in \{(\mathrm{S},\mathrm{S}_1), (\mathrm{S}_K, \mathrm{D})\}$.

By variable substitution and applying Kummer's transform, the PDF and cumulative density function (CDF) of $\lambda_{ij}$ can be presented as

$$f_{\lambda_{ij}}(x) = \alpha \sum_{k=0}^{m-1} \frac{\varsigma(k)}{\bar{\lambda}_i^{k+1}} x^k \exp\left(-\frac{\beta-\delta}{\bar{\lambda}_i} x\right) \tag{5.16}$$

and

$$F_{\lambda_{ij}}(x) = 1 - \alpha \sum_{k=0}^{m-1} \frac{\varsigma(k)}{\bar{\lambda}_i^{k+1}} \sum_{p=0}^{k} \frac{k!}{p!} \left(\frac{\beta-\delta}{\bar{\lambda}_i}\right)^{-(k+1-p)} x^p \exp\left(-\frac{\beta-\delta}{\bar{\lambda}_i} x\right), \tag{5.17}$$

respectively, where $\varsigma(k) = \frac{(-1)^k (1-m)_k \delta^k}{(k!)^2}$ and $(t)_k = t(t+1)\cdots(t+k-1)$ is the Pochhammer symbol [2].

#### 5.1.3.1 Transmission Delay

Applying (5.9), the average transmission delay between $i$ and $j$ (where $(i, j) \in \{(\mathrm{S}, \mathrm{S}_1), (\mathrm{S}_K, \mathrm{D})\}$) can be calculated as

$$\begin{aligned} \bar{\tau}_{ij} &= \int_{H}^{\sqrt{H^2+L^2}} \frac{x}{c} \cdot f_{d_{ij}}(x)\, dx \\ &= \frac{2}{cL^2} \int_{H}^{\sqrt{H^2+L^2}} x^2 dx \\ &= \frac{2}{3cL^2} \left[\left(H^2+L^2\right)^{\frac{3}{2}} - H^3\right]. \end{aligned} \tag{5.18}$$

[6] When $m = 0$, it denotes the case without LOS; when $0 < m < \infty$, it stands for the case of both small-scale fading and LOS; when $m = \infty$, it represents the LOS case.

#### 5.1.3.2 Outage Performance

Using (5.11), the OP over the link between $i$ and $j$ (where $(i, j) \in \{(\mathrm{S}, \mathrm{S}_1), (\mathrm{S}_K, \mathrm{D})\}$) can be written as

$$\begin{aligned}
P_{\text{out}}^{i,j} &= \Pr\left\{\gamma_{ij} \le \gamma_{\text{th}}\right\} \\
&= \Pr\left\{\frac{\lambda_{ij}}{d_{ij}^n} \le \gamma_{\text{th}}\right\} \\
&= 1 - \alpha \sum_{k=0}^{m-1} \frac{\varsigma(k)}{\bar{\lambda}_i^{k+1}} \sum_{p=0}^{k} \gamma_{\text{th}}^p \frac{k!}{p!} \left(\frac{\beta-\delta}{\bar{\lambda}_i}\right)^{-(k+1-p)} d_{ij}^{np} \exp\left(-\frac{\beta-\delta}{\bar{\lambda}_i}\gamma_{\text{th}} d_{ij}^n\right),
\end{aligned} \tag{5.19}$$

where $\gamma_{\text{th}}$ is the predefined threshold.[7]

Considering the randomness of the position of the terrestrial terminals, OP can be further calculated as (5.20), shown on the top of this page, where $Y_{\max,i} = \frac{\gamma_{\text{th}}(\beta-\delta)}{\bar{\lambda}_i}(H^2+L^2)^{\frac{n}{2}}$, $Y_{\min,i} = \frac{\gamma_{\text{th}}(\beta-\delta)}{\bar{\lambda}_i}H^n$, and $\gamma(a, x) = \int_0^x t^{a-1}\exp(-t)\,dt$ is the lower incomplete gamma function.

$$\begin{aligned}
P_{\text{out}}^{i,j} &= 1 - \int_H^{\sqrt{H^2+L^2}} \alpha \sum_{k=0}^{m-1} \frac{\varsigma(k)}{\bar{\lambda}_i^{k+1}} \sum_{p=0}^{k} \gamma_{\text{th}}^p \frac{k!}{p!} \left(\frac{\beta-\delta}{\bar{\lambda}_i}\right)^{-(k+1-p)} x^{np} \exp\left(-\frac{\beta-\delta}{\bar{\lambda}_i}\gamma_{\text{th}} x^n\right) f_{d_{ij}}(x)\,dx \\
&= 1 - \frac{2\alpha}{L^2} \sum_{k=0}^{m-1} \frac{\varsigma(k)}{\bar{\lambda}_i^{k+1}} \sum_{p=0}^{k} \gamma_{\text{th}}^p \frac{k!}{p!} \left(\frac{\beta-\delta}{\bar{\lambda}_i}\right)^{-(k+1-p)} \int_H^{\sqrt{H^2+L^2}} x^{np+1} \exp\left(-\frac{\beta-\delta}{\bar{\lambda}_i}\gamma_{\text{th}} x^n\right) dx \\
&= 1 - \frac{2\alpha}{nL^2\gamma_{\text{th}}^{\frac{2}{n}}} \sum_{k=0}^{m-1} \frac{\varsigma(k)}{\bar{\lambda}_i^{k+1}} \sum_{p=0}^{k} \frac{k!}{p!} \left(\frac{\beta-\delta}{\bar{\lambda}_i}\right)^{-\left(k+\frac{2}{n}+1\right)} \\
&\quad \left[\gamma\left(p+\frac{2}{n}, Y_{\max,i}\right) - \gamma\left(p+\frac{2}{n}, Y_{\min,i}\right)\right]
\end{aligned} \tag{5.20}$$

### 5.1.4 *Terrestrial Relay Assisted Transmissions*

In this subsection, we will investigate the transmission quality while the data transmission between S and D is carried out via terrestrial cooperative $\mathrm{S}_p$-G-$\mathrm{S}_q$ link shown in Fig. 5.1.

Obviously, G can assist the transmission between $\mathrm{S}_p$ and $\mathrm{S}_q$, only when G locates in the intersection of the coverage areas of $\mathrm{S}_p$ and $\mathrm{S}_q$, as presented in Fig. 5.4.

[7] In this section, we assume that the outage threshold at the terrestrial terminal and the LEO satellite is the same one for simplification.

Because it requires G to be able to bi-directionally communicate with $S_p$ and $S_q$ simultaneously to set up the cooperative $S_p$-G-$S_q$ link.

As suggested in Fig. 5.4, the coverage areas of $S_p$ and $S_q$ are treated as circles with equal radius $L$, and original $O_p$ and $O_q$, respectively. The distance between $O_p$ and $O_q$ is denoted as $T$.

Furthermore, as observed from Fig. 5.4, if there exists terrestrial relay for $S_p$ and $S_q$, the area of the intersection of the coverage areas of $S_p$ and $S_q$ should not be 0. Thus, we can have

$$L < T \leq 2L. \tag{5.21}$$

To characterize the statistics of the distance between G and $S_p$ and $S_q$, we will first derive the PDF of the square of $d_{O_p\mathrm{G}}/d_{O_q\mathrm{G}}$.

**Theorem 5.1** *As presented in Fig. 5.4, assuming that the distance between the originals of two circles with equal radius, L, is l ($L \leq T \leq 2L$), and a node, G, is uniformly distributed in the intersection of the two circles, the PDF of the distance between G and $O_p/O_q$ (namely, $d_{O_p\mathrm{G}}/d_{O_q\mathrm{G}}$ shown in Fig. 5.4) can be written as*

$$f_{d_{O_{p/q}\mathrm{G}}}(x) = \frac{2x}{S_{\mathrm{in}}} \cdot \arccos \frac{T^2 - L^2 + x^2}{2Tx}, \tag{5.22}$$

*where $T - L \leq x \leq L$, $S_{\mathrm{in}} = 2L^2 \cos^{-1}\left(\frac{T}{2L}\right) - \frac{T}{2}\sqrt{4L^2 - T^2}$ is the area of the intersection of the two circles.*

***Proof*** Please refer to Appendix I. □

Similar to the previous section, for tractability purpose, we consider the cases that the projection of the LEO satellites is the the centre of their coverage areas. Then, by using Theorem 1, the PDF of the square of the distance between G and $S_p/S_q$, $d^2_{S_p\mathrm{G}} = H^2 + d^2_{O_p\mathrm{G}}/d^2_{\mathrm{GS}_q} = H^2 + d^2_{O_q\mathrm{G}}$, can be derived as

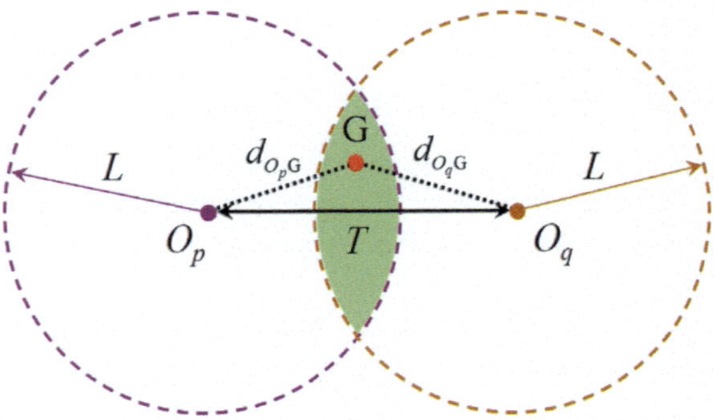


**Fig. 5.4** Terrestrial relay assisted transmissions

$$f_{d^2_{S_{p/q}G}}(x) = \frac{1}{S_{\text{in}}} \cdot \arccos \frac{T^2 - L^2 - H^2 + x}{2T\sqrt{x - H^2}}, \tag{5.23}$$

where $(T - L)^2 + H^2 = Y_{\text{min}} \leq x \leq Y_{\text{max}} = L^2 + H^2$.

Thus, the PDF of the distance between G and $S_p/S_q$, $d_{S_pG} = \sqrt{H^2 + d^2_{O_pG}}/d_{GS_q} = \sqrt{H^2 + d^2_{O_qG}}$, can be derived as

$$f_{d_{S_{p/q}G}}(x) = \frac{2x}{S_{\text{in}}} \cdot \arccos \frac{T^2 - L^2 - H^2 + x^2}{2T\sqrt{x^2 - H^2}}, \tag{5.24}$$

where $\sqrt{(T - L)^2 + H^2} = Z_{\text{min}} \leq x \leq Z_{\text{max}} = \sqrt{L^2 + H^2}$.

Moreover, it is easy to obtain the received SNR over satellite-terrestrial relay links in Fig. 5.1 as $\gamma_{ij} = \frac{\lambda_{ij}}{d^n_{ij}} = \frac{\lambda_{ij}}{d^2_{ij}}$, where $(i, j) \in \left\{\left(S_p, G\right), \left(G, S_q\right)\right\}$, and the path-loss factor is set as 2 for LoS propagation scenarios.

Obviously, the PDF and CDF of $\lambda_{ij}$ $(i \in \left\{S_p, G\right\})$ can be obtained as (5.16) and (5.17), respectively.

#### 5.1.4.1 Transmission Delay

By adopting (5.11) and (5.24), and employing Chebyshev-Gauss quadrature, the average transmission delay for terrestrial relay assisted transmissions can be derived as

$$\begin{aligned}
\bar{\tau}_{ij} &= \frac{2}{cS_{\text{in}}} \int_{Z_{\text{min}}}^{Z_{\text{max}}} x^2 \arccos \frac{T^2 - L^2 - H^2 + x^2}{2T\sqrt{x^2 - H^2}} dx \\
&= \frac{2}{cS_{\text{in}}} \cdot \frac{Z_{\text{max}} - Z_{\text{min}}}{2} \int_{-1}^{1} (a_1 t + a_2)^2 \\
&\quad \times \arccos \frac{T^2 - L^2 - H^2 + (a_1 t + a_2)^2}{2T\sqrt{(a_1 t + a_2)^2 - H^2}} dt \\
&\approx A_{TR} \sum_{i=1}^{P_G} w_i \sqrt{1 - \varepsilon_i^2} (a_1 \varepsilon_i + a_2)^2 \\
&\quad \times \arccos \frac{T^2 - L^2 - H^2 + (a_1 \varepsilon_i + a_2)^2}{2T\sqrt{(a_1 \varepsilon_i + a_2)^2 - H^2}},
\end{aligned} \tag{5.25}$$

where $A_{TR} = \frac{Z_{\text{max}} - Z_{\text{min}}}{cS_{\text{in}}}$, $a_1 = \frac{Z_{\text{max}} - Z_{\text{min}}}{2}$, $a_2 = \frac{Z_{\text{max}} + Z_{\text{min}}}{2}$, $w_i = \frac{\pi}{P_G}$, and $\varepsilon_i = \cos\left(\frac{2i-1}{2P_G}\pi\right)$.

#### 5.1.4.2 Outage Performance

Considering the randomness of the position of the terrestrial relay and using (5.23), the OP between $i$ and $j$ (where $(i, j) \in \{(\mathrm{S}_p, \mathrm{G}), (\mathrm{G}, \mathrm{S}_q)\}$) for terrestrial relay assisted transmissions can be derived as

$$P_{\mathrm{out}}^{i,j} = 1 - \frac{\alpha}{S_{\mathrm{in}}} \sum_{k=0}^{m-1} \frac{\varsigma(k)}{\bar{\lambda}_i^{k+1}} \sum_{p=0}^{k} \gamma_{\mathrm{th}}^{p} \frac{k!}{p!} \left(\frac{\beta-\delta}{\bar{\lambda}_i}\right)^{-(k+1-p)} \times \int_{Y_{\min}}^{Y_{\max}} y^p \exp\left(-\frac{\beta-\delta}{\bar{\lambda}_i}\gamma_{\mathrm{th}} y\right) \arccos \frac{T^2 - L^2 - H^2 + y}{2T\sqrt{y - H^2}} dy. \tag{5.26}$$

Observing the integral involved in last equation, it is impossible to derive a closed-form analytical solution for such a complicated integral. Thus, by employing Chebyshev-Gauss quadrature, the approximated OP for terrestrial relay assisted transmissions can be obtained as (5.27), shown on top of next page, where $B_{TR} = \frac{\alpha}{S_{\mathrm{in}}} \cdot \frac{Y_{\max}-Y_{\min}}{2}$, $a_3 = \frac{Y_{\max}-Y_{\min}}{2}$, $a_4 = \frac{Y_{\max}+Y_{\min}}{2}$, $\upsilon_{\mathrm{G}i} = \cos\left(\frac{2i-1}{2Q_{\mathrm{G}}}\pi\right)$, and $u_{\mathrm{G}i} = \frac{\pi}{Q_{\mathrm{G}}}$.

$$\begin{aligned} P_{\mathrm{out}}^{i,j} &= 1 - B_{TR} \sum_{k=0}^{m-1} \frac{\varsigma(k)}{\bar{\lambda}_i^{k+1}} \sum_{p=0}^{k} \gamma_{\mathrm{th}}^{p} \frac{k!}{p!} \left(\frac{\beta-\delta}{\bar{\lambda}_i}\right)^{-(k+1-p)} \\ &\quad \times \int_{-1}^{1} (a_3 t + a_4)^p \exp\left(-\frac{\gamma_{\mathrm{th}}(\beta-\delta)}{\bar{\lambda}_i}(a_3 t + a_4)\right) \\ &\quad \times \arccos \frac{T^2 - L^2 - H^2 + a_3 t + a_4}{2T\sqrt{a_3 t + a_4 - H^2}} dt \\ &\approx 1 - B_{TR} \sum_{k=0}^{m-1} \frac{\varsigma(k)}{\bar{\lambda}_i^{k+1}} \sum_{p=0}^{k} \gamma_{\mathrm{th}}^{p} \frac{k!}{p!} \left(\frac{\beta-\delta}{\bar{\lambda}_i}\right)^{-(k+1-p)} \\ &\quad \times \sum_{i=1}^{Q_{\mathrm{G}}} u_{\mathrm{G}i} \sqrt{1-\upsilon_{\mathrm{G}i}^2} (a_3 \upsilon_{\mathrm{G}i} + a_4)^p \exp\left(-\frac{\gamma_{\mathrm{th}}(\beta-\delta)}{\bar{\lambda}_i}(a_3 \upsilon_{\mathrm{G}i} + a_4)\right) \\ &\quad \times \arccos \frac{T^2 - L^2 - H^2 + a_3 \upsilon_{\mathrm{G}i} + a_4}{2T\sqrt{a_3 \upsilon_{\mathrm{G}i} + a_4 - H^2}} \end{aligned} \tag{5.27}$$

### *5.1.5 Aerial Relay Assisted Transmissions*

In this subsection, we will study the time delay and outage performance when aerial relay-assisted transmissions are considered to replace ISLs in Fig. 5.1.

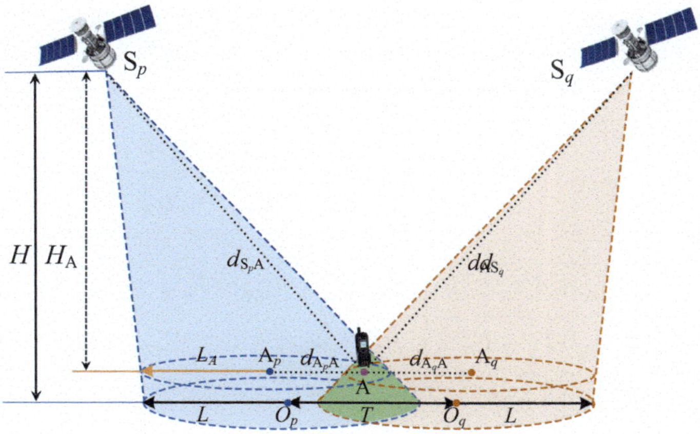


**Fig. 5.5** Aerial relay assisted transmissions

As observed in Fig. 5.5, only the aerial relay, which is distributed in the intersection space of the coverage spaces of LEO satellites $\mathrm{S}_p$ and $\mathrm{S}_q$ (the green-colored space shown in Fig. 5.5), can serve as the relay to replace the inter-satellite transmissions between LEO satellites $\mathrm{S}_p$ and $\mathrm{S}_q$. Furthermore, in this section, it is assumed that A is uniformly distributed in the green-colored space shown in Fig. 5.5 to study the impact of the randomness of its position on the performance of aerial relay assisted transmissions.

In practice, A operates in a fixed height range due to its design or physical limitations. In this section, we assume that the height of A ranges from $A_{\min}$ to $A_{\max}$, where $0 < A_{\min} < A_{\max}$. Then, one can easily have $H - A_{\max} \leq H_{\mathrm{A}} \leq H - A_{\min}$.

To characterise the statistical characteristics of the distances between A and $\mathrm{S}_p/\mathrm{S}_q$ (namely, $d_{\mathrm{S}_p\mathrm{A}}/d_{\mathrm{S}_q\mathrm{A}}$ shown in Fig. 5.5), two circles with centre $\mathrm{A}_p$ and $\mathrm{A}_q$, and radius $L_{\mathrm{A}}$, which are parallel with the ground coverage areas of $\mathrm{S}_p$ and $\mathrm{S}_q$, namely, the two circles with centre $O_p$ and $O_q$, and radius $L$. Then, one can see that A is uniformly distributed in the intersection area of the two circles with centre $\mathrm{A}_p$ and $\mathrm{A}_q$, and radius $L_{\mathrm{A}}$.

Observing Fig. 5.5, one can derive $L_{\mathrm{A}}$ as

$$L_{\mathrm{A}} = \frac{H_{\mathrm{A}}}{H} L, \tag{5.28}$$

where $\frac{H-A_{\max}}{H} L \leq L_{\mathrm{A}} \leq \frac{H-A_{\min}}{H} L$.

We denote the distance between $\mathrm{A}_p$ and $\mathrm{A}_q$ is $T_{\mathrm{A}}$ ($L_A < T_{\mathrm{A}} \leq 2L_A$). As depicted in Fig. 5.5, by using the elementary knowledge of geometry, one can easily obtain

$$T_{\mathrm{A}} = T + (H - H_{\mathrm{A}}) \left( \cot \omega_{O_p} + \cot \omega_{O_q} \right), \tag{5.29}$$

where $\omega_{O_p}$ and $\omega_{O_q}$ are the angle between line $\mathrm{S}_p\mathrm{O}_p/\mathrm{S}_q\mathrm{O}_q$ and the horizon, which is decided by the beam direction of the satellite and the range of which is $(0, \pi/2]$.

**Theorem 5.2** *As presented in Fig. 5.5, assuming that A is uniformly distributed in the intersection of the circles with centre* $A_p$ *and* $A_q$*, and radius* $L_A$*, the PDF of the distance between A and* $S_p/S_q$ *can be written as*

$$f_{d_{S_{p/q}A}}(x) = \frac{2x}{S_{in,A}} \cdot \arccos \frac{T_A^2 - L_A^2 - H_A^2 + x^2}{2T_A\sqrt{x^2 - H_A^2}}, \tag{5.30}$$

*where* $d_{S_pA} = \sqrt{H_A^2 + d_{A_pA}^2}$, $d_{AS_q} = \sqrt{H_A^2 + d_{A_qA}^2}$, $S_{in,A} = 2L_A^2\cos^{-1}\left(\frac{T_A}{2L_A}\right) - \frac{T_A}{2}\sqrt{4L_A^2 - T_A^2}$*, and* $\sqrt{(T_A - L_A)^2 + H_A^2} = W_{min} \le x \le W_{max} = \sqrt{L_A^2 + H_A^2}$.

***Proof*** Please refer to the Appendix II. □

Moreover, it is easy to obtain the received SNR over satellite-aerial relay links in Fig. 5.5 as $\gamma_{ij} = \frac{\lambda_{ij}}{d_{ij}^n} = \frac{\lambda_{ij}}{d_{ij}^2}$, where $(i, j) \in \{(S_p, A), (A, S_q)\}$, and the path-loss factor is set as 2 for LoS propagation scenarios.

#### 5.1.5.1 Transmission Delay

By using (5.11) and (5.30), and employing Chebyshev-Gauss quadrature, the average transmission delay for aerial relay assisted transmissions can be derived as

$$\begin{aligned}
\bar{\tau}_{ij} &= \frac{2}{cS_{in,A}} \int_{W_{min}}^{W_{max}} x^2 \arccos \frac{T_A^2 - L_A^2 - H_A^2 + x^2}{2T_A\sqrt{x^2 - H_A^2}} dx \\
&= \frac{W_{max} - W_{min}}{cS_{in,A}} \cdot \int_{-1}^{1} (a_5 t + a_6)^2 \arccos \frac{T_A^2 - L_A^2 - H_A^2 + (a_5 t + a_6)^2}{2T_A\sqrt{(a_5 t + a_6)^2 - H_A^2}} dt \\
&= A_{AR1} \sum_{i=1}^{P_{A1}} w_{Ai}\sqrt{1 - \varepsilon_{Ai}^2}(a_5\varepsilon_{Ai} + a_6)^2 \arccos \frac{T_A^2 - L_A^2 - H_A^2 + (a_5\varepsilon_{Ai} + a_6)^2}{2T_A\sqrt{(a_5\varepsilon_{Ai} + a_6)^2 - H_A^2}},
\end{aligned} \tag{5.31}$$

where $A_{AR1} = \frac{W_{max} - W_{min}}{cS_{in,A}}$, $a_5 = \frac{W_{max} - W_{min}}{2}$, $a_6 = \frac{W_{max} + W_{min}}{2}$, $w_{Ai} = \frac{\pi}{P_{A1}}$, and $\varepsilon_{Ai} = \cos\left(\frac{2i-1}{2P_{A1}}\pi\right)$.

As A is uniformly distributed in the green-colored space shown in Fig. 5.5, by considering the randomness of $H_A$ which is uniformly distributed in the region $[H - A_{max}, H - A_{min}]$, one can finally derive the average transmission delay for aerial relay assisted transmissions. However, as $A_{AR1}$, $a_5$, and $a_6$ are functions of $H_A$, one can imagine that the final approximated analytical expression for the average transmission delay, which is obtained by using the Chebyshev-Gauss quadrature to deal with the randomness of $H_A$, will be very complicated. Then, in this section, we will discuss the impact of the randomness of $H_{\mathrm{A}}$ on the average transmission via simulations in Sect. 5.1.8, instead of presenting such a complicated expression.

#### 5.1.5.2 Outage Performance

In this section, we also consider SR fading to model the channel gain between $i$ and $j$ (where $(i,j) \in \{(\mathrm{S}_p, \mathrm{A}), (\mathrm{A}, \mathrm{S}_q)\}$), because SR channel model can cover LoS propagation scenarios by setting $m = 0$ in (5.16).

Considering the randomness of the position of the aerial relay and using (11) in Appendix II, the OP between $i$ and $j$ (where $(i,j) \in \{(\mathrm{S}_p, \mathrm{A}), (\mathrm{A}, \mathrm{S}_q)\}$) for aerial relay assisted transmissions can be derived as

$$P_{\text{out}}^{i,j} = 1 - \frac{\alpha}{S_{\text{in,A}}} \sum_{k=0}^{m-1} \frac{\varsigma(k)}{\bar{\lambda}_i^{k+1}} \sum_{p=0}^{k} \gamma_{\text{th}}^{p} \frac{k!}{p!} \left( \frac{\beta - \delta}{\bar{\lambda}_i} \right)^{-(k+1-p)} \times \int_{T_{\min}}^{T_{\max}} y^p \exp\left( -\frac{\beta - \delta}{\bar{\lambda}_i} \gamma_{\text{th}} y \right) \arccos \frac{T_{\mathrm{A}}^2 - L_{\mathrm{A}}^2 - H_{\mathrm{A}}^2 + y}{2T_{\mathrm{A}}\sqrt{y - H_{\mathrm{A}}^2}} dy. \tag{5.32}$$

Similar to the derivation of (5.27), by employing Chebyshev-Gauss quadrature, the approximated OP for aerial relay assisted transmissions can be obtained as (5.33), shown on top of next page, where $B_{AR1} = \frac{\alpha}{S_{\text{in,A}}} \cdot \frac{T_{\max} - T_{\min}}{2}$, $a_7 = \frac{T_{\max} - T_{\min}}{2}$, $a_8 = \frac{T_{\max} + T_{\min}}{2}$, $\upsilon_{\mathrm{A}i} = \cos\left(\frac{2i-1}{2Q_{\mathrm{A1}}}\pi\right)$, and $u_{\mathrm{A}i} = \frac{\pi}{Q_{\mathrm{A1}}}$.

$$\begin{aligned} P_{\text{out}}^{i,j} &= 1 - B_{AR1} \sum_{k=0}^{m-1} \frac{\varsigma(k)}{\bar{\lambda}_i^{k+1}} \sum_{p=0}^{k} \gamma_{\text{th}}^{p} \frac{k!}{p!} \left( \frac{\beta - \delta}{\bar{\lambda}_i} \right)^{-(k+1-p)} \\ &\times \int_{-1}^{1} (a_7 t + a_8)^p \exp\left( -\frac{\gamma_{\text{th}}(\beta - \delta)}{\bar{\lambda}_i} (a_7 t + a_8) \right) \\ &\times \arccos \frac{T_{\mathrm{A}}^2 - L_{\mathrm{A}}^2 - H_{\mathrm{A}}^2 + a_7 t + a_8}{2T_{\mathrm{A}}\sqrt{a_7 t + a_8 - H_{\mathrm{A}}^2}} dt \\ &\approx 1 - B_{AR1} \sum_{k=0}^{m-1} \frac{\varsigma(k)}{\bar{\lambda}_i^{k+1}} \sum_{p=0}^{k} \gamma_{\text{th}}^{p} \frac{k!}{p!} \left( \frac{\beta - \delta}{\bar{\lambda}_i} \right)^{-(k+1-p)} \\ &\times \sum_{i=1}^{Q_{\mathrm{A1}}} u_{\mathrm{A}i} \sqrt{1 - \upsilon_{\mathrm{A}i}^2} (a_7 \upsilon_{\mathrm{A}i} + a_8)^p \exp\left( -\frac{\gamma_{\text{th}}(\beta - \delta)}{\bar{\lambda}_i} (a_7 \upsilon_{\mathrm{A}i} + a_8) \right) \\ &\times \arccos \frac{T_{\mathrm{A}}^2 - L_{\mathrm{A}}^2 - H_{\mathrm{A}}^2 + a_7 \upsilon_{\mathrm{A}i} + a_8}{2T_{\mathrm{A}}\sqrt{a_7 \upsilon_{\mathrm{A}i} + a_8 - H_{\mathrm{A}}^2}} \end{aligned} \tag{5.33}$$

Then, considering the randomness of $H_A$, the OP for aerial relay-assisted transmissions can be finally approximated. However, we can see that $T_{max}$, $T_{min}$, $a_7$, and $a_8$ are functions of $H_A$, thus the final approximated analytical expression for the OP, which is obtained by using Chebyshev-Gauss quadrature to deal with the randomness of $H_A$, will be too complicated. Then, in this section, we will investigate the impact of the randomness of $H_A$ on the OP via simulations in Sect. 5.1.8, in place of presenting such a complicated expression.

### 5.1.6 Satellite Relay Assisted Transmissions

In this subsection, the time delay and outage performance of satellite relay-assisted transmissions will be studied, respectively.

As depicted in Fig. 5.6, only the relay satellite, which is located in the intersection of the coverage space of LEO satellite $S_p$ and $S_q$ (namely, the intersection of the two hemispheres with centre $S_p$ and $S_q$, and radius $L_S$), can serve as the relay satellite between $S_p$ and $S_q$.

To characterise the statistical characteristics of the distances between $S_0$ and $S_p$/$S_q$ (namely, $d_{S_pS_0}$/$d_{S_qS_0}$ shown in Fig. 5.6), two circles with centre $U_p$ and $U_q$, and radius $R_S$, which are parallel with the the two circles with $S_p$ and $S_q$, and radius $L_S$). Then, one can see that $S_0$ is uniformly distributed in the intersection area of the two circles with centre $U_p$ and $U_q$, and radius $R_S$.

Practically, the orbit height of $S_0$ is much higher than the one of LEO satellites. Then, in this section, we assume that the orbit height of $S_0$, $H_{0,min} > H$, where $H_{0,min}$ the minimum orbit height of the relay satellite, e.g., normally, geostationary Earth orbit (GEO) satellites orbit at 35,786 km from the earth and medium Earth orbit (MEO) satellites operate in the space between the two most popular orbital regimes (namely, LEO and GEO oribis), while the orbit height of LEO satellites ranges from 160 to 2000 km. Therefore, to facilitate the following modeling and analysis, in this section, it is assumed that the interface between the lower side of the distributed space of $S_0$ and the intersection of the coverage space of LEO satellite $S_p$ and $S_q$ is a plane, which is parallel to the plane consisting of the two circles with $S_p$ and $S_q$, and radius $L_S$, as presented in Fig. 5.6. So, the distribution space of $S_0$ is the green-colored space in Fig. 5.6. Furthermore, in this section, it is assumed that $S_0$ is uniformly distributed in the green-colored space shown in Fig. 5.6 to study the impact of the randomness of its position on the performance of satellite relay assisted transmissions.

In this subsection, we also denote $H_{S,min}$ as the distance between the bottom of the distribution space of $S_0$ and the plane consisting two circles with with $S_p$ and $S_q$, and radius $L_S$. Then, let $H_S$ ($H_{S,min} \leq H_S \leq H_{S,max}$) be the distance of $S_p$ and $U_p$ or $S_q$ and $U_q$, and $T_S$ be the distance between $S_p$ and $S_q$, or $U_p$ and $U_q$. Furthermore, as observed from Fig. 5.6, $H_{S,min}$ is decided by the orbit height of $S_0$, which is constant in a specific practical system.

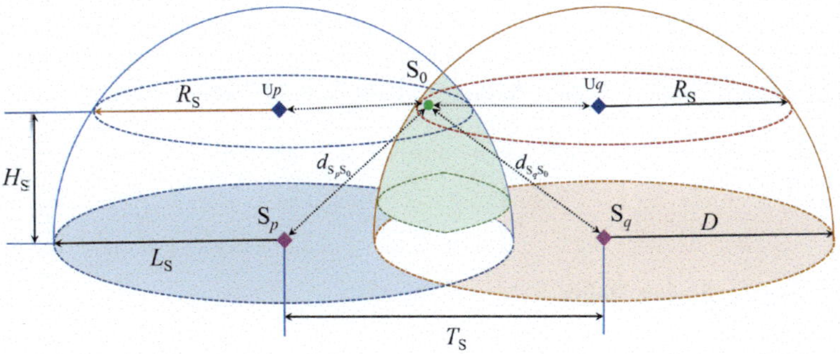


**Fig. 5.6** The distribution of satellite relay

Observing Fig. 5.2 and using $\theta = \pi/N$, $T_S$ can be calculated as

$$T_S = 2(R_E + H)\sin\frac{(q-p)\theta}{2}$$
$$= 2(R_E + H)\sin\frac{(q-p)\pi}{2N}. \tag{5.34}$$

By using the knowledge of the elementary geometry, the maximum value of $H_S$, $H_{S,\max}$, and $R_S$ can be obtained as

$$H_{S,\max} = \sqrt{L_S^2 - \left(\frac{T_S}{2}\right)^2}$$
$$= \sqrt{L_S^2 - (R_E + H)^2 \sin^2\frac{(q-p)\pi}{2N}} \tag{5.35}$$

and

$$R_S = \sqrt{L_S^2 - H_S^2}, \tag{5.36}$$

respectively.

**Theorem 5.3** *As presented in Fig. 5.6, assuming that* $S_0$ *is uniformly distributed in the intersection of the circles with centre* $U_p$ *and* $U_q$*, and radius* $R_S$*, the PDF of the distance between* $S_0$ *and* $S_p/S_q$ *can be derived as*

$$f_{d_{S_{p/q}S_0}}(x) = \frac{2x}{S_{\text{in},S}} \cdot \arccos\frac{T_S^2 - R_S^2 - H_S^2 + x^2}{2T_S\sqrt{x^2 - H_S^2}}$$
$$= \frac{2x}{S_{\text{in},S}} \cdot \arccos\frac{T_S^2 - L_S^2 + x^2}{2T_S\sqrt{x^2 - H_S^2}}, \tag{5.37}$$

*where* $d_{\mathrm{S}_p\mathrm{S}_0} = \sqrt{H_\mathrm{S}^2 + d_{\mathrm{S}_p\mathrm{S}_0}^2}$, $d_{\mathrm{S}_q\mathrm{S}_0} = \sqrt{H_\mathrm{S}^2 + d_{\mathrm{S}_q\mathrm{S}_0}^2}$, $S_{\mathrm{in,S}} = 2R_\mathrm{S}^2\cos^{-1}\left(\frac{T_\mathrm{S}}{2R_\mathrm{S}}\right) - \frac{T_\mathrm{S}}{2}\sqrt{4R_\mathrm{S}^2 - T_\mathrm{S}^2}$, *and* $\sqrt{H_\mathrm{S}^2 + (T_\mathrm{S} - L_\mathrm{S})^2} = M_{\min} \le x \le M_{\max} = L_\mathrm{S}$.

***Proof*** The proof of Theorem 3 is similar to that of Theorem 2 given in Appendix II. □

#### 5.1.6.1 Transmission Delay

By using (5.11) and (5.37) and employing Chebyshev-Gauss quadrature, the average transmission delay for satellite relay assisted transmissions can be derived as

$$\begin{aligned}
\bar{\tau}_{ij} &= \frac{2}{cS_{\mathrm{in,S}}} \int_{M_{\min}}^{M_{\max}} x^2 \arccos \frac{T_\mathrm{S}^2 - L_\mathrm{S}^2 + x^2}{2T_\mathrm{S}\sqrt{x^2 - H_\mathrm{S}^2}} dx \\
&= \frac{M_{\max} - M_{\min}}{cS_{\mathrm{in,S}}} \int_{-1}^{1} (a_9 t + a_{10})^2 \arccos \frac{T_\mathrm{S}^2 - L_\mathrm{S}^2 + (a_9 t + a_{10})^2}{2T_\mathrm{S}\sqrt{(a_9 t + a_{10})^2 - H_\mathrm{S}^2}} dt \\
&\approx A_{SR1} \sum_{i=1}^{P_{\mathrm{S}1}} w_{\mathrm{S}i} \sqrt{1 - \varepsilon_{\mathrm{S}i}^2} (a_9 \varepsilon_{\mathrm{S}i} + a_{10})^2 \arccos \frac{T_\mathrm{S}^2 - L_\mathrm{S}^2 + (a_9 \varepsilon_{\mathrm{S}i} + a_{10})^2}{2T_\mathrm{S}\sqrt{(a_9 \varepsilon_{\mathrm{S}i} + a_{10})^2 - H_\mathrm{S}^2}},
\end{aligned} \tag{5.38}$$

where $A_{SR1} = \frac{M_{\max} - M_{\min}}{cS_{\mathrm{in,S}}}$, $a_9 = \frac{M_{\max} - M_{\min}}{2}$, $a_{10} = \frac{M_{\max} + M_{\min}}{2}$, $w_{\mathrm{S}i} = \frac{\pi}{P_{\mathrm{S}1}}$, and $\varepsilon_{\mathrm{S}i} = \cos\left(\frac{2i-1}{2P_{\mathrm{S}1}}\pi\right)$.

As $\mathrm{S}_0$ is uniformly distributed in the green-colored space shown in Fig. 5.6, by considering the randomness of $H_\mathrm{S}$ which is uniformly distributed in the region $\left[H_{\mathrm{S,min}}, H_{\mathrm{S,max}}\right]$, one can finally derive the average transmission delay for satellite relay assisted transmissions. However, we can see that $M_{\min}$, $A_{SR1}$, $a_9$, and $a_{10}$ are functions of $H_\mathrm{S}$, thus the final approximated analytical expression for the average transmission delay, which is obtained by using Chebyshev-Gauss quadrature to treat the randomness of $H_\mathrm{S}$, will be too complicated. Then, in this section, we will investigate the impact of the randomness of $H_\mathrm{S}$ on the average transmission delay via simulations in Sect. 5.1.8, in place of presenting such a complicated expression.

#### 5.1.6.2 Outage Performance

In this subsubsection, outage analysis will be given for inter-satellite relay links while RF and FSO communication techniques are considered, respectively.

Besides traditional RF communication techniques, FSO has been proved and applied as a promising technique to set up inter-satellite relay links, due to its inherent merits, e.g., high bandwidth, high data rate, small size, etc. In the following, we will present the analysis models for the outage performance over the inter-satellite relay links between $S_0$ and $S_p/S_q$, while considering RF and FSO communication techniques, respectively.

Then, it is easy to obtain the received SNR over inter satellite relay links in Fig. 5.6 as [3–5]

$$\gamma_{ij} = \begin{cases} \frac{\bar{\lambda}_i G_{FSO}}{d_{ij}^4}, & \text{When FSO technique is adopted;} \\ \frac{\bar{\lambda}_i}{d_{ij}^2}, & \text{When RF technique is adopted} \end{cases}, \tag{5.39}$$

where $G_{FSO}$ is the e2e signal power gain over the FSO channel from the transmitter to the receiver, $(i, j) \in \{(S_p, S_0), (S_0, S_q)\}$, and the path-loss factor is set as 2 for RF transmissions in LoS propagation scenarios.

*(1) RF Transmission over Inter Satellite Relay Links:* When RF transmission techniques are adopted for inter satellite relay links, free-space transmissions are obtained, as there is no multi-path transmission with no reflections, diffraction, and scattering. Then, the OP under this case can be derived as

$$\begin{aligned} P_{\text{out}}^{i,j} &= \Pr\{\gamma_{ij} \le \gamma_{\text{th}}\} = \Pr\left\{\frac{\bar{\lambda}_i}{d_{ij}^2} \le \gamma_{\text{th}}\right\} \\ &= 1 - F_{d_{S_{p/q}S_0}}\left(\sqrt{\frac{\bar{\lambda}_i}{\gamma_{\text{th}}}}\right), \end{aligned} \tag{5.40}$$

where $F_{d_{S_{p/q}S_0}}(x)$ is the CDF of $d_{S_pS_0}/d_{S_qS_0}$.

Using (5.37), the CDF of $d_{S_pS_0}/d_{S_qS_0}$ can be derived as

$$\begin{aligned} F_{d_{S_{p/q}S_0}}(x) &= \frac{2}{S_{\text{in,S}}} \int_{M_{\min}}^{x} x \cdot \arccos \frac{T_S^2 - L_S^2 + x^2}{2T_S\sqrt{x^2 - H_S^2}} dx \\ &= \frac{x - M_{\min}}{S_{\text{in,S}}} \int_{-1}^{1} (c_1 t + c_2) \arccos \frac{T_S^2 - L_S^2 + (c_1 t + c_2)^2}{2T_S\sqrt{(c_1 t + c_2)^2 - H_S^2}} dt \\ &\approx \frac{x - M_{\min}}{S_{\text{in,S}}} \sum_{i=1}^{Q_{S1}} \delta_{Si}\sqrt{1 - \tau_{Si}^2}(c_1\tau_{Si} + c_2) \arccos \frac{T_S^2 - L_S^2 + (c_1\tau_{Si} + c_2)^2}{2T_S\sqrt{(c_1\tau_{Si} + c_2)^2 - H_S^2}}, \end{aligned} \tag{5.41}$$

where $c_1 = \frac{x - M_{\min}}{2}$, $c_2 = \frac{x + M_{\min}}{2}$, $\delta_{Si} = \frac{\pi}{Q_{S1}}$, and $\tau_{Si} = \cos\left(\frac{2i-1}{2Q_{S1}}\pi\right)$.

Considering the randomness of of $H_S$, the CDF of $d_{S_pS_0}/d_{S_qS_0}$ can be further achieved. However, it is easy to see that $M_{\min}$, $S_{\text{in,S}}$, $c_1$, and $c_4$ are functions of $H_S$, thus the final approximated analytical expression for the OP, which is obtained by using Chebyshev-Gauss quadrature to treat the randomness of $H_S$, will be too complicated. Then, in this part, we will investigate the impact of the randomness of $H_S$ on the OP via simulations in Sect. 5.1.8, rather than presenting such a complicated expression.

Then, using (5.41) into (5.40), one can finally achieve the OP for the case of adopting RF transmissions for inter-satellite relay links.

*(2) FSO Transmission over Inter Satellite Relay Links:* If FSO transmission techniques are adopted for inter-satellite relay delivery, $G_{FSO}$, the e2e signal power gain over the FSO channel from the transmitter to the receiver, can be written as [3–5]

$$G_{FSO} = \eta_T^2 \eta_R^2 G_T^2 G_R^2 L_T^2 L_R^2 \left(\frac{\lambda}{4\pi}\right)^4, \tag{5.42}$$

where $\eta_T$ and $\eta_R$ are the efficiencies of the transmitter and the receiver optics, $G_T$ and $G_R$ are the gains of transmitting and receiving telescopes, respectively, $\lambda$ is the wavelength, and $L_T$ and $L_R$ are the pointing loss factor at the transmitter and receiver, respectively.

Thus, the OP under this case can be derived as

$$\begin{aligned} P_{\text{out}}^{i,j} &= \Pr\left\{\frac{\bar{\lambda}_i G_{FSO}}{d_{ij}^4} \leq \gamma_{\text{th}}\right\} \\ &= 1 - F_{d_{S_{p/q}S_0}}(\Lambda), \end{aligned} \tag{5.43}$$

where $\Lambda = \left(\frac{\bar{\lambda}_i G_{FSO}}{\gamma_{\text{th}}}\right)^{\frac{1}{4}}$.

Substituting (5.41) in (5.43), the OP of adopting FSO transmissions for inter satellite relay links can be derived as

$$\begin{aligned} P_{\text{out}}^{i,j} \approx{}& 1 - \frac{\Lambda - M_{\min}}{S_{\text{in,S}}} \sum_{i=1}^{Q_{S1}} \delta_{Si} \sqrt{1 - \tau_{Si}^2} (c_3 \tau_{Si} + c_4) \\ &\times \arccos \frac{T_S^2 - L_S^2 + (c_3 \tau_{Si} + c_4)^2}{2T_S \sqrt{(c_3 \tau_{Si} + c_4)^2 - H_S^2}}, \end{aligned} \tag{5.44}$$

where $c_3 = \frac{\Lambda - M_{\min}}{2}$ and $c_4 = \frac{\Lambda + M_{\min}}{2}$.

Similarly, the impact of the randomness of $H_S$ on the OP will be studied via simulations in Sect. 5.1.8.

### 5.1.7 *Numerical Results*

In this subsection, numerical results will be provided to study the performance of the considered system, as well as to verify the proposed analytical models. In the following, we use $P_{\mathrm{S}}$ instead of $P_{\mathrm{S}_k}$ $(0 \le k \le K)$ for simplification, and path-loss factors for all links are set as 2. Unless otherwise explicitly specified, the parameters are set as $H = 550$ km, $N = 25$, $K = 15$, $p = 8$, $q = 11$, $R_{\mathrm{E}} = 6371$ km, $L = 100$ km, $T = 110$ km, $A_{\max} = 10$ km, $m = 2$, $b = \Omega = 15$ dB, $\omega_l = 80°$ $(l \in \{\mathrm{S}, \mathrm{D}, O_k \,|0 \le k \le K\})$, $\gamma_{th} = 0$ dB, $\eta_T = \eta_R = 0.95$, $\lambda = 1.55 \times 10^{-6}$, $G_T = G_R = 5.03 \times 10^{11}$, $L_T = L_R = 0.0145$, $P_{\mathrm{S}} = 25$ dBW, $\epsilon_i = 50$ ms (where $i \in \{\mathrm{G}, \mathrm{A}, \mathrm{S}_k \,|0 \le k \le K\}$), and $\sigma^2 = -64$ dBm.

Furthermore, we run $1 \times 10^6$ times of the realizations of the considered system and $1 \times 10^5$ trials of Monte-Carlo simulations, to model the randomness of the positions of terrestrial terminals and terrestrial/aerial/satellite relays, and channel gains over each link. In other words, the channel gains are randomly generated during each trial of Monte-Carlo simulation and the positions of terrestrial terminals and the three types of relays are uniformly selected in their corresponding distribution space during each realization of the considered system. Therefore, under the Monte-Carlo method, all presented results in the following are statistical mean values.

#### 5.1.7.1 Outage Performance

In this subsubsection, the outage performance of the links between satellites and terrestrial terminal, and terrestrial/aerial/satellite relay links will be investigated versus the transmit power at the satellite, respectively, while considering the randomness of the positions of the terrestrial terminal, and terrestrial/aerial/satellite relay. Moreover, in this section, we only consider the transmission process from the satellite to terrestrial terminal, and terrestrial/aerial/satellite relay to demonstrate the impacts of the randomness of the positions of these nodes on the outage performance, and the outage performance of the transmission over the inverse direction can be easily studied via the same analysis method, due to the reciprocity of the wireless channels.

In Fig. 5.7, the outage performance over the satellite-to-terrestrial link is presented for various kinds of $L$. One can see that the OP with small $L$ outperforms that with large $L$, which suggests that the large coverage area of the satellite leads to degraded outage performance. Because a large $L$ means a large transmission distance between the satellite and terrestrial terminal, and then results in large path loss. Furthermore, we can also observe that the impact of $L$ on the OP is very slight since the orbit height of the satellite plays a major role in the path loss, the value of which here we consider is 550 km.

Figure 5.8 presents the outage performance over terrestrial relay link while considering various $T$ and $Q_G = 200$. Obviously, similar to $L$ in Fig. 5.7, $T$, the distance between the centers of the satellites' coverage area, also exhibits a weak effect on the outage performance. Furthermore, $T$ shows a negative influence on the OP, which

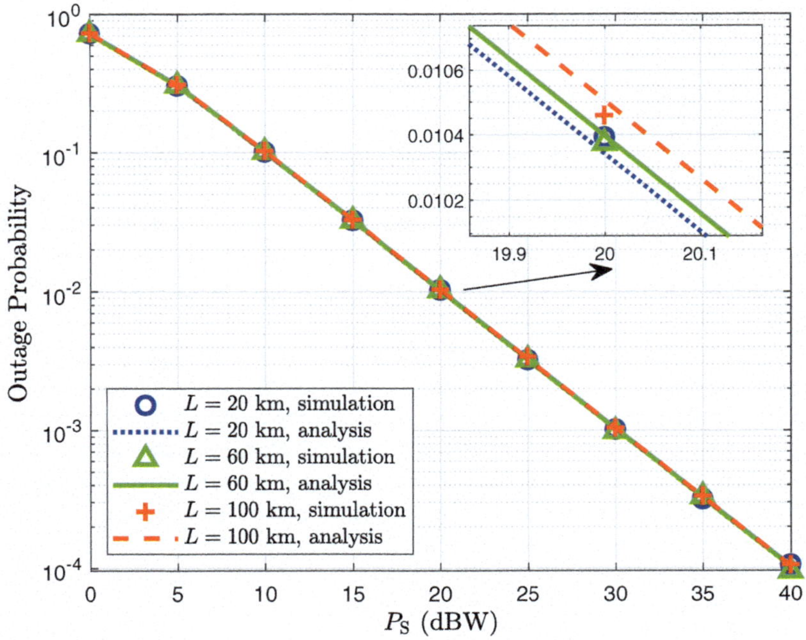


**Fig. 5.7** OP of satellite-to-terrestrial transmission link

can be explained by the fact that a large $T$ represents a large transmission distance and large path loss in return.

In Fig. 5.9, we study the outage performance over the aerial relay link for different $A_{\max}$. Here, we set $H_A = H - A_{\max}$ for simplification and $Q_{A1} = 300$. As shown in Fig. 5.9, it is clear that large $A_{\max}$ shows a positive impact on the OP, because a large $A_{\max}$ means a short transmission distance for aerial relay link. Also, similar to the previous two figures, the impact of $A_{\max}$ on the outage performance is not strong, which can be explained by a similar reason.

In Figs. 5.10 and 5.11, the outage performance over satellite relay links is investigated for RF and FSO, while $Q_{S1} = 500$ and 200, respectively. In detail, Fig. 5.10 shows that, when $P_S$ is larger than 31 dBW, $L_S$ show an obvious negative impact on the OP and opposite observation can be achieved when $P_S$ is less than 31 dBW. In Fig. 5.11, $H_S$ exhibits a negative effect on the OP, which becomes very slight when $P_S$ is larger than 40.5 dBW.

Furthermore, as presented in Figs. 5.7, 5.8, 5.9 and 5.11, the OP can be improved while the transmit power at the satellite increases, because a large $P_S$ means a large transmit SNR.

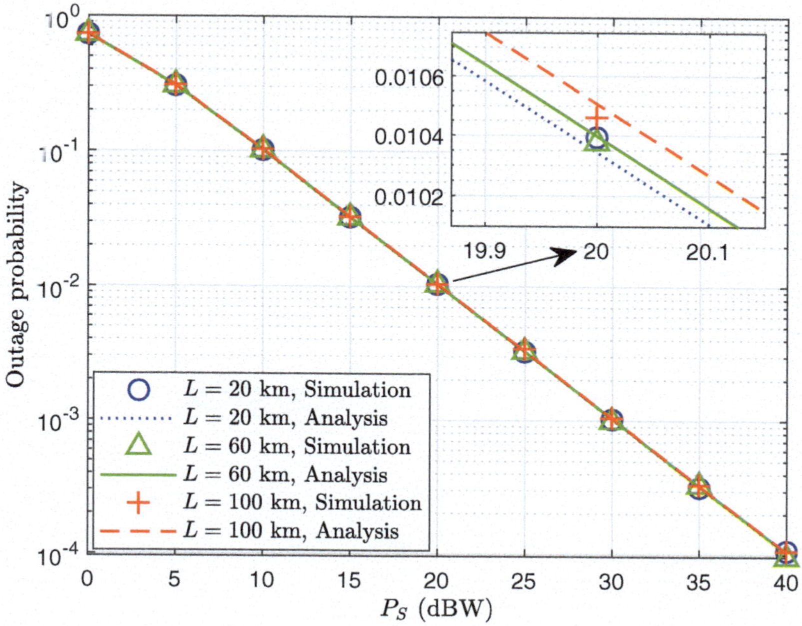


**Fig. 5.8** OP of terrestrial relay link

#### 5.1.7.2 Latency Performance

In this subsubsection, we will investigate the latency performance for the links between the satellite and terrestrial terminal, and terrestrial/aerial/satellite relay.

As shown in Fig. 5.12, the radius of the coverage area of the satellite exhibits a negative impact on the transmission delay. Because a large $T$ represents a large transmission distance, leading to a long transmission time.

Figure 5.13 presents the transmission delay for terrestrial relay link, while $P_G = 50$. Clearly, $T$ plays a negative role on the transmission delay, as a large $T$ incurs a large transmission distance.

An interesting finding can be obtained that $L$ in Fig. 5.15 and $T$ in Fig. 5.13 both show a very weak influence on the transmission delay, which also be uncovered by the reasons given for the observations in Figs. 5.7 and 5.8.

In Fig. 5.14, we investigate the transmission delay for various $A_{\max}$, while $P_{A1} = 50$. It is observed that $A_{\max}$ shows an oppositive impact on the transmission like it does in Fig. 5.9. This finding can also be demonstrated by the same explanation given for Fig. 5.9.

The transmission delay over satellite relay link is presented in Fig. 5.15, while $P_{S1} = 200$. We can easily see that the transmission delay with a large $L_S$ is worse than that with a small $L_S$, since large $L_S$ leads to large transmission distance.

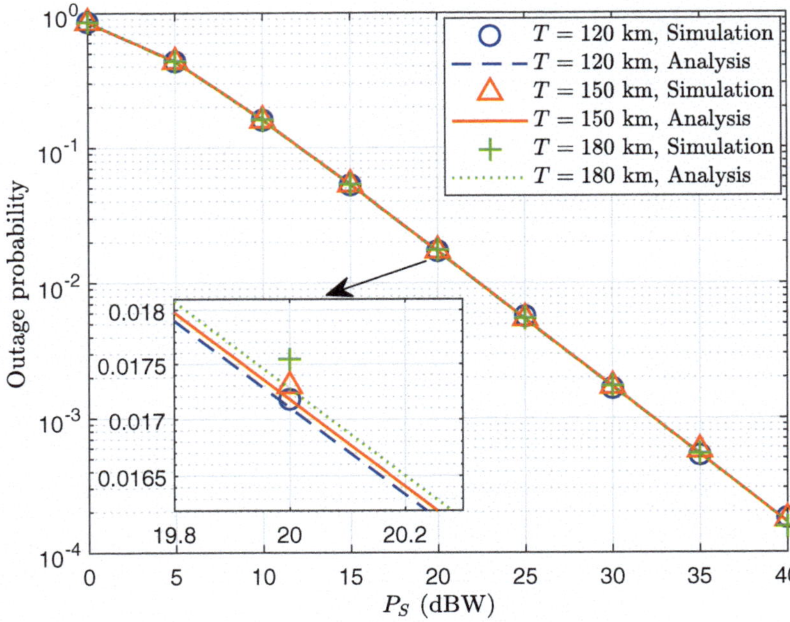


**Fig. 5.9** OP of aerial relay link

Moreover, as given in Figs. 5.12, 5.13, 5.14 and 5.15, the higher the orbit height is, the larger the transmission delay achieves, due to the increased transmission distance.

#### 5.1.7.3 E2e Performance

In this subsubsection, we will study the e2e outage and latency performance of the considered system while three different relays are considered for comparison purpose, where $m = 16$ for aerial relay links, $P_{S_k} = 35$ dBW, $P_A = 15$ dBW, $P_G = 61$ dBW, $L_S = 2500$ km, and $H_S = 1600$ km.

As shown in Fig. 5.16, the e2e time delay over the inter-satellite link is the largest one among the four considered cases, because of the frequent forwarding behaviors at the LEO satellites. Though the satellite relay can save the forwarding delay at the LEO satellites, the transmission delay is still larger than that over two other relay links, due to its long transmission distance. Also, we can see that aerial and satellite relay links show similar e2e time delay performance, as the transmission distances over these kinds of relay links are much shorter than those over two other links.

Moreover, Fig. 5.17 presents the e2e OP versus the orbit height of the LEO satellite, $H$, for the four relay links. In the low $H$ region ($H < 800$ km), the OP of the aerial relay link outperforms the three other ones, and while $H > 800$ km, the OP of the satellite relay link exhibits the worst e2e outage performance among the four ones.

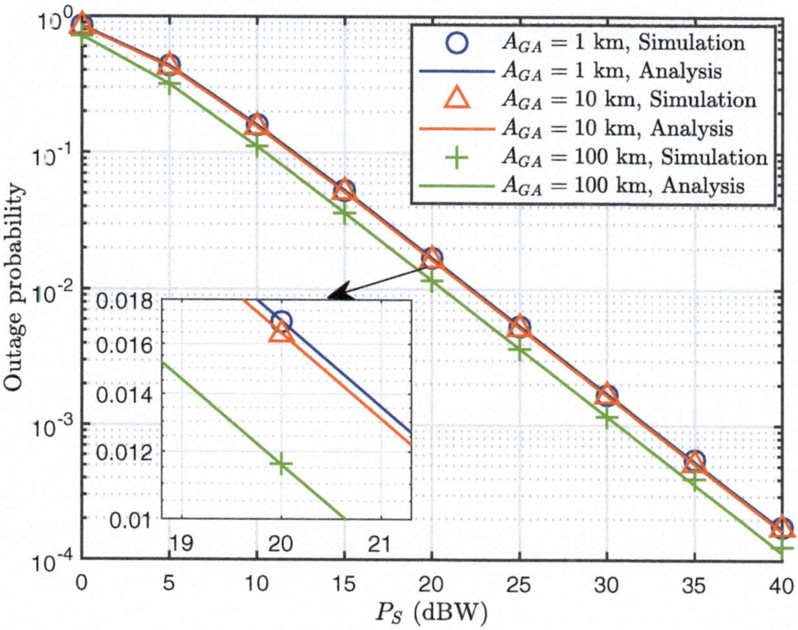


**Fig. 5.10** OP of satellite relay RF link

Because the transmission distance of the relay link tightly depends on the orbit height of the LEO satellite. One can also easily observe from Figs. 5.16 and 5.17 that the e2e delay and OP for the four relay link degrade while $H$ enlarges, because of the increased path loss.

In Fig. 5.18, $q - p$ shows a strong impact on the e2e delay of terrestrial, aerial, and satellite relay links. Because $q - p$ denotes the number of inter-satellite hops which can be reduced by introducing relay links. Moreover, in Fig. 5.19 $q - p$ exhibits a weak influence on the e2e OP of terrestrial and aerial relay links but strong impacts on that of satellite relay links. The reason for this observation is that the increment of the transmission distance of satellite relay links is much larger compared to those of terrestrial and aerial relay links while $q - p$ increases.

To study the impact of the forwarding delay at each node, in Fig. 5.20 we plot the e2e time delay of inter LEO satellite link and the considered three kinds of relay links under different cases of $\epsilon_{\mathrm{A}}$, $\epsilon_{\mathrm{S}_0}$, and $\epsilon_{\mathrm{G}}$. It can be seen that $\epsilon_{\mathrm{A}}$, $\epsilon_{\mathrm{S}_0}$, and $\epsilon_{\mathrm{G}}$ show strong negative effects on the e2e time delay, and the optimal relay link in term of the e2e time delay strongly depends on the forwarding delay.

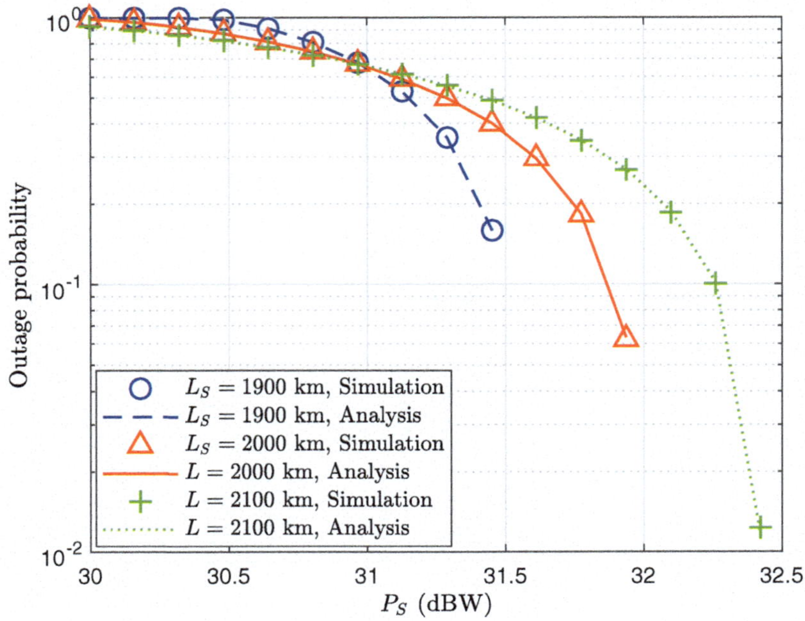


**Fig. 5.11** OP of satellite relay FSO link

## *5.1.8 Conclusion*

In this part, by considering an LEO mega-constellation system, we first studied the outage and transmission delay over the links between the LEO satellite and terrestrial terminal, and terrestrial/aerial/satellite relay, while considering the randomness of the positions of the terrestrial terminal, and terrestrial/aerial/satellite relay. These proposed analytical models are verified by simulation and numerical results and one can see that the computational complexity of the proposed models only depends on the times of summations as suggested by the proposed analytical expressions given in Sects. 5.1.3–5.1.7, which is quite acceptable.

Some insightful conclusions can be obtained as follows.

1. The size of the coverage area of the LEO satellites, namely, the radius of the coverage area on the ground, exhibits a weak negative impact on both outage and transmission delay performance;

2. The size of the intersection coverage area/space of two LEO satellites, which can be inversely represented by $T$ in Fig. 5.4, $T_A$ in Fig. 5.5, and $T_S$ in Fig. 5.6, shows a slight negative influence on both outage and transmission delay performance;

3. The orbit height of the LEO satellites shows negative impacts on outage and transmission delay performance for terrestrial and aerial relay links, and the height of aerial relay shows a positive influence on both outage and transmission delay performance because of the shortened transmission distance;

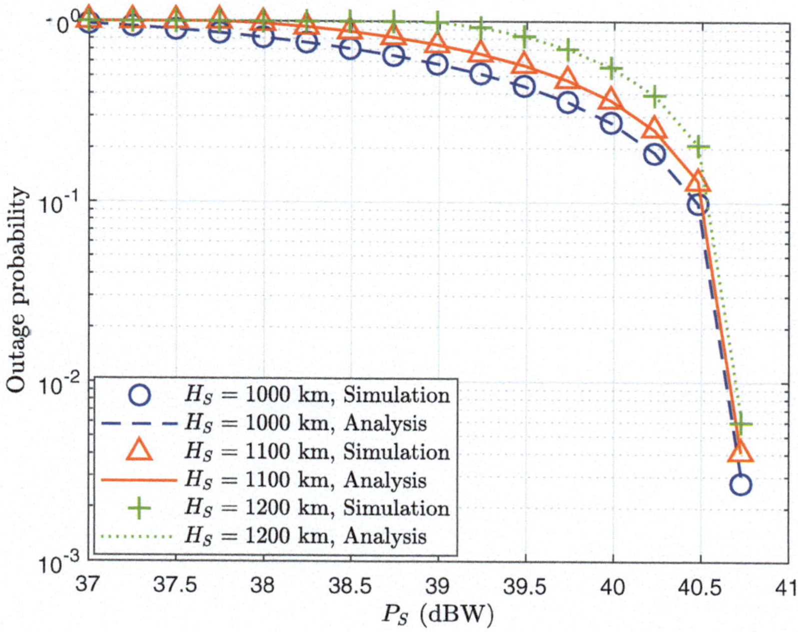


**Fig. 5.12** The transmission delay over satellite-to-terrestrial transmission link

4. The e2e time delay can be improved by introducing terrestrial/aerial/satellite relay, which can greatly save the forwarding delay at each LEO satellite;

5. In piratical, when relay link should be introduced to LEO mega-constellation systems and which kind of relay links is the optimal one depend on the details of the system setting, e.g., the forwarding delay, the orbit height of LEO/relay satellite, the height of aerial relay, and the distance between two neighboring LEO satellites, etc.

Therefore, for practical LEO mega-constellation systems, choosing which kind of relays and how to place them depends on the real application scenarios and requirements, as each kind of relays unavoidably suffers its inherent limitations and possesses unique merits simultaneously. For example, terrestrial relay suffers the worst small-scale fading but has a continual power supply; though there is no small-scale fading over aerial/satellite relay links, the stable and sustainable power supply is a tough problem; the long transmission distance and high mobile speed will bring huge challenges to satellite relay link.

Moreover, one can see that we carry out the analysis by using the classic analysis method for wireless communication systems, which ignores the timeline. In future works, we will study the delivery process with the consideration of the transmission sequence of the information bits/packets via the analysis methods adopted for upper layers (like random theory for the medium access control layer). Thus, the impacts of the moving speed and direction of the nodes (terrestrial terminals, satellites, and relays) and the transmission/queuing time will be directly modeled and demonstrated.

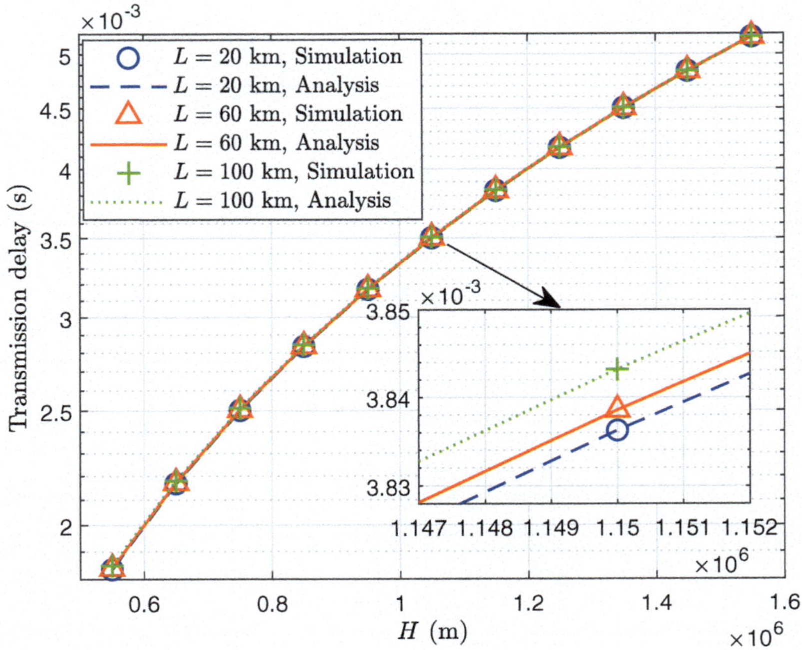


**Fig. 5.13** The transmission delay over terrestrial relay link

Finally, the adopted analysis method and technical road-map in this work can serve as a useful reference for the researchers who work in analyzing other networks, e.g., Internet of Things, Power Grids, etc.

## 5.2 Cooperative Satellite-Aerial-Terrestrial Systems in Presence of Spatially Aerial Random Interference

### 5.2.1 System Model

In this section, we consider a cooperative satellite-aerial-terrestrial network (SATN), which includes a satellite ($S$), a number of aerial relays ($R_n, n \in \mathbb{Z}^+, n \in [1, \ldots, N]$, $N \geq 1$), a group of aerial transmitters ($A_m$, $m \in \mathbb{Z}^+$, $m \in [1, \ldots, M]$, e.g., UAVs,

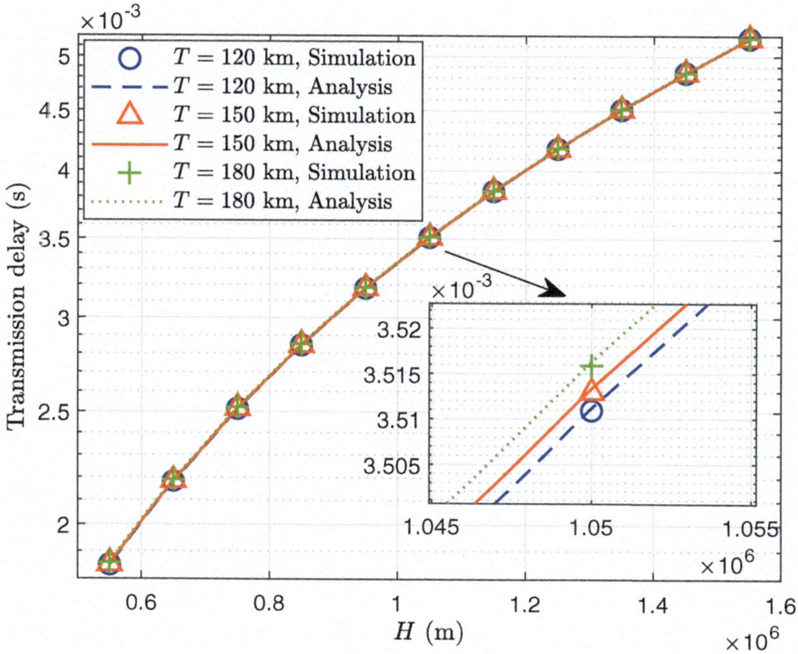


**Fig. 5.14** The transmission delay over aerial relay link

and $M \geq 1$), and a destination ($D$),[8] as shown in Fig. 5.21. The aerial transmitters, e.g., UAVs, aim to transmit their information to a remote terrestrial destination $D$. However, as the distances between $A_m$ and $D$ are quite large, the direct links from the transmitters to the terrestrial receiver are unavailable. Therefore, satellite communications are exploited as a solution to aid the targeted remote wireless transmission.

Additionally, there is an alternative relaying link via a series of aerial relays $R_n$ to link the aerial transmitters and $D$. The $N$ aerial relays are ordered from one to $N$ to perform cooperative transmission with $N + 1$ hops[9] Therefore, the received signals at $D$ can be enhanced to achieve enlarged performance. We split our system into a downlink, an uplink, and inter-aerial relaying links,[10] shown in Fig. 5.22.

[8] In this section, we consider two types of relaying links to assist the aerial source nodes and the terrestrial destination for comparison purpose. Because both of them are possible solutions in reality to bridge the source and destination nodes which are far from each other and the two kinds of relaying links show advantages and disadvantages in transmission performance (e.g., power consumption and transmission delay) due to their transmission characteristics [6].

[9] The received signal at the other aerial relays except for the first one will not suffer multi-user interference (MUI) which exists at the first one, as the aerial relays will transmit. the information of different users on orthogonal resources, e.g., time slots, to improve the transmission quality.

[10] Here, the effect of Doppler shifts at the receiver has been ignored to make the analysis more concise, as it has been well studied and is out of the main focus of our work.

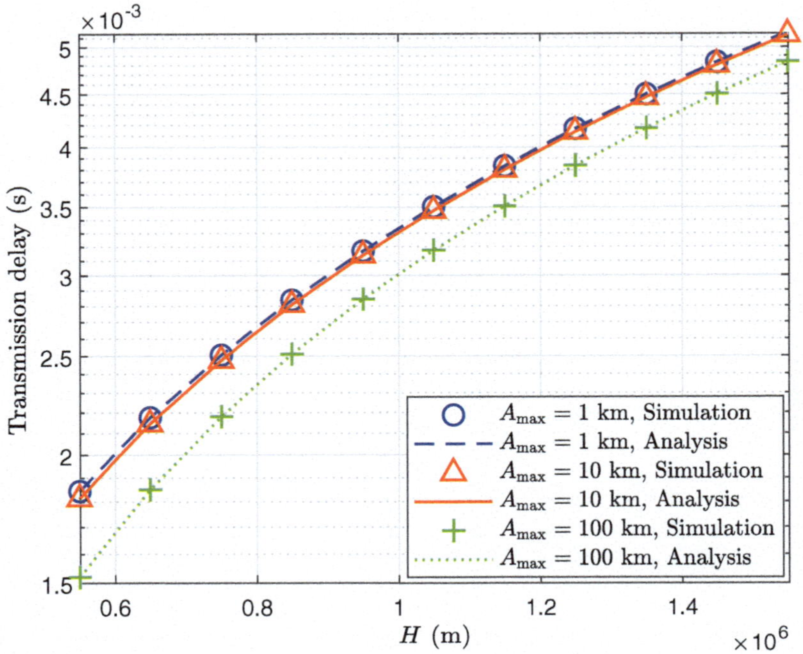


**Fig. 5.15** The transmission delay over satellite relay link

#### 5.2.1.1 The Uplink

For the uplink, we consider two types of channels, i.e., the channels between the aerial transmitters and aerial relays and the channels between aerial transmitters and the satellite.

*(1) Multi-Access Mechanism at* $S/R_1$*:* As there are a group of aerial transmitters trying to deliver their information to $D$ via the help of $S/R_1$, the multi-access mechanism adopted at $S/R_1$ is a key factor affecting the transmission performance of the uplink.

In this section, we assume that these aerial transmitters share the same frequency by using CDMA. In this case, the received signal at $S/R_1$ will suffer from the MUI arising from other aerial transmitters, which will be discussed in the following analysis.

*(2) Deployment of Aerial Transmitters:* In this section, we exploit the Matérn hard-core point process (MHCPP) type-II to mimic and reflect the random deployment of practical aerial scenarios for the aerial transmitters.[11] As considered, the aerial transmitters are randomly deployed in the ball space, denoted as $\mathcal{V}$. Each candidate

[11] The MHCPP is adopted here is to accurately reflect and model the practical deployment of aerial terminals, as the aerial terminals cannot be too close to each other to promise their safety or to hold their own serving space.

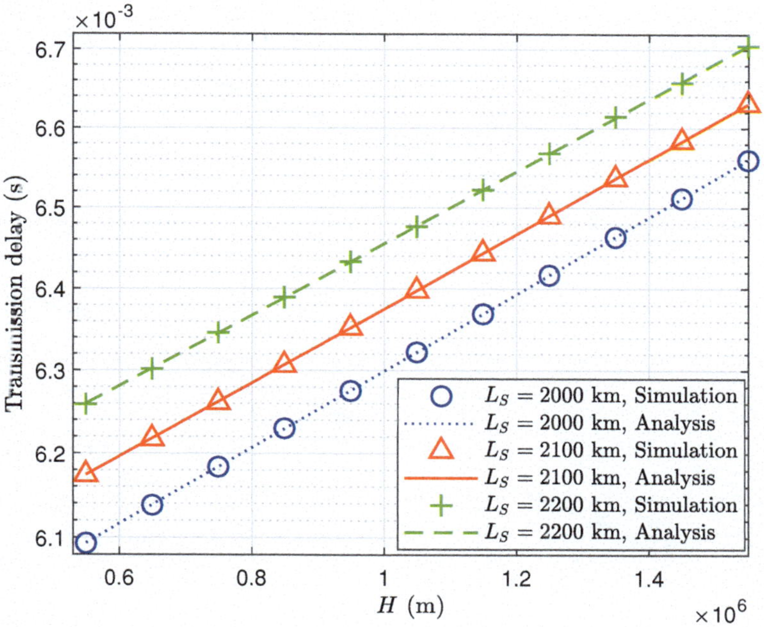


**Fig. 5.16** The e2e delay versus $H$

point has a minimum distance $D_{\min}$ to maintain aviation safety. Before presenting the detailed distributions for the candidate points, we introduce the concept of the MHCPP type-II first as follows.

We define that the locations of the candidate points have a density $\lambda_A$ and each candidate point has the minimum distance $D_{\min}$. Then, we present the processes to build up a MHCPP, denoted as $\Phi_A$, within three steps.

- In the first step, we follows the homogeneous Poisson point processes (HPPP) to initially generate the candidate points. Hence, the candidate points are randomly chosen from the space $\mathcal{V}$ with the density $\lambda_P$. If we denote the number of the candidate points as $N_P$, the probability mass function of the Poisson distribution with a mean $\lambda_P V$, when the number of the candidate points equals to $s$, can be expressed as $\Pr\{N_P = s\} = \frac{(\lambda_P V)^s}{s!} \exp(-\lambda_P V)$, where $V$ is the volume of space $\mathcal{V}$.
- In the second step, we generate an independent mark for each candidate point. The value of the marks is chosen based on a uniform distribution with a range from [0, 1].
- In the final step, we only choose the point with the smallest mark within a small space with the distance $D_{\min}$, while the others are eliminated. More specifically for a certain point $K$, we choose $K$ as the center of a small spherical repulsion space with the radius $D_{\min}$. If there are other candidate points in the spherical repulsion space, we compare the mark value from one to another. We only keep the point

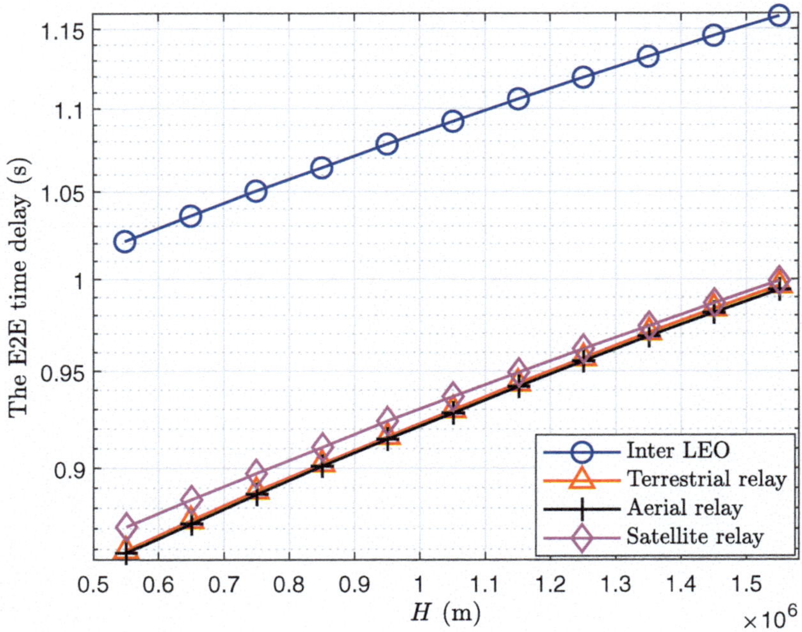


**Fig. 5.17** The e2e OP versus $H$

with the smallest mark in each spherical repulsion space. Then, we move to another candidate point to do the same processes until each spherical repulsion space has only one point. Based on this process, we have the relationship of densities, i.e., $\lambda_A$ and $\lambda_P$, which is expressed as $\lambda_A = \frac{1-\exp(-\frac{4}{3}\pi D_{\min}^3 \lambda_P)}{\frac{4}{3}\pi D_{\min}^3}$.

*(3) Channel Fading:* Though LoS propagation plays a leading role over air-to-air/space links, the path impairments in the aeronautical mobile-satellite communications include surface reflection (multi-path) effects [10]. Thus, without loss of generality and facilitating the following analysis, we assume that all uplink channels suffer independent and identically distributed (i.i.d.) Nakagami-$m$ fading, which covers the statistical characteristics of LoS propagation. Hence, the expressions of the PDF and CDF are presented as

$$f_{|h_{A_m X}|^2}(x) = \left(\frac{m_n}{\Omega}\right)^{m_n} \frac{x^{m_n-1}}{(m_n-1)!} \exp\left(-\frac{m_n}{\Omega}x\right) \tag{5.45}$$

and

$$F_{|h_{A_m X}|^2}(x) = 1 - \sum_{i=0}^{m_n-1} \left(\frac{m_n}{\Omega}\right)^{m_n-i-1} \frac{x^{m_n-i-1}}{(m_n-i-1)!} \times \exp\left(-\frac{m_n}{\Omega}x\right), \tag{5.46}$$

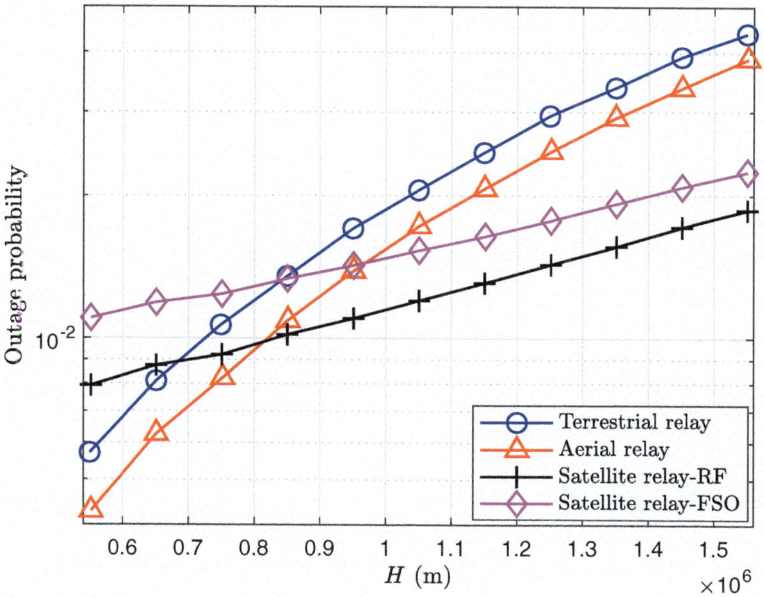


**Fig. 5.18** The e2e delay versus $q - p$

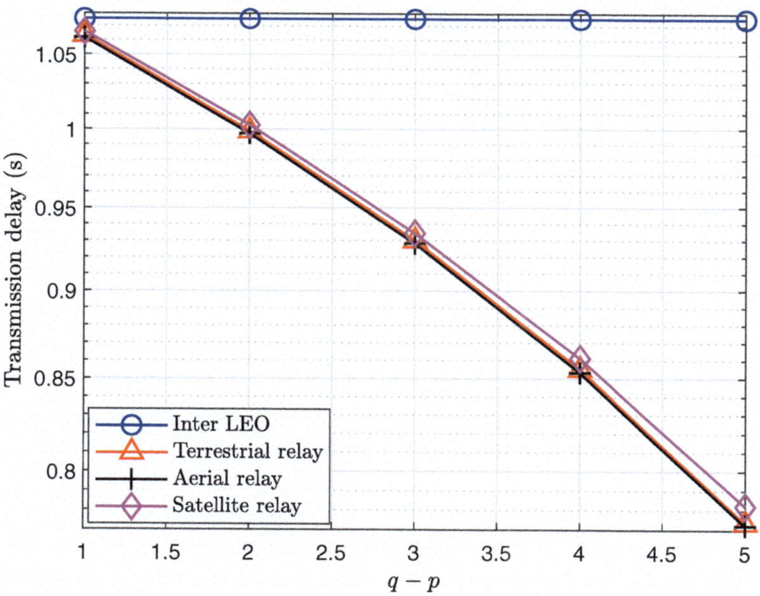


**Fig. 5.19** The e2e OP versus $q - p$

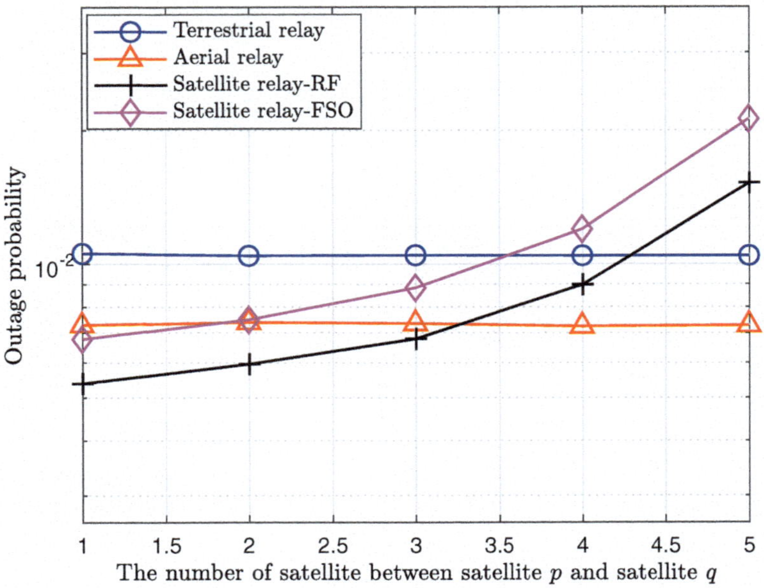


**Fig. 5.20** The e2e delay versus $\epsilon_A$

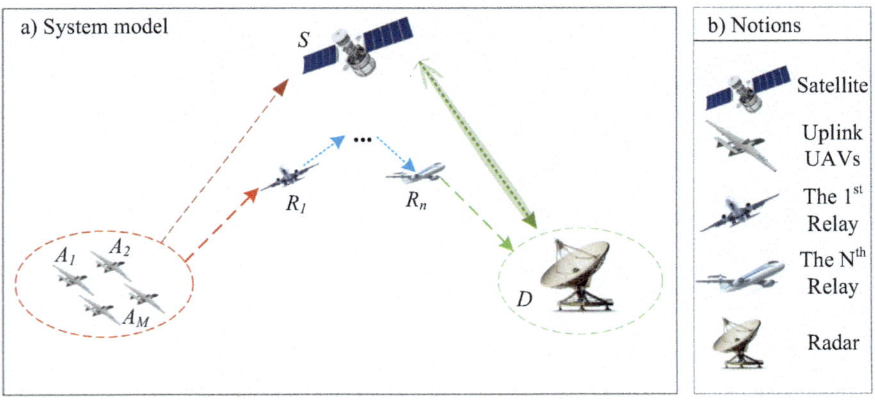


**Fig. 5.21** Illustration of the cooperative SATN system: **a** The cooperative communications with the aid of a series of aerial relays and a satellite; and **b** Notions

respectively, where $X \in \{S, R_1\}$ to present the satellite and the first aerial relay, respectively, $|h_{A_m X}|^2$ is the channel gain of the channel from the aerial transmitter $A_m$ to the relay or the satellite, $X$, and $m_n$ and $\Omega$ is the parameters of Nakagami-$m$ fading channels.

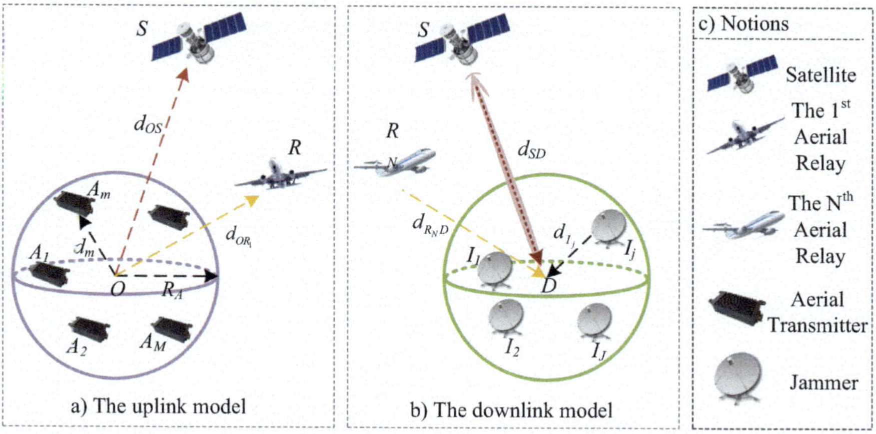


**Fig. 5.22** Illustration of the cooperative SATN system: **a** The uplink channels from the aerial sources to the satellite/the aerial relays; **b** The downlink channels from the satellite/ the aerial relays to the terrestrial destination; and **c** Notions

*(4) Signal Model:* We consider a practical satellite-aerial-terrestrial scenario where the aerial transmitters suffer the interference from each other. Hence, the signal-to-interference-plus-noise ratio (SINR) for an aerial transmitter $A_m$ to $X \in \{S, R_1\}$ can be expressed as

$$\gamma_{A_m X} = \frac{P_{A_m} \left|h_{A_m X}\right|^2 d_{A_m X}^{-\alpha_X}}{\sigma^2 + \sum\limits_{k=1\&k\neq m}^{M} P_{A_k} \left|h_{A_k X}\right|^2 d_{A_k X}^{-\alpha_X}} \approx \frac{P_{A_m} \left|h_{A_m X}\right|^2 d_{A_m X}^{-\alpha_X}}{\sum\limits_{k=1\&k\neq m}^{M} P_{A_k} \left|h_{A_k X}\right|^2 d_{A_k X}^{-\alpha_X}}, \tag{5.47}$$

where $P_{A_m}$ is the transmit power at $A_m$, $d_{A_m X}$ is the distance from $A_m$ to the relay or the satellite $X$, $\alpha_X$ is the path loss exponent for the channel to the relay or the satellite, and $\sigma^2$ is the strength of additive white Gaussian noise (AWGN).

#### 5.2.1.2 The Downlink

In this work, we assume that we exploit the DF relay scheme to process and transmit the signals. Over the downlink, the signals from the aerial transmitters have been decoded and re-transmitted by the satellite or the aerial relays. At the terrestrial destination, we consider that there are a group of near-ground aerial jammers, denoted as $I_j$ with $j \in \mathbb{Z}^+$ and $j \in [1, \ldots, J]$ $(J \geq 1)$, interfering the signals transmitted from the satellite or the aerial relay.

*(1) Deployment of Aerial Jammers:* We consider that these aerial jammers follow a Poisson point process (PPP),[12] which are uniformly distributed in a hemisphere, $\mathcal{V}_h$, whose bottom is on the ground with the radius $R_I$ and the original, $D$. The number of the aerial jammers is Poisson distributed with density $\lambda_J$, i.e., $\Pr\{N = k\} = \left(\mu_{\mathcal{V}_h}^k / k!\right) \exp\left(-\mu_{\mathcal{V}_h}\right)$, where $\mu_{\mathcal{V}_h} = \frac{2\pi {R_I}^3}{3}\lambda_J$ is the mean measure.

Define the distance from a jammer $I_j$ to the center of the hemisphere, $D$, as $d_{I_j}$. The PDF of $d_{I_j}$ is expressed as

$$f_{d_{I_j}}(x) = \begin{cases} 0, & x < 0; \\ \frac{3x^2}{R_I^3}, & 0 \leqslant x \leqslant R_I; \\ 0, & x > R_I \end{cases}. \tag{5.48}$$

*(2) Channel Fading:* For the $S - D$ and $R_N - D$ links, we consider the Shadowed-Rician model, which has been widely utilized by the researches in satellite/aerial-terrestrial communication area. We denote the channel gains between $S$ and $D$, and $R_N$ and $D$ as $|h_{iD}|^2$ ($i \in \{S, R_N\}$), thus the PDF of $|h_{iD}|^2$ is expressed as [1]

$$f_{|h_{iD}|^2}(x) = \chi \exp(-\beta x)\, {}_1F_1(m_n; 1; \delta x), \tag{5.49}$$

where $\chi = \left(\frac{2bm_n}{2bm_n+\Theta}\right)^{m_n} \Big/ (2b)$, $\beta = \frac{1}{2b}$, $\delta = \frac{\Theta}{2b(2bm_n+\Theta)}$, $\Theta$ and $2b$ are the average power for the line of sight component and the multi-path component, respectively, $m_n$ is the fading parameters based on the Nakagami-$m$ fading channels, and ${}_1F_1(\cdot;\cdot;\cdot)$ is the confluent hypergeometric function of the first kind.[13]

Moreover, for the jamming links between the aerial jammers and $D$, without loss of generality, it is also assumed that all jamming channels suffer i.i.d. Nakagami-$m$ fading to facilitate the following analysis since LoS propagation always plays a key role in aerial-to-ground transmission scenarios.

*(3) Signal Model:* For the $S - D$ and $R_N - D$ links, the SINR of $|h_{iD}|^2$ ($i \in \{S, R_N\}$) is

$$\gamma_{iD} = \frac{P_S |h_{iD}|^2 d_{iD}^{-\alpha_S}}{\sum_{j=1}^{J} P_j |h_{jD}|^2 d_{I_j}^{-\alpha_S} + \sigma^2} \approx \frac{P_S |h_{iD}|^2 d_{iD}^{-\alpha_S}}{\sum_{j=1}^{J} P_j |h_{jD}|^2 d_{I_j}^{-\alpha_S}}, \tag{5.50}$$

---

12 Here, for the downlink, the interference may not only come from the hostile neighboring terminals distributed in the considered space but also be the ones which are non-hostile and do not locate in the considered interfering space. For the later ones, the interferer may be far away from the receiver, but we can cast it into the considered hemisphere region by varying the transmission power. In this way, the interferer may be coincide in the hemisphere. Thus, the aerial jammers are modeled with normal PPP without restriction on the inter-node distance to reflect the worst case possibly existing in the practice .

13 The altitude dependent shadowing can be analysed by converting the fading parameters. Here, the effect of elevation angle on the Shadowed-Rician model has been ignored for simplification purpose, which have been well studied and is out of the main focus of our work.

where $P_i$ is the transmit power at $i$, $P_j$ is the transmit power at the $j$th aerial jammer, $d_{iD}$ is the distance between $i$ and $D$, and $\alpha_S$ is the path-loss exponent.

#### 5.2.1.3 Inter-Aerial Relaying Links

*(1) Channel Fading:* In this work, it is assumed that all inter-aerial relaying links suffer Nakagami-$m$ fading channels, which covers typical LoS propagation scenarios. Hence, the expressions of PDF and the CDF of the channel power gain between the $n$th ($n = 1, \ldots, N-1$) and the $(n+1)$th aerial relays, $\left|h_{R_n R_{n+1}}\right|^2$, are presented as

$$f_{\left|h_{R_n R_{n+1}}\right|^2}(x) = \left(\frac{m_n}{\Omega}\right)^{m_n} \frac{x^{m_n-1}}{(m_n-1)!} \exp\left(-\frac{m_n}{\Omega}x\right). \tag{5.51}$$

*(2) Signal Model:* Next, we present the SNR over the link between the $n$th ($n = 1, \ldots, N-1$) and the $(n+1)$th aerial relays as $\gamma_{R_n R_{n+1}} = \frac{P_{R_n}\left|h_{R_n R_{n+1}}\right|^2}{d_{R_n R_{n+1}}^{\alpha_{R_N}}\sigma^2}$, where $P_{R_n}$ is the transmit power at the $n$th aerial relay, $d_{R_n R_{n+1}}$ is the distance between the $n$th and the $(n+1)$th aerial relays, and $\alpha_{R_n}$ is the path-loss exponent.

### 5.2.2 *Outage Analysis of the Uplink*

For simplification, assume that all the aerial transmitters except the target transmitter share the same transmit power with $P_{A_k} = P_{A_u}$ and denote $m_i = m_n - i - 1$.

**Lemma 5.1** *The OP for the received signal at $X$ ($X \in \{S, R_1\}$), which is transmitted by the mth aerial transmitter, can be derived as*

$$P_{out}^{A_m X} = 1 - \sum_{i=0}^{m_n-1} \frac{(-1)^{m_i}}{(m_i)!} \mathbb{E}_{d_{A_m X}} \left\{ s^{m_i} \frac{\mathrm{d}^{m_i} \mathbb{E}_I \left\{\exp(sI)\right\}}{\mathrm{d}s^{m_i}} \right\}, \tag{5.52}$$

*where* $I = \sum\limits_{k=1\&k\neq m}^{M} \left|h_{A_k X}\right|^2 d_{A_k X}^{-\alpha_X}$ *and* $s = -\frac{m_n d_{A_m X}^{\alpha_X} \gamma_{out} P_{A_u}}{P_{A_m}\Omega}$.

***Proof*** Please refer to Appendix A. □

To obtain a closed-form expression for (5.52), $\frac{\mathrm{d}^{m_i}\mathbb{E}_I\{\exp(sI)\}}{\mathrm{d}s^{m_i}}$ should be calculated. Thus, we need to first derive $\mathbb{E}_I\left\{\exp(sI)\right\}$.

**Lemma 5.2** $\mathbb{E}_I\left\{\exp\left(sI\right)\right\}$ *can be expressed as*

$$\mathbb{E}_I\left\{\exp\left(sI\right)\right\} = \exp\left[\lambda_A V_1 (D_2 - 1)\right], \tag{5.53}$$

*where* $V_1 = \frac{4\pi}{3}(R_A^3 - D_{\min}^3)$ *and* $D_2 = \mathbb{E}_{d_{A_k X}}\left\{\left(\frac{\frac{m_n}{\Omega s} d_{A_k X}^{-\alpha_X}}{\frac{m_n}{\Omega s} d_{A_k X}^{-\alpha_X} - 1}\right)^{m_n}\right\}$.

***Proof*** Please refer to Appendix B □

**Lemma 5.3** *The OP is*

$$\begin{aligned} P_{out}^{A_m X} &= 1 - \sum_{i=0}^{m_n - 1} \frac{1}{(m_i)!} \mathbb{E}_{d_{A_m X}} \left\{ \left( \frac{m_n d_{A_m X}^{\alpha_X} \gamma_{out} P_{A_u}}{P_{A_m} \Omega} \right)^{m_i} \right. \\ &\quad \times \exp\left[\lambda_A V_1 (D_2 - 1)\right] \left[ 1 + \mathbb{1}\{m_i > 0\} \sum_{k=1}^{m_i} B_{m_i, k} \right. \\ &\quad \left. \left. \times \left( \lambda_A V_1 D_2^{(1)}, \ldots, \lambda_A V_1 D_2^{(m_i - k + 1)} \right) \right] \right\}, \end{aligned} \tag{5.54}$$

*where* $\mathbb{1}\{\cdot\}$ *is the indicator function and* $D_2^{(k)}$ $(k = 0, 1, \ldots, m_i)$ *is*

$$D_2^{(k)} = \frac{(m_n + k - 1)!}{(m_n - 1)!} \mathbb{E}_{d_{A_k X}} \left\{ \left( \frac{\Omega}{m_n d_{A_k X}^{\alpha_X}} \right)^k \left( \frac{d_{A_k X}^{\alpha_X}}{d_{A_k X}^{\alpha_X} + \frac{d_{A_m X}^{\alpha_X} \gamma_{out} P_{A_u}}{P_{A_m}}} \right)^{m_n + k} \right\}. \tag{5.55}$$

***Proof*** Please refer to Appendix C. □

#### 5.2.2.1 Uplink with the Satellite

When the aerial transmitters communicate with the satellite, we assume that all the aerial transmitters have the same transmission distance $d_{OS}$.[14] By substituting $d_{A_m X} = d_{OS}$ and $d_{A_k X} = d_{OS}$ into (5.54), we can get the OP of the uplink from the aerial transmitter cluster to the satellite as

[14] This assumption is reasonable since the distance to the satellite (normally on the order of hundreds/thousands of km) is quite larger than that to each other (normally on the order of km to tens of km).

$$P_{out}^{OS} = 1 - \sum_{i=0}^{m_n-1} \frac{1}{(m_i)!} \left( \frac{m_n d_{OS}^{\alpha_S} \gamma_{out} P_{A_u}}{P_{A_m} \Omega} \right)^{m_i} \exp[\lambda_A V_1 (D_2 - 1)]$$
$$\times \left[ 1 + \mathbb{1}\{m_i > 0\} \sum_{k=1}^{m_i} B_{m_i,k} \left( \lambda_A V_1 D_2^{(1)}, ..., \lambda_A V_1 D_2^{(m_i-k+1)} \right) \right], \quad (5.56)$$

where $D_2^{(k)}$ $(k = 0, 1, ..., m_i)$ is $D_2^{(k)} = \frac{(m_n+k-1)!}{(m_n-1)!} \left( \frac{\Omega}{m_n d_{OS}^{\alpha_S}} \right)^k \left( \frac{1}{1+\frac{\gamma_{out} P_{A_u}}{P_{A_m}}} \right)^{m_n+k}$.

#### 5.2.2.2 Uplink with the Aerial Relay

In this case, we suppose the transmitter is located in the center of the aerial transmitter cluster and the distance from the cluster center to the aerial relay is a constant, namely, $d_{A_m R_1} = d_{O R_1}$.

To get $D_2^{(k)}$, we should derive the PDF of $d_{A_k R_1}$.

**Lemma 5.4** *The PDF of* $d_{A_k R_1}^2$ *is*

$$f_{d_{A_k R_1}^2}(x) = \frac{\pi [R_A^2 - \tau(x)^2]}{2 V_1 d_{O R_1}}, \quad (5.57)$$

*where* $\tau(x) = \max\{D_{\min}, |\sqrt{x} - d_{O R_1}|\}$ *and* $d_{\max}^d = (d_{O R_1} - R_A)^2 \le x \le (d_{O R_1} + R_A)^2 = d_{\max}^u$.

***Proof*** Please refer to Appendix D. □

Then, by denoting $\Lambda = \frac{d_{O R_1}^{\alpha_{R_1}} \gamma_{out} P_{A_u}}{P_{A_m}}$, $D_2^{(k)}$ in (5.90) can be achieved as

$$D_2^{(k)} = \frac{(m_n + k - 1)!}{(m_n - 1)!} \mathbb{E}_{d_{A_k X}} \left\{ \left( \frac{\Omega}{m_n d_{A_k X}^{\alpha_{R_1}}} \right)^k \left( \frac{d_{A_k X}^{\alpha_{R_1}}}{d_{A_k X}^{\alpha_{R_1}} + \frac{d_{A_m X}^{\alpha_{R_1}} \gamma_{out} P_{A_u}}{P_{A_m}}} \right)^{m_n+k} \right\}$$
$$= \frac{(m_n + k - 1)!}{(m_n - 1)!} \left( \frac{\Omega}{m_n} \right)^k \int_{d_{\max}^d}^{d_{\max}^u} \frac{x^{\frac{\alpha_{R_1} m_n}{2}}}{\left( x^{\frac{\alpha_{R_1}}{2}} + \Lambda \right)^{m_n+k}} f_{d_{A_k R_1}^2}(x) dx$$
$$= \frac{\pi (m_n + k - 1)!}{2 V_1 d_{O R_1} (m_n - 1)!} \left( \frac{\Omega}{m_n} \right)^k \int_{d_{\max}^d}^{d_{\max}^u} \frac{x^{\frac{\alpha_{R_1} m_n}{2}} [R_A^2 - \tau(x)^2]}{\left( x^{\frac{\alpha_{R_1}}{2}} + \Lambda \right)^{m_n+k}} dx \quad (5.58)$$

When $\tau(x) = D_{\min}$ which means $D_{\min} > |\sqrt{x} - d_{O R_1}|$, we can get $d_{\min}^d = (d_{O R_1} - D_{\min})^2 < x < (d_{O R_1} + D_{\min})^2 = d_{\min}^u$ and $R_A^2 - \tau(x)^2 = R_A^2 - D_{\min}^2$.

When $\tau(x) = |\sqrt{x} - d_{OR_1}|$ which indicates $|\sqrt{x} - d_{OR_1}| > D_{\min}$, thus $(d_{OR_1} - R_A)^2 < x < (d_{OR_1} - D_{\min})^2$, $(d_{OR_1} + D_{\min})^2 < x < (d_{OR_1} + R_A)^2$, and $R_A^2 - \tau(x)^2 = R_A^2 - d_{OR_1}^2 - x + 2d_{OR_1}\sqrt{x}$.

By using [2, Eq. (3.194.1)], $D_2^{(k)}$ can be obtained as

$$\begin{aligned} D_2^{(k)} =& \frac{\pi(m_n+k-1)!}{2V_1 d_{OR_1}(m_n-1)!}\left(\frac{\Omega}{m_n}\right)^k \Lambda^{-m_n-k}\Bigg\{(R_A^2 - D_{\min}^2)\mathcal{F}\left(\frac{2}{\alpha_{R_1}}, d_{\min}^u, d_{\min}^d, k\right) \\ &+ (R_A^2 - d_{OR_1}^2)\left[\mathcal{F}\left(\frac{2}{\alpha_{R_1}}, d_{\min}^d, d_{\max}^d, k\right) + \mathcal{F}\left(\frac{2}{\alpha_{R_1}}, d_{\max}^u, d_{\min}^u, k\right)\right] \\ &- \left[\mathcal{F}\left(\frac{4}{\alpha_{R_1}}, d_{\min}^d, d_{\max}^d, k\right) + \mathcal{F}\left(\frac{4}{\alpha_{R_1}}, d_{\max}^u, d_{\min}^u, k\right)\right] \\ &+ 2d_{OR_1}\left[\mathcal{F}\left(\frac{3}{\alpha_{R_1}}, d_{\min}^d, d_{\max}^d, k\right) + \mathcal{F}\left(\frac{3}{\alpha_{R_1}}, d_{\max}^u, d_{\min}^u, k\right)\right]\Bigg\}, \end{aligned} \tag{5.59}$$

where $\mathcal{F}(a,b,c,k) = \frac{2}{(m_n+a)\alpha_{R_1}}\Bigg[b^{\frac{(m_n+a)\alpha_{R_1}}{2}} {}_2F_1\left(m_n+k, m_n+a; m_n+a+1; -\frac{b^{\frac{\alpha_{R_1}}{2}}}{\Lambda}\right)$ $-c^{\frac{(m_n+a)\alpha_{R_1}}{2}} {}_2F_1\left(m_n+k, m_n+a; m_n+a+1; -\frac{c^{\frac{\alpha_{R_1}}{2}}}{\Lambda}\right)\Bigg]$ and the detailed derivation of (5.59) is given in Appendix E.

Substituting (5.59) and $d_{A_m X} = d_{OR_1}$ into (5.54), we can get the OP over the uplink from the aerial transmitter to the first aerial relay as

$$\begin{aligned} P_{out}^{OR_1} =& 1 - \sum_{i=0}^{m_n-1} \frac{1}{(m_i)!}\left(\frac{m_n d_{OR_1}^{\alpha_X} \gamma_{out} P_{A_u}}{P_{A_m}\Omega}\right)^{m_i} \exp\left[\lambda_A V_1(D_2 - 1)\right] \\ &\times \left[1 + \mathbb{K}\{m_i > 0\} \sum_{k=1}^{m_i} B_{m_i,k}\left(\lambda_A V_1 D_2^{(1)}, \ldots, \lambda_A V_1 D_2^{(m_i-k+1)}\right)\right]. \end{aligned} \tag{5.60}$$

Interestingly, comparing (5.56) and (5.60), we can find that the closed-form analytical expression for the OP over the uplink from the aerial transmitter to the satellite is similar to that for the OP over the uplink from the aerial transmitter to the aerial relay. Thus, in Sect. 5.2.6, we will study the outage performance of the uplink by merging these two cases into one for simplicity.

#### 5.2.2.3 Asymptotic Outage Performance of the Uplink

In this part, we will present the asymptotic analysis of the outage performance of the uplink of the considered system. In the following, we derive the approximate expressions assuming $\lambda = \frac{P_{A_m X}}{N_0} \to \infty$.

Adopting the series representations of the exponential function, $\exp\left(-\frac{m_n x}{\Omega}\right) = \sum_{n=0}^{\infty} \frac{\left(-\frac{m_n x}{\Omega}\right)^n}{n!}$, and keeping the first two terms while ignoring the higher order term, the asymptotic CDF of $\lambda\left|h_{A_m X}\right|^2$ can be obtained as

$$F^{\infty}_{\lambda\left|h_{A_m X}\right|^2}(x) = 1 - \sum_{i=0}^{m_n-1} \left(\frac{m_n}{\Omega}\right)^{m_n-i-1} \frac{x^{m_n-i-1}}{(m_n-i-1)!}\left(1-\frac{m_n x}{\Omega\lambda}\right). \tag{5.61}$$

**Lemma 5.5** *The asymptotic OP can be presented as*

$$P^{\infty}_{out} = 1 - \sum_{i=0}^{m_n-1} \frac{(-1)^{m_i}}{(m_i)!}\left\{\mathbb{E}_{d_{A_m X}}\left\{s^{m_i}\right\}\mathbb{E}_I\left\{I^{m_i}\right\} - \mathbb{E}_{d_{A_m X}}\left\{s^{m_i+1}\right\}\mathbb{E}_I\left\{I^{m_i+1}\right\}\right\}, \tag{5.62}$$

*where*

$$\mathbb{E}_I\{I\} = \sum_{M=0}^{\infty} \frac{(\lambda_A V_1)^M}{(M-1)!}\exp(-\lambda_A V_1)\left\{1 - \sum_{i=0}^{m_n-1}\left(\frac{m_n}{\Omega}\right)^{m_n-i-1} \times \frac{x^{m_n-i}}{(m_n-i-1)!}\exp\left(-\frac{m_n}{\Omega}x\right)\right\}\mathbb{E}_{d_{A_k X}}\left\{d^{-\alpha_X}_{A_k X}\right\} \tag{5.63}$$

*and*

$$\mathbb{E}_{d_{A_m X}}\{s\} = \mathbb{E}_{d_{A_m X}}\left\{-\frac{m_n d^{\alpha_X}_{A_m X}\gamma_{out} P_{A_u}}{P_{A_m}\Omega}\right\}. \tag{5.64}$$

***Proof*** Please refer to Appendix F. □

When the aerial transmitters communicate with the satellite, $\mathbb{E}_I\{I\}$ can be further derived as

$$\mathbb{E}_I\{I\} = \sum_{M=0}^{\infty} \frac{(\lambda_A V_1)^M}{(M-1)!}\exp(-\lambda_A V_1)\left\{1 - \sum_{i=0}^{m_n-1}\left(\frac{m_n}{\Omega}\right)^{m_n-i-1} \times \frac{x^{m_n-i}}{(m_n-i-1)!}\exp\left(-\frac{m_n}{\Omega}x\right)\right\} d^{-\alpha_X}_{A_k X}. \tag{5.65}$$

Thus, $P^{\infty}_{out}$ can be rewritten as

$$P_{out}^{\infty} = 1 - \sum_{i=0}^{m_n-1} \frac{(-1)^{m_i}}{(m_i)!} \Big\{ s^{m_i} \{\mathbb{E}_I\{I\}\}^{m_i} - s^{m_i+1} \{\mathbb{E}_I\{I\}\}^{m_i+1} \Big\}, \tag{5.66}$$

When the aerial transmitters communicate with the aerial relay, we can obtain $\mathbb{E}_{d_{A_k X}}\left\{d_{A_k X}^{-\alpha_X}\right\}$ as

$$\begin{aligned}
\mathbb{E}_{d_{A_k X}}\left\{d_{A_k X}^{-\alpha_X}\right\} &= \int_{d_{\max}^d}^{d_{\max}^u} x^{-\frac{\alpha_X}{2}} f_{d_{A_k R_1}^2}(x)\mathrm{d}x \\
&= \frac{\pi}{2V_1 d_{OR_1}} \int_{d_{\max}^d}^{d_{\max}^u} x^{-\frac{alpha_X}{2}} \left[R_A^2 - \tau^2(x)\right] \mathrm{d}x \\
&= \frac{\pi}{2V_1 d_{OR_1}} \Bigg\{ \left(R_A^2 - D_{\min}^2\right) \int_{d_{\min}^d}^{d_{\min}^u} x^{-\frac{\alpha_X}{2}} \mathrm{d}x \\
&\quad + \int_{d_{\max}^d}^{d_{\min}^d} \left(R_A^2 - d_{OR_1}^2\right) x^{-\frac{\alpha_X}{2}} - x^{-\frac{\alpha_X}{2}+1} + 2d_{OR_1} x^{-\frac{\alpha_X-1}{2}} \mathrm{d}x \\
&\quad + \int_{d_{\min}^u}^{d_{\max}^u} \left(R_A^2 - d_{OR_1}^2\right) x^{-\frac{\alpha_X}{2}} - x^{-\frac{\alpha_X}{2}+1} + 2d_{OR_1} x^{-\frac{\alpha_X-1}{2}} \mathrm{d}x \Bigg\} \\
&= \frac{\pi}{2V_1 d_{OR_1}} \Bigg\{ \frac{\left(R_A^2 - D_{\min}^2\right)}{-\frac{\alpha_X}{2}+1} \left({d_{\min}^u}^{-\frac{\alpha_X}{2}+1} - {d_{\min}^d}^{-\frac{\alpha_X}{2}+1}\right) \\
&\quad + \frac{\left(R_A^2 - d_{OR_1}^2\right)}{-\frac{\alpha_X}{2}+1} \left({d_{\min}^d}^{-\frac{\alpha_X}{2}+1} - {d_{\max}^d}^{-\frac{\alpha_X}{2}+1}\right) \\
&\quad - \frac{1}{-\frac{\alpha_X}{2}+2} \left({d_{\min}^d}^{-\frac{\alpha_X}{2}+2} - {d_{\max}^d}^{-\frac{\alpha_X}{2}+2}\right) \\
&\quad + \frac{2d_{OR_1}}{-\frac{\alpha_X-1}{2}+1} \left({d_{\min}^d}^{-\frac{\alpha_X-1}{2}+1} - {d_{\max}^d}^{-\frac{\alpha_X-1}{2}+1}\right) \\
&\quad + \frac{\left(R_A^2 - d_{OR_1}^2\right)}{-\frac{\alpha_X}{2}+1} \left({d_{\max}^u}^{-\frac{\alpha_X}{2}+1} - {d_{\min}^u}^{-\frac{\alpha_X}{2}+1}\right) \\
&\quad - \frac{1}{-\frac{\alpha_X}{2}+2} \left({d_{\max}^u}^{-\frac{\alpha_X}{2}+2} - {d_{\min}^u}^{-\frac{\alpha_X}{2}+2}\right) \\
&\quad + \frac{2d_{OR_1}}{-\frac{\alpha_X-1}{2}+1} \left({d_{\max}^u}^{-\frac{\alpha_X-1}{2}+1} - {d_{\min}^u}^{-\frac{\alpha_X-1}{2}+1}\right) \Bigg\}
\end{aligned} \tag{5.67}$$

Thus, we can obtain the asymptotic OP over the uplink from the aerial transmitter to the satellite by substituting (5.67) into (5.62).

### 5.2.3 Outage Analysis of the Downlink

The OP of the downlink depicted in Fig. 5.22b can be given as

$$P_{out}^{iD} = \Pr\{\gamma_{iD} \leq \gamma_{out}\}$$
$$\approx \Pr\left\{\frac{P_i|h_{iD}|^2 d_{iD}^{-\alpha_S}}{\sum_{j=1}^{J} P_j\left|h_{jD}\right|^2 d_{I_j}^{-\alpha_S}} \leq \gamma_{out}\right\}, \quad i \in \{S, R_N\}. \tag{5.68}$$

Moreover, as the focus of this section is to investigate the impacts of these spatial distribution aerial jammers on the outage performance of the downlink, here we also assume that the distance between the satellite/the last aerial relay and $D$ is fixed for the feasibility of the analysis. Namely, $d_{iD}$ $(i \in \{S, R_N\})$ is with a given value. Thus, we can rewrite the previous equation as

$$P_{out}^{iD} = \Pr\left\{\frac{Z}{Y} \leq \frac{d_{iD}^{\alpha_S}}{P_i}\gamma_{out}\right\}, \tag{5.69}$$

where $i \in \{S, R_N\}$, $Z = |h_{iD}|^2$, and $Y = \sum_{j=1}^{J} P_j\left|h_{jD}\right|^2 d_{I_j}^{-\alpha_S}$.

Recalling (5.49) and using the Kummer's transform of the hypergeometric function, we easily rewrite the PDF of $|h_{iD}|^2$ as

$$f_{|h_{iD}|^2}(x) = \sum_{k=0}^{m_n-1} \Psi(k)\, x^k \exp\left(-(\beta-\delta)\, x\right), \tag{5.70}$$

where $\Psi(k) = \delta^k \frac{(-1)^k}{(k!)^2} \chi (1-m_n)_k$ and $(\cdot)_n$ is the Pochhammer symbol. Then, the CDF of $|h_{iD}|^2$ can be presented as

$$F_{|h_{iD}|^2}(x) = \sum_{k=0}^{m_n-1} \Psi(k) \int_0^x t^k e^{-(\beta-\delta)t} dt$$
$$= \sum_{k=0}^{m_n-1} \frac{\Psi(k)}{(\beta-\delta)^{k+1}} \gamma\left(k+1, (\beta-\delta)\, x\right). \tag{5.71}$$

Further, using (5.69) and (5.71) and denote $\kappa = \frac{\gamma_{out} d_{iD}^{\alpha_S}}{P_i}(\beta - \delta)$, we can be obtain

$$\begin{aligned} P_{out}^{iD} &= \Pr\{\gamma_{iD} \le \gamma_{out}\} \\ &= \sum_{k=0}^{m_n-1} \frac{\Psi(k)}{(\beta-\delta)^{k+1}} \gamma\left(k+1, \frac{\gamma_{out} d_{iD}^{\alpha_S}}{P_i}(\beta-\delta) Y\right) \\ &= \sum_{k=0}^{m_n-1} \frac{\Psi(k)}{(\beta-\delta)^{k+1}} \gamma(k+1, \kappa Y). \end{aligned} \tag{5.72}$$

As shown in [8], $\gamma(k+1, x) < \Gamma(k+1)(1 - \exp(-\zeta x))^{k+1}$, $\zeta = (\Gamma(k+2))^{-\frac{1}{k+1}}$. Then, $P_{out}^{iD}$ can be approximated as

$$\begin{aligned} P_{out}^{iD} &\approx \sum_{k=0}^{m_n-1} \frac{\Psi(k)\Gamma(k+1)}{(\beta-\delta)^{k+1}} (1 - \exp(-\zeta\kappa Y))^{k+1} \\ &= \sum_{k=0}^{m_n-1} \frac{\Psi(k)\Gamma(k+1)}{(\beta-\delta)^{k+1}} \sum_{t=0}^{k+1} \binom{k+1}{t} (-1)^t \exp(-\zeta\kappa t Y). \end{aligned} \tag{5.73}$$

Furthermore, considering the randomness of $Y$, we can obtain

$$\begin{aligned} P_{out}^{iD} &= \sum_{k=0}^{m_n-1} \frac{\Psi(k)\Gamma(k+1)}{(\beta-\delta)^{k+1}} \sum_{t=0}^{k+1} \binom{k+1}{t} (-1)^t \mathbb{E}\left[\exp(-\zeta\kappa t Y)\right] \\ &= \sum_{k=0}^{m_n-1} \frac{\Psi(k)\Gamma(k+1)}{(\beta-\delta)^{k+1}} \sum_{t=0}^{k+1} \binom{k+1}{t} (-1)^t \mathscr{L}_Y(\zeta\kappa t), \end{aligned} \tag{5.74}$$

where $\mathscr{L}_Y(s)$ is the Laplace transform of random variable $Y$.

**Lemma 5.6** *Considering that all jamming channels between the aerial jammers and the terrestrial receiver, D, suffer i.i.d. Nakagmi-m fading, namely,* $f_{|h_{jD}|^2}(x) = \left(\frac{m_n}{\Omega}\right)^{m_n} \frac{x^{m_n-1}}{(m_n-1)!} \exp\left(-\frac{m_n}{\Omega}x\right)$, *$\mathscr{L}_Y(\zeta\kappa t)$ can be calculated as*

$$\begin{aligned} \mathscr{L}_Y(\zeta\kappa t) = &\exp\left(-\pi R_I \lambda_I \sum_{i=1}^{V} \sqrt{1-t_i^2}\,\omega_i \upsilon_i^2 \right. \\ &\left. \times \left(1 - \left(\frac{m_n}{\Omega}\right)^{m_n} \left(\frac{m_n}{\Omega} + \zeta\kappa t P_j \upsilon_i^{-\alpha_S}\right)^{-m_n}\right)\right), \end{aligned} \tag{5.75}$$

*where* $\upsilon_i = \frac{R_I(t_i+1)}{2}$, $t_i = \cos\left(\frac{2i-1}{2V}\pi\right)$, *and* $\omega_i = \frac{\pi}{V}$.

***Proof*** *Please refer to Appendix G.*

Therefore, the OP over the downlink can be achieved via inserting (5.75) into (5.74).

Adopting the series representations of the exponential function, $\exp\left(-\beta\frac{x}{\lambda_{dn}}\right) = \sum_{n=0}^{\infty}\frac{\left(-\beta\frac{x}{\lambda_{dn}}\right)^n}{n!}$, and keeping the first two terms while ignoring the higher order term, the asymptotic CDF and PDF of $\lambda_{dn} = P_i|h_{iD}|^2$ can be expressed as

$$f_{\lambda_{dn}}^{\infty}(x) = \sum_{k=0}^{m_n-1}\Psi(k)\,x^k\left(1-\frac{\beta-\delta}{\lambda_{dn}}x\right) \tag{5.76}$$

and

$$F_{\lambda_{dn}}^{\infty}(x) = \sum_{k=0}^{m_n-1}\frac{\Psi(k)}{k+1}x^{k+1} - \frac{\Psi(k)(\beta-\delta)}{(k+2)\lambda_{dn}}x^{k+2}. \tag{5.77}$$

Then, we can obtain $P_{out}^{\infty}$ as

$$P_{out}^{\infty} = \sum_{k=0}^{m_n-1}\frac{\Psi(k)}{k+1}(\kappa\mathbb{E}_Y\{Y\})^{k+1} - \frac{\Psi(k)(\beta-\delta)}{(k+2)\lambda_{dn}}(\kappa\mathbb{E}_Y\{Y\})^{k+2}, \tag{5.78}$$

where

$$\begin{aligned}
\mathbb{E}_Y\{Y\} &= \mathbb{E}_{|h_{jD}|^2, d_{I_j}}\left[\sum_{j=1}^{J}\frac{P_j|h_{jD}|^2}{d_{I_j}^{\alpha_S}}\right] \\
&= \sum_{j=1}^{J}\mathbb{E}_{d_{I_j}}\left[\underbrace{\int_0^{\infty}\frac{P_j|h_{jD}|^2}{d_{I_j}^{\alpha_S}}f_{|h_{jD}|^2}(x)\,dx}_{I_2}\right] \\
&= \sum_{j=1}^{J}\left(\lambda_I\int_{\mathbb{R}^3}I_2\left(d_{I_j}\right)dd_{I_j}\right),
\end{aligned} \tag{5.79}$$

where

$$\begin{aligned}
I_2 &= \left(\frac{m_n}{\Omega}\right)^{m_n}\frac{P_j}{d_{I_j}^{\alpha_S}(m_n-1)!}\int_0^{\infty}x^{m_n-1}\exp\left(-\frac{m_n}{\Omega}x\right)dx \\
&= \left(\frac{m_n}{\Omega}\right)^{-1}\frac{P_j}{d_{I_j}^{\alpha_S}(m_n-1)!}\Gamma(m_n+1),
\end{aligned} \tag{5.80}$$

and $\mathbb{R}^3$ is the 3D distribution space for these aerial jammers presented in Fig. 5.22b.

Then, $\mathbb{E}_Y\{Y\}$ can be finally derived as

$$\begin{aligned}\mathbb{E}_Y\{Y\} &= \sum_{j=1}^{J}\left(\lambda_I \int_{\mathbb{R}^3} I_2\left(d_{I_j}\right) dd_{I_j}\right)\\ &= \sum_{j=1}^{J}\left(-\pi R_I \lambda_I \sum_{i=1}^{V} \sqrt{1-t_i^2}\omega_i \upsilon_i^{2}\left(\frac{m_n}{\Omega}\right)^{-1} \frac{P_j \Gamma\left(m_n+1\right)}{\upsilon_i^{\alpha_S}\left(m_n-1\right)!}\right).\end{aligned} \tag{5.81}$$

Finally, the asymptotic OP over the downlink can be achieved by substituting (5.81) into (5.78).

## 5.2.4 *Outage of Inter-Aerial Relaying and e2e Links*

### 5.2.4.1 Outage Analysis of Inter-Aerial Relaying Links

Denoting $\Delta_{n,n+1} = \frac{d_{R_n R_{n+1}}^{\alpha_{R_N}} \sigma^2 \gamma_{out}}{P_{R_n}}$, one easily has the OP over the link between the $n$th ($n = 1, \ldots, N-1$) and the $(n+1)$th aerial relays as

$$\begin{aligned}P_{out}^{R_n R_{n+1}} &= \Pr\{\gamma_{R_n R_{n+1}} \le \gamma_{out}\}\\ &= \Pr\left\{\frac{P_{R_n}\left|h_{R_n R_{n+1}}\right|^2}{d_{R_n R_{n+1}}^{\alpha_{R_N}}\sigma^2} \le \gamma_{out}\right\}\\ &= 1 - \sum_{i=0}^{m_n-1}\left(\frac{m_n}{\Omega}\right)^{m_n-i-1} \frac{\Delta_{n,n+1}^{m_n-i-1}}{\left(m_n-i-1\right)!}\exp\left(-\frac{m_n}{\Omega}\Delta_{n,n+1}\right).\end{aligned} \tag{5.82}$$

Moreover, here we also assume that the transmission distance over each aerial relay hop is fixed to facilitate the analysis, as the main concern of this work is the impacts of the distribution of the aerial sources and jammers.

### 5.2.4.2 e2e OP Analysis

Using the analysis offered in previous sections, the e2e OP of the target system will be presented by considering the adopted relay types.

*(1) Satellite Relay Link:* As observed from Fig. 5.21 and considering DF scheme is adopted at the satellite, the e2e OP over satellite relay link can be easily written as

$$P_{out}^{SR} = 1 - \left(1 - P_{out}^{OS}\right)\left(1 - P_{out}^{SD}\right). \tag{5.83}$$

*(2) Aerial Relay Link:* Similar to the previous subsection, we can present the e2e OP over the aerial relay link as

$$P_{out}^{AR} = 1 - \left(1 - P_{out}^{OR_1}\right)\left(1 - P_{out}^{R_N D}\right)\prod_{n=1}^{N-1}\left(1 - P_{out}^{R_n R_{n+1}}\right). \tag{5.84}$$

### 5.2.5 *Numerical Results*

In this section, numerical results will be presented to investigate the performance of the considered cooperative satellite/aerial-terrestrial communication system. We run $10^6$ times of the realizations of the considered system and $10^6$ trials of Monte-Carlo simulations to model the randomness of the positions of the aerial transmitters and jammers and channel gains over each link.

Following the transmit power and the orbit altitude shown in [9, 10], in this subsection, unless otherwise explicitly specified, the parameters are set as follows: (1) Uplink: $d_{OS} = 300$ km, $R_A = 10$ km, $D_{\min} = 1$ km, $d_{OR_1} = 20$ km, $\alpha_S = 2$, $\alpha_X = 2$, $\gamma_{out} = -1$ dB, $\lambda_A = 1 \times 10^{-4}$, $m_n = 2$, and $\Omega = 1$; (2) Downlink: $d_{SD} = 300$ km, $R_I = 10$ km, $d_{R_N D} = 20$ km, $\alpha_S = 2$, $\gamma_{out} = -1$ dB, $\lambda_I = 1 \times 10^{-11}$, $m_n = 2$, $\Omega = 1$, $b = 1$ dB, $P_S = 30$ dBW, and $P_{A_u} = 30$ dBW.

#### 5.2.5.1 The Outage Performance Over the Uplink

Figure 5.23 presents the influence of $R_A$ on the outage performance of the uplink. One can see that the OP over the uplink gets worse as the size of the distribution space of the aerial transmitters increases. This observation can be easily understood, since a large distribution space of the aerial transmitters indicates that there will be more aerial transmitters accessing the satellite/the aerial relay simultaneously, resulting in increased MUI accordingly.

Furthermore, the relationship between the OP of the uplink and the transmission distances of the uplink is investigated in Fig. 5.25. As expected, the transmission distances of the uplink do not have a large impact on the outage performance over the uplink. Because the target transmission signal and corresponding self-interfering signal emitted by other aerial transmitters suffer from the same path loss when traveling over the same uplink.

Finally, the effects of the minimum safety distance among the aerial transmitters, $D_{\min}$, on the outage performance over the uplink are shown in Fig. 5.26. We can observe that $D_{\min}$ exhibits a positive impact on the OP over the uplink. In other words, a large $D_{\min}$ stands for a small OP over the uplink, which can be interpreted by the fact that for the given distribution space, a large $D_{\min}$ implies a low distribution density of the aerial transmitters and further low MUI at the satellite/the aerial relay. However, one can also see that such a positive effect is very weak as the OP curves

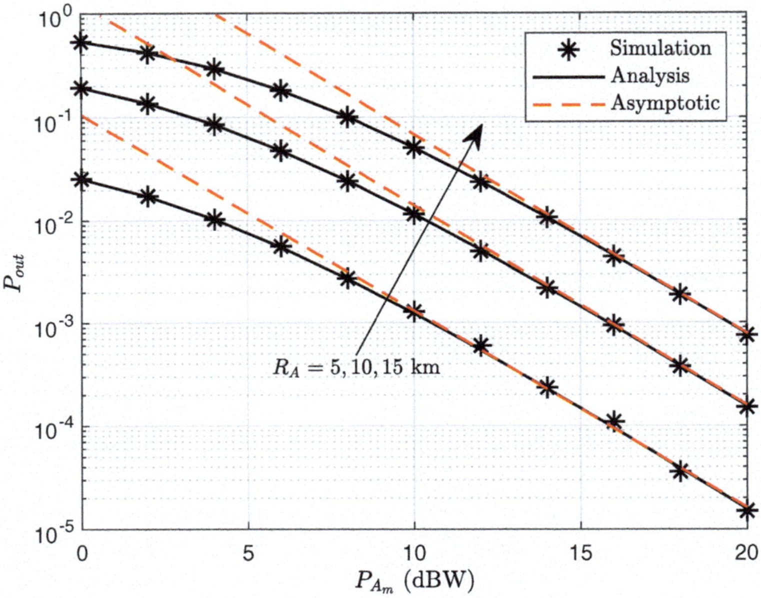


**Fig. 5.23** OP over the uplink for various $R_A$

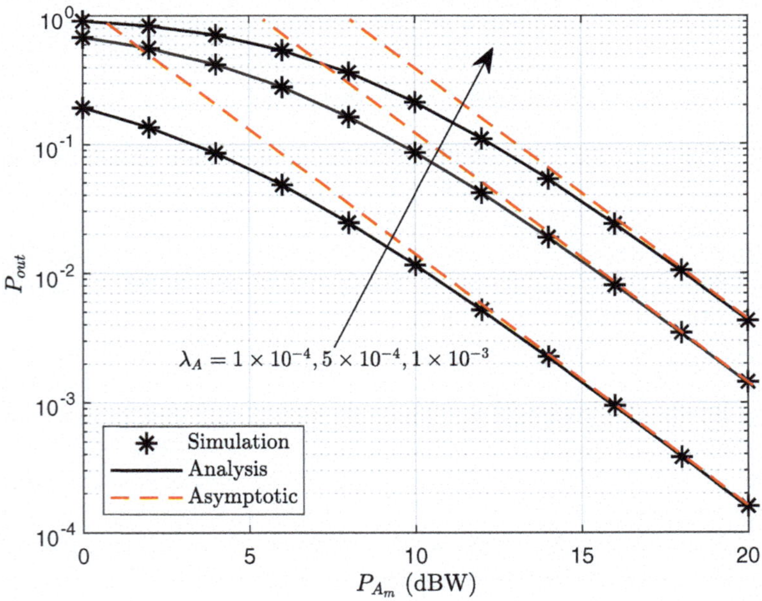


**Fig. 5.24** OP over the uplink for various $\lambda_A$

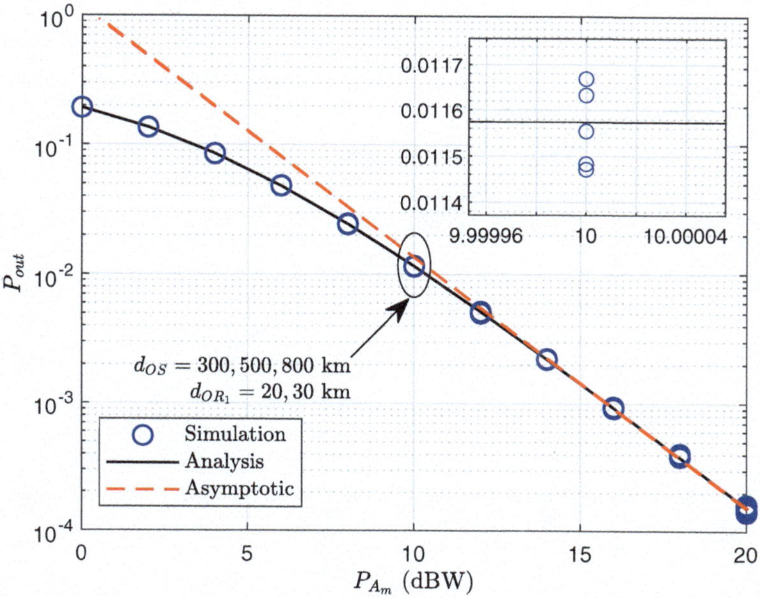


**Fig. 5.25** OP over the uplink for various $d_{OS}$ and $d_{OR_1}$

plotted in this figure overlap each other from a macroscopic view. Because the range of $D_{\min}$ to promise the safety of the aerial transmitters is relatively quite smaller than the size of their distribution space and the length of the uplinks.

Next, in Fig. 5.24, the impacts of the distribution density of the aerial transmitters on the OP over the uplink are investigated. It is noted from this figure that a large distribution density of the aerial transmitter incurs a reduced outage performance. Because more aerial transmitters will be brought by a large $\lambda_A$ in the given distribution space, causing large MUI at the satellite/the aerial relay.

#### 5.2.5.2 The Outage Performance Over the Downlink

In Figs. 5.27, 5.28 and 5.29, the OP over the downlink between $S/R_N$ and $D$ is depicted. $P_J$ exhibits an obvious negative impact on the outage performance over the downlink. Because a large $P_J$ represents a large interfering power at $D$.

In Fig. 5.27, the OP with different $d_{iD}$ ($i \in \{S, R_N\}$) is presented. The outage performance over the downlink degrades while the transmission distance over the downlink enlarges. More specifically, one can see that the OP over the downlink between the satellite and the terrestrial receiver gets worse when the link length increases from 300 km to 800 km. Similar observations can be achieved for the downlink between $R_N$ and $D$ when $d_{R_N D}$ reaches 30 km from 20 km.

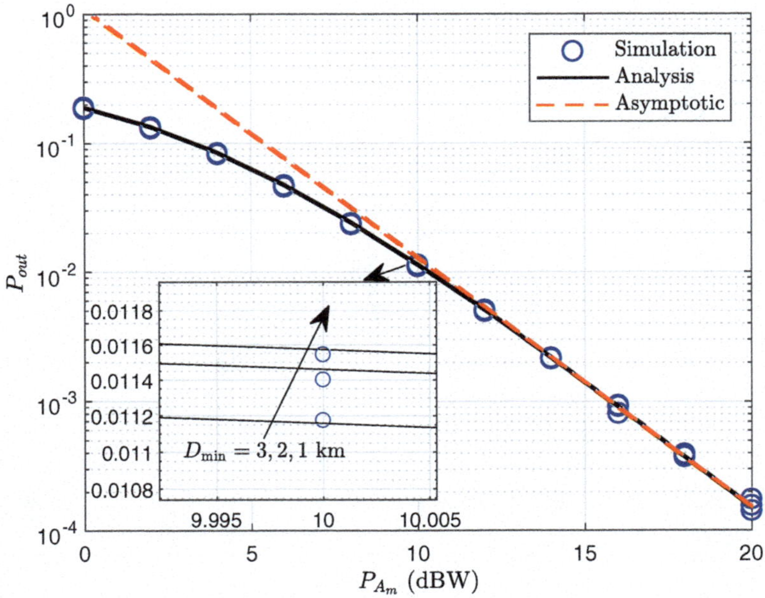


**Fig. 5.26** OP over the uplink for various $D_{\min}$

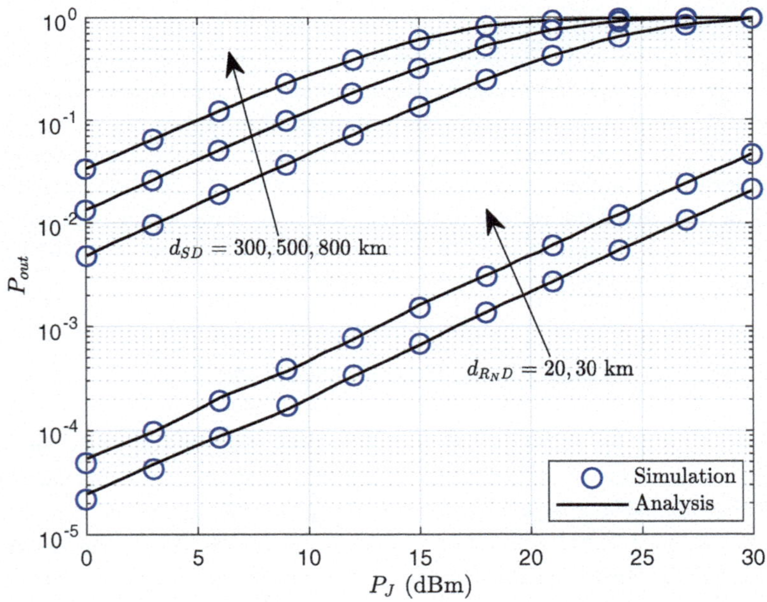


**Fig. 5.27** OP over the downlink for various $d_{iD}$

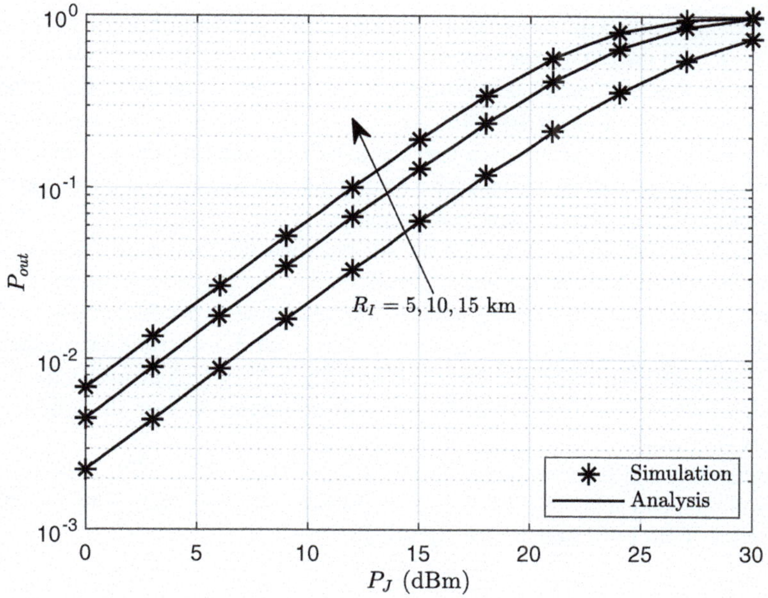


**Fig. 5.28** OP over the downlink for various $R_I$

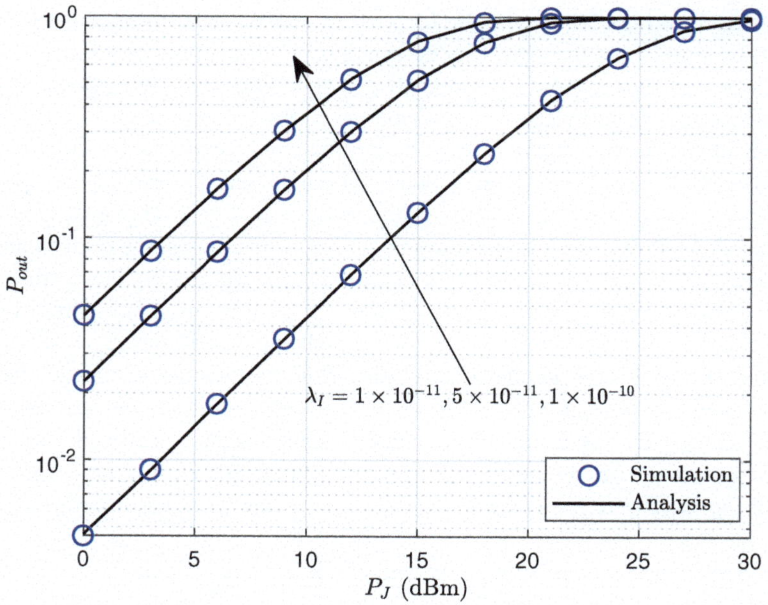


**Fig. 5.29** OP over the downlink for various $\lambda_I$

Figure 5.28 studied the influence of the size of the distribution space of the aerial jammers, $R_I$, on the outage performance over the downlink. The OP over the downlink enlarges while $R_I$ increases. For example, when $P_J = 10$ dBm, the OP of the downlink is on the order of $10^{-3}$ for $R_I = 5$ km, and that is on the order of $10^{-2}$ for $R_I = 15$ km. This observation can be easily explained by the fact that a large $R_I$ means a large distribution space and further denotes more aerial jammers, which finally incurs the degraded outage performance.

In Fig. 5.29, we studied the relationship between the outage performance and the distribution density of the aerial jammers, $\lambda_I$. For a given distribution space for the aerial jammers, a large $\lambda_I$ leads to worse outage performance since a large $\lambda_I$ implies more aerial jammers operate in the given 3D distribution space. This observation is similar to the one achieved from Fig. 5.23, which uncovers the impact of the distribution density of the aerial transmitters on the OP over the uplink in the previous subsection.

Moreover, it can be seen from Figs. 5.23, 5.24, 5.25, 5.26, 5.27, 5.28 and 5.29 that simulation and analysis results agree with each other very well, which verify the correctness of our proposed analytical models.

#### 5.2.5.3 The E2e Outage Performance

As illustrated in Figs. 5.30 and 5.31, for a given transmission distance between the aerial transmitters and the terrestrial destination, $N$ offers a positive impact on the e2e outage performance of the considered system via the aerial relay path, but such a positive influence gets weak while $N$ enlarges as flat bottoms can be seen from the OP lines plotted in both figures. This observation can be explained as follows: The transmission distance between every two neighboring aerial relays gets smaller and smaller when more aerial relays are introduced into the aerial relay link, leading to the improved OP over each aerial relay hop; then, when the outage performance of each aerial relay hop is improved to a certain degree, the e2e outage performance of the considered system will be only decided by the transmission quality over the uplink and the downlink shown in Fig. 5.22, which agrees with the expression of the e2e OP with aerial relay addressed by (5.84). Therefore, we can conclude: Though the e2e outage performance can be enhanced somewhat, bringing more aerial relays into the aerial relay link is not always a good choice for the considered system and optimal system design should be carried out to achieve a trade-off between the e2e OP and the increased system resource overhead arisen from the introduced aerial relays.

Moreover, another interesting finding here is that the e2e OP via the satellite relay link does not always outperform that via the aerial relay links because the e2e outage performance under these two cases varies with numerous system factors, e.g., the transmit power at the aerial transmitter/the satellite/the aerial relays, the number of aerial relays, the link distance among all terminals included in the considered system, etc. Thus, to realize an optimal considered system for a practical scenario setting, the selection of the relay link depends on the feasibility of achieving a balance between

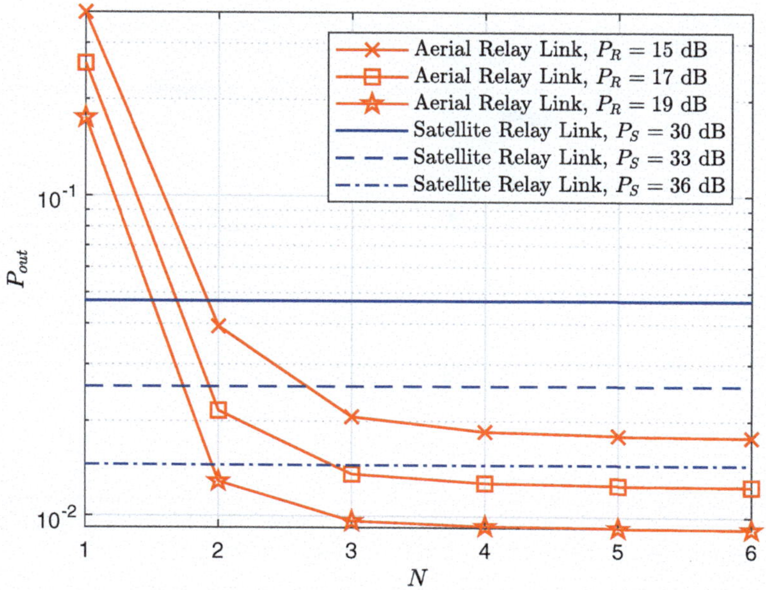


**Fig. 5.30** The e2e OP for various $P_S$ and $P_R$

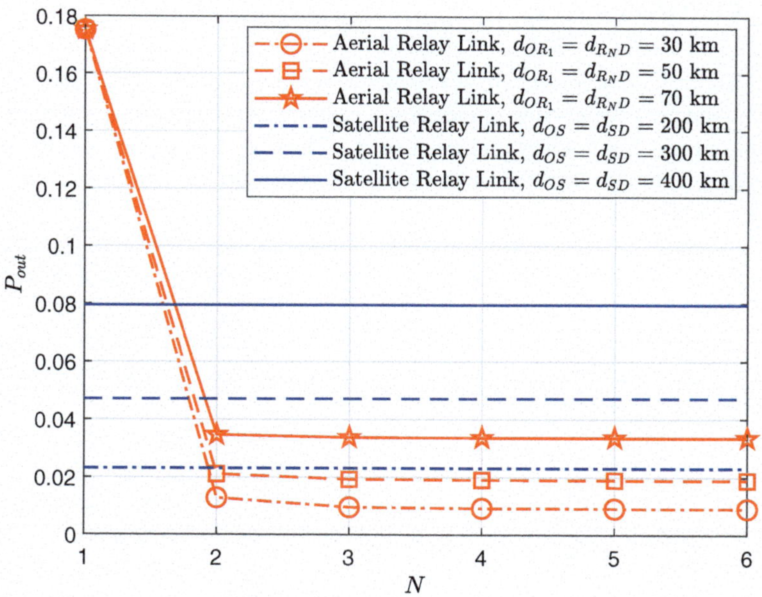


**Fig. 5.31** The e2e OP for various $d_{OS}/d_{OR_1}$ and $d_{SD}/d_{R_ND}$

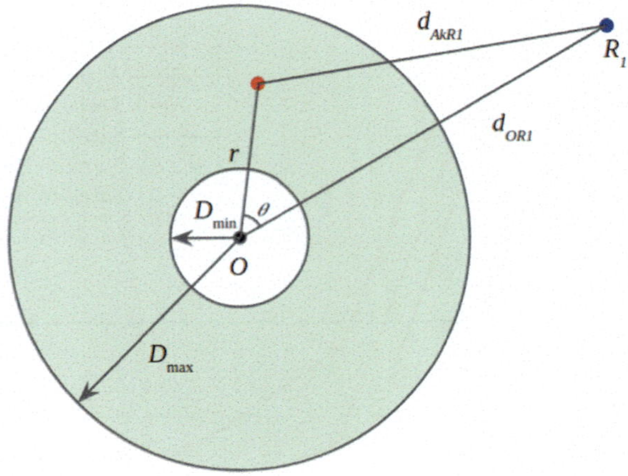


**Fig. 5.32** Aerial transmitter-aerial relay link model

the e2e OP and other performance indices, for example, the time delay incurred by the multi-hop forwarding over the aerial relay link.

### *5.2.6 Conclusion*

In this work, we have investigated the e2e outage performance of a cooperative SATN, in which there are two relaying choices for a group of aerial sources to forward their information to a remote terrestrial destination. The e2e outage performance of the considered system via two different relay links has been studied and compared. Some remarks are obtained:

1. In presence of co-channel interference among the aerial sources, the transmission distances over the uplinks have little impact on the outage performance of the uplinks;
2. The distribution density and the size of the distribution space of the aerial sources and the aerial jammers play a negative role on the OP over the uplink and the downlink, respectively;
3. The safety distance among the aerial sources exhibits a positive impact on the OP over the uplink;
4. For the aerial relay link, the number of aerial relays can be optimized to realize an optimal trade-off between the e2e OP and system resource overhead.
5. The choice of the relay link for the considered system depends on the practical scenario settings. In other words, both two kinds of relay links can provide optimal trade-offs between the e2e outage performance and the other performance indices/constraints, and then people should choose the relay type to satisfy the given system settings.

## Appendix A: Proof of Lemma 1

According to the definition of OP, the OP over the link from the $m$th aerial transmitter to $X$ ($X \in \{S, R_1\}$) can be written as

$$\begin{aligned}
P_{out}^{A_m X} &= \Pr\{\gamma_{A_m X} \leq \gamma_{out}\} \\
&= \Pr\left\{ \frac{P_{A_m} \left|h_{A_m X}\right|^2 d_{A_m X}^{-\alpha_X}}{\sum\limits_{k=1\&k\neq m}^{M} P_{A_k} \left|h_{A_k X}\right|^2 d_{A_k X}^{-\alpha_X}} \leq \gamma_{out} \right\} \\
&= \Pr\left\{ \left|h_{A_m X}\right|^2 \leq \frac{d_{A_m X}^{\alpha_X} \gamma_{out} P_{A_u}}{P_{A_m}} \sum_{k=1\&k\neq m}^{M} \left|h_{A_k X}\right|^2 d_{A_k X}^{-\alpha_X} \right\} \\
&= 1 - \sum_{i=0}^{m_n-1} \frac{(-1)^{m_i}}{(m_i)!} \mathbb{E}_{I, d_{A_m X}} \left\{ (sI)^{m_i} \exp(sI) \right\} \\
&= 1 - \sum_{i=0}^{m_n-1} \frac{(-1)^{m_i}}{(m_i)!} \mathbb{E}_{d_{A_m X}} \left\{ s^{m_i} \underbrace{\mathbb{E}_I \left\{ I^{m_i} \exp(sI) \right\}}_{D_1} \right\},
\end{aligned} \tag{5.85}$$

where $I = \sum\limits_{k=1\&k\neq m}^{M} \left|h_{A_k X}\right|^2 d_{A_k X}^{-\alpha_X}$ and $s = -\frac{m_n d_{A_m X}^{\alpha_X} \gamma_{out} P_{A_u}}{P_{A_m} \Omega}$.

It is easy to get

$$D_1 = \frac{\mathrm{d}^{m_i} \mathbb{E}_I \{\exp(sI)\}}{\mathrm{d}s^{m_i}}. \tag{5.86}$$

Thus, we can obtain (5.52) via substituting (5.86) into (5.85).

## Appendix B: Proof of Lemma 2

As $\left|h_{A_k X}\right|^2$ and $d_{A_k X}$ are i.i.d random variables, we can get

$$\begin{aligned}
\mathbb{E}_I \{\exp(sI)\} &= \mathbb{E}_M \left\{ \mathbb{E}_{\left|h_{A_k X}\right|^2, d_{A_k X}} \left\{ \exp\left( s \sum_{k=1\&k\neq m}^{M} \frac{\left|h_{A_k X}\right|^2}{d_{A_k X}^{\alpha_X}} \right) \right\} \right\} \\
&= \mathbb{E}_M \left\{ \prod_{k=1\&k\neq m}^{M} \mathbb{E}_{\left|h_{A_k X}\right|^2, d_{A_k X}} \left\{ \exp\left( \frac{s\left|h_{A_k X}\right|^2}{d_{A_k X}^{\alpha_X}} \right) \right\} \right\}
\end{aligned}$$

$$
\begin{aligned}
&= \sum_{M=0}^{\infty} \frac{(\lambda_A V_1)^M}{M!} \exp(-\lambda_A V_1) \\
&\quad \times \left\{ \underbrace{\mathbb{E}_{|h_{A_k X}|^2, d_{A_k X}} \left\{ \exp\left( \frac{s |h_{A_k X}|^2}{d_{A_k X}^{\alpha_X}} \right) \right\}}_{D_2} \right\}^M \\
&= \sum_{M=0}^{\infty} \frac{(\lambda_A V_1 D_2)^M}{M!} \exp(-\lambda_A V_1) \\
&= \exp[\lambda_A V_1 (D_2 - 1)],
\end{aligned}
\tag{5.87}
$$

where $V_1 = \frac{4\pi}{3}(R_A^3 - D_{\min}^3)$.

Using the moment generating function (MGF) of $|h_{A_k X}|^2$, it deduces

$$
D_2 = \mathbb{E}_{d_{A_k X}} \left\{ \left( \frac{\frac{m_n}{\Omega s} d_{A_k X}^{\alpha_X}}{\frac{m_n}{\Omega s} d_{A_k X}^{\alpha_X} - 1} \right)^{m_n} \right\}. \tag{5.88}
$$

Then, (5.53) can be reached by combining (5.87) and (5.88).

## Appendix C: Proof of Lemma 3

After achieving $\mathbb{E}_I\{\exp(sI)\}$, we will derive its $(m_i)$th derivative. We know that the 0th derivative of any function is itself. When $m_i > 0$, according to the Faàdi Bruno's formula, we can get

$$
\frac{\mathrm{d}^{m_i} \mathbb{E}_I\{\exp(sI)\}}{\mathrm{d}s^{m_i}} = \exp[\lambda_A V_1 (D_2 - 1)] \sum_{k=1}^{m_i} B_{m_i,k}\left(\lambda_A V_1 D_2^{(1)}, \ldots, \lambda_A V_1 D_2^{(m_i - k + 1)}\right), \tag{5.89}
$$

where $B_{m,k}(\cdot)$ is the Bell polynomials and $D_2^{(k)}$ is the $k$th derivative of $D_2$ w.r.t. $s$.

It is easy to get $D_2^{(k)}$ after substituting $s = -\frac{m_n d_{A_m X}^{\alpha_X} \gamma_{out} P_{A_u}}{P_{A_m} \Omega}$ as

$$
\begin{aligned}
D_2^{(k)} &= \frac{(m_n + k - 1)!}{(m_n - 1)!} \mathbb{E}_{d_{A_k X}} \left\{ \left( \frac{\Omega}{m_n d_{A_k X}^{\alpha_X}} \right)^k \left( \frac{m_n d_{A_k X}^{\alpha_X}}{m_n d_{A_k X}^{\alpha_X} - \Omega s} \right)^{m_n + k} \right\} \\
&= \frac{(m_n + k - 1)!}{(m_n - 1)!} \mathbb{E}_{d_{A_k X}} \left\{ \left( \frac{\Omega}{m_n d_{A_k X}^{\alpha_X}} \right)^k \left( \frac{d_{A_k X}^{\alpha_X}}{d_{A_k X}^{\alpha_X} + \frac{d_{A_m X}^{\alpha_X} \gamma_{out} P_{A_u}}{P_{A_m}}} \right)^{m_n + k} \right\}.
\end{aligned}
\tag{5.90}
$$

Substituting (5.89), (5.90), and $s = -\frac{m_n d_{A_m X}^{\alpha X} \gamma_{out} P_{A_u}}{P_{A_m} \Omega}$ into (5.52), the OP can be presented as (5.54).

## Appendix D: Proof of Lemma 4

As aerial transmitters are assumed to be independently and uniformly distributed in $\mathcal{V}$, the joint CDF of $r$ and $\theta$ can be expressed as

$$F_{r,\theta}(x, y) = \frac{\int_0^{2\pi} d\varphi \int_0^{y} \sin\theta d\theta \int_{D_{\min}}^{x} r^2 dr}{V_1} = \frac{2\pi(1 - \cos y)(x^3 - D_{\min}^3)}{3V_1}. \tag{5.91}$$

Then, the joint PDF of $r$ and $\theta$ can be written as

$$f_{r,\theta}(x, y) = \frac{\partial^2 F_{r,\theta}(x, y)}{\partial x \partial y} = \frac{2\pi x^2 \sin y}{V_1}. \tag{5.92}$$

From Fig. 5.32, the relationships between $r$, $\theta$, and $d_{A_k R_1}^2$ can be represented as $d_{A_k R_1}^2 = r^2 + d_{OR_1}^2 - 2rd_{OR_1} \cos\theta$. It can be easily seen that $d_{OR_1} - R_A \leq d_{A_k R_1} \leq d_{OR_1} + R_A$. To obtain the PDF of $d_{A_k R_1}^2$, we first derive the joint PDF of $d_{A_k R_1}^2$ and $r$.

According to the multivariate change of variables formula, the Jacobian determinant of matrix $\partial(d_{A_k R_1}^2, r)/\partial(r, \theta)$ is

$$\left| \frac{\partial(d_{A_k R_1}^2, r)}{\partial(r, \theta)} \right| = \begin{vmatrix} 2r - 2d_{OR_1} \cos\theta & 2rd_{OR_1} \sin\theta \\ 1 & 0 \end{vmatrix} = 2rd_{OR_1} \sin\theta. \tag{5.93}$$

Then, the joint PDF of $d_{A_k R_1}^2$ and $r$ can be achieved as

$$f_{d_{A_k R_1}^2, r}(x, y) = \frac{f_{r,\theta}(x, y)}{\left| \frac{\partial(d_{A_k R_1}^2, r)}{\partial(r,\theta)} \right|} = \frac{\pi y}{V_1 d_{OR_1}}, \tag{5.94}$$

where $D_{\min} \leq y \leq R_A$ and $-1 \leq \cos\theta = \frac{y^2 + d_{OR_1}^2 - x}{2d_{OR_1} y} \leq 1$.

From $-1 \leq \cos\theta = \frac{y^2 + d_{OR_1}^2 - x}{2d_{OR_1} y} \leq 1$, we can get $x \leq (d_{OR_1} + y)^2$ and $(d_{OR_1} - y)^2 \leq x$. As $R_A \ll d_{OR_1}$, one can obtain $\sqrt{x} - d_{OR_1} \leq y$ and $d_{OR_1} - \sqrt{x} \leq y$, namely, $|\sqrt{x} - d_{OR_1}| \leq y$. Observing $D_{\min} \leq y \leq R_A$, the range of $y$ is $\max\{D_{\min}, |\sqrt{x} - d_{OR_1}|\} \leq y \leq R_A$.

The PDF of $d^2_{A_kR_1}$ can be acquired through the integration of $f_{d^2_{A_kR_1},r}(x,y)$ in (5.94) according to $r$ and finally shown in (5.57).

## Appendix E: Proof of (5.59)

By using [2, Eq. (3.194.1)], $D_2^{(k)}$ can be obtained as

$$
\begin{aligned}
D_2^{(k)} &= \frac{\pi(m_n+k-1)!}{2V_1 d_{OR_1}(m_n-1)!}\left(\frac{\Omega}{m_n}\right)^k \Bigg\{(R_A^2-D_{\min}^2)\int_{d_{\min}^d}^{d_{\min}^u} \frac{x^{\frac{\alpha_{R_1} m_n}{2}}}{\left(x^{\frac{\alpha_{R_1}}{2}}+\Lambda\right)^{m_n+k}}dx \\
&+ \int_{d_{\max}^d}^{d_{\min}^d} \frac{(R_A^2-d_{OR_1}^2)x^{\frac{\alpha_{R_1} m_n}{2}} - x^{\frac{\alpha_{R_1} m_n}{2}+1} + 2d_{OR_1}x^{\frac{\alpha_{R_1} m_n+1}{2}})}{\left(x^{\frac{\alpha_{R_1}}{2}}+\Lambda\right)^{m_n+k}}dx \\
&+ \int_{d_{\min}^u}^{d_{\max}^u} \frac{(R_A^2-d_{OR_1}^2)x^{\frac{\alpha_{R_1} m_n}{2}} - x^{\frac{\alpha_{R_1} m_n}{2}+1} + 2d_{OR_1}x^{\frac{\alpha_{R_1} m_n+1}{2}})}{\left(x^{\frac{\alpha_{R_1}}{2}}+\Lambda\right)^{m_n+k}}dx\Bigg\} \\
&= \frac{\pi(m_n+k-1)!}{2V_1 d_{OR_1}(m_n-1)!}\left(\frac{\Omega}{m_n}\right)^k \Lambda^{-m_n-k}\frac{2}{\alpha_{R_1}}\Bigg\{(R_A^2-D_{\min}^2)\int_{d_{\min}^d}^{d_{\min}^u} \frac{x^{\frac{\alpha_{R_1}}{2}\left(m_n-1+\frac{2}{\alpha_{R_1}}\right)}}{\left(\frac{x^{\frac{\alpha_{R_1}}{2}}}{\Lambda}+1\right)^{m_n+k}}dx^{\frac{\alpha_{R_1}}{2}} \\
&+ \int_{d_{\max}^d}^{d_{\min}^d} \frac{(R_A^2-d_{OR_1}^2)x^{\frac{\alpha_{R_1}}{2}\left(m_n-1+\frac{2}{\alpha_{R_1}}\right)} - x^{\frac{\alpha_{R_1}}{2}\left(m_n-1+\frac{4}{\alpha_{R_1}}\right)} + 2d_{OR_1}x^{\frac{\alpha_{R_1}}{2}\left(m_n-1+\frac{3}{\alpha_{R_1}}\right)}}{\left(\frac{x^{\frac{\alpha_{R_1}}{2}}}{\Lambda}+1\right)^{m_n+k}}dx^{\frac{\alpha_{R_1}}{2}} \\
&+ \int_{d_{\min}^u}^{d_{\max}^u} \frac{(R_A^2-d_{OR_1}^2)x^{\frac{\alpha_{R_1}}{2}\left(m_n-1+\frac{2}{\alpha_{R_1}}\right)} - x^{\frac{\alpha_{R_1}}{2}\left(m_n-1+\frac{4}{\alpha_{R_1}}\right)} + 2d_{OR_1}x^{\frac{\alpha_{R_1}}{2}\left(m_n-1+\frac{3}{\alpha_{R_1}}\right)}}{\left(\frac{x^{\frac{\alpha_{R_1}}{2}}}{\Lambda}+1\right)^{m_n+k}}dx^{\frac{\alpha_{R_1}}{2}}\Bigg\} \\
&= \frac{\pi(m_n+k-1)!}{2V_1 d_{OR_1}(m_n-1)!}\left(\frac{\Omega}{m_n}\right)^k \Lambda^{-m_n-k}\Bigg\{(R_A^2-D_{\min}^2)\mathcal{F}\left(\frac{2}{\alpha_{R_1}}, d_{\min}^u, d_{\min}^d, k\right) \\
&+ (R_A^2-d_{OR_1}^2)\left[\mathcal{F}\left(\frac{2}{\alpha_{R_1}}, d_{\min}^d, d_{\max}^d, k\right)+\mathcal{F}\left(\frac{2}{\alpha_{R_1}}, d_{\max}^u, d_{\min}^u, k\right)\right] \\
&- \left[\mathcal{F}\left(\frac{4}{\alpha_{R_1}}, d_{\min}^d, d_{\max}^d, k\right)+\mathcal{F}\left(\frac{4}{\alpha_{R_1}}, d_{\max}^u, d_{\min}^u, k\right)\right] \\
&+ 2d_{OR_1}\left[\mathcal{F}\left(\frac{3}{\alpha_{R_1}}, d_{\min}^d, d_{\max}^d, k\right)+\mathcal{F}\left(\frac{3}{\alpha_{R_1}}, d_{\max}^u, d_{\min}^u, k\right)\right]\Bigg\},
\end{aligned}
\tag{5.95}
$$

where $\mathcal{F}(a,b,c,k) = \left[b^{\frac{(m_n+a)\alpha_{R_1}}{2}}{}_2F_1\left(m_n+k, m_n+a; m_n+a+1; -\frac{b^{\frac{\alpha_{R_1}}{2}}}{\Lambda}\right)\right.$
$\left. -c^{\frac{(m_n+a)\alpha_{R_1}}{2}}{}_2F_1\left(m_n+k, m_n+a; m_n+a+1; -\frac{c^{\frac{\alpha_{R_1}}{2}}}{\Lambda}\right)\right] \times \frac{2}{(m_n+a)\alpha_{R_1}}$.

## Appendix F: Proof of Lemma 5

By substituting (5.61) into (5.85), we can obtain that

$$\begin{aligned} P_{out}^{\infty} &= 1 - \sum_{i=0}^{m_n-1} \frac{(-1)^{m_i}}{(m_i)!} \mathbb{E}_{I,d_{A_m X}} \left\{ (sI)^{m_i} (1 - sI) \right\} \\ &= 1 - \sum_{i=0}^{m_n-1} \frac{(-1)^{m_i}}{(m_i)!} \mathbb{E}_{d_{A_m X}} \left\{ s^{m_i} \mathbb{E}_I \left\{ I^{m_i} (1 - sI) \right\} \right\} \\ &= 1 - \sum_{i=0}^{m_n-1} \frac{(-1)^{m_i}}{(m_i)!} \left\{ \mathbb{E}_{d_{A_m X}} \left\{ s^{m_i} \right\} \mathbb{E}_I \left\{ I^{m_i} \right\} - \mathbb{E}_{d_{A_m X}} \left\{ s^{m_i+1} \right\} \mathbb{E}_I \left\{ I^{m_i+1} \right\} \right\}. \end{aligned} \tag{5.96}$$

As $\left|h_{A_k X}\right|^2$ and $d_{A_k X}$ are i.i.d random variables, $\mathbb{E}_I \{I\}$ can be derived as

$$\begin{aligned} \mathbb{E}_I \{I\} &= \mathbb{E}_M \left\{ \mathbb{E}_{\left|h_{A_k X}\right|^2, d_{A_k X}} \left\{ \sum_{k=1\&k\neq m}^{M} \frac{\left|h_{A_k X}\right|^2}{d_{A_k X}^{\alpha_X}} \right\} \right\} \\ &= \sum_{M=0}^{\infty} \frac{(\lambda_A V_1)^M}{(M-1)!} \exp(-\lambda_A V_1) \\ &\quad \times \mathbb{E}_{\left|h_{A_k X}\right|^2} \left\{ \left|h_{A_k X}\right|^2 \right\} \mathbb{E}_{d_{A_k X}} \left\{ d_{A_k X}^{-\alpha_X} \right\} \\ &= \sum_{M=0}^{\infty} \frac{(\lambda_A V_1)^M}{(M-1)!} \exp(-\lambda_A V_1) \\ &\quad \times \left\{ 1 - \sum_{i=0}^{m_n-1} \left( \frac{m_n}{\Omega} \right)^{m_n-i-1} \frac{x^{m_n-i}}{(m_n-i-1)!} \right. \\ &\quad \left. \times \exp\left( -\frac{m_n}{\Omega} x \right) \right\} \mathbb{E}_{d_{A_k X}} \left\{ d_{A_k X}^{-\alpha_X} \right\}. \end{aligned} \tag{5.97}$$

Then, (5.62) can be reached by combining (5.96) and (5.97).

## Appendix G: Proof of Lemma 6

As it has been assumed that the channels between the aerial jammers and the terrestrial receiver, $D$, suffer i.i.d. Nakagmi-$m$ fading, we have $f_{\left|h_{jD}\right|^2}(x) = \left(\frac{m_n}{\Omega}\right)^{m_n} \frac{x^{m_n-1}}{(m_n-1)!} \exp\left(-\frac{m_n}{\Omega}x\right)$. Thus, we derive $\mathscr{L}[Y](s)$ as

$$
\begin{aligned}
\mathscr{L}_Y(s) &= \mathbb{E}\left[\exp(-sY)\right] = \mathbb{E}\left[\exp\left(-s\sum_{j=1}^{J}\frac{P_j\left|h_{jD}\right|^2}{d_{I_j}^{\alpha_S}}\right)\right] \\
&= \mathbb{E}\left[\prod_{j=1}^{J}\exp\left(-\frac{sP_j\left|h_{jD}\right|^2}{d_{I_j}^{\alpha_S}}\right)\right] \\
&= \mathbb{E}\left[\prod_{j=1}^{J}\underbrace{\int_0^\infty \exp\left(-\frac{sP_j\left|h_{jD}\right|^2}{d_{I_j}^{\alpha_S}}\right) f_{\left|h_{jD}\right|^2}(x)\,dx}_{I_1}\right] \\
&= \exp\left(-\lambda_I \int\limits_{\mathbb{R}^3} \left(1 - I_1\left(d_I\right)\right) dd_I\right),
\end{aligned}
\tag{5.98}
$$

where $\mathbb{R}^3$ is the 3D distribution space for these aerial jammers presented in Fig. 5.22b,

$$
\begin{aligned}
I_1 &= \left(\frac{m_n}{\Omega}\right)^{m_n}\frac{1}{(m_n-1)!}\int_0^\infty x^{m_n-1}\exp\left(-\left(\frac{m_n}{\Omega}+\frac{sP_j}{d_{I_j}^{\alpha_S}}\right)x\right)dx \\
&= \left(\frac{m_n}{\Omega}\right)^{m_n}\left(\frac{m_n}{\Omega}+\frac{sP_j}{d_{I_j}^{\alpha_S}}\right)^{-m_n},
\end{aligned}
\tag{5.99}
$$

and the last step follows the probability generating functional of the PPP, which means for function $f(x)$ that $\mathbb{E}\left[\prod_{\mathbb{R}} f(x)\right] = \exp\left(-\lambda\int_0^\infty (1-f(x))\,dx\right)$.

Then, thinking that $\mathbb{R}^3$ is a hemisphere space depicted in Fig. 5.22b, one can finally derive $\mathscr{L}_Y(s)$ as (5.100) presented on the top of the page after next page,

$$
\begin{aligned}
\mathscr{L}_Y(s) &= \exp\left(-\lambda_I\int_0^{R_I}\int_0^{\frac{\pi}{2}}\int_0^{2\pi}\left(1-\left(\frac{m_n}{\Omega}\right)^{m_n}\left(\frac{m_n}{\Omega}+\frac{sP_j}{r^{\alpha_S}}\right)^{-m_n}\right)r^2\sin\phi\, d\psi\, d\phi\, dr\right) \\
&= \exp\left(-2\pi\lambda_I\int_0^{R_I} r^2\left(1-\left(\frac{m_n}{\Omega}\right)^{m_n}\left(\frac{m_n}{\Omega}+\frac{sP_j}{r^{\alpha_S}}\right)^{-m_n}\right)dr\int_0^{\frac{\pi}{2}}\sin\phi\, d\phi\right) \\
&= \exp\left(-2\pi\lambda_I\int_0^{R_I} r^2\left(1-\left(\frac{m_n}{\Omega}\right)^{m_n}\left(\frac{m_n}{\Omega}+\frac{sP_j}{r^{\alpha_S}}\right)^{-m_n}\right)dr\right) \\
&= \exp\left(-\pi R_I\lambda_I\sum_{i=1}^{V}\sqrt{1-t_i^2}\,\omega_i{\upsilon_i}^2\left(1-\left(\frac{m_n}{\Omega}\right)^{m_n}\left(\frac{m_n}{\Omega}+\frac{sP_j}{{\upsilon_i}^{\alpha_S}}\right)^{-m_n}\right)\right)
\end{aligned}
\tag{5.100}
$$

where $\upsilon_i = \frac{R_I(t_i+1)}{2}$, $t_i = \cos\left(\frac{2i-1}{2V}\pi\right)$, and $\omega_i = \frac{\pi}{V}$.

Finally, replacing $s$ in (5.100) by $\zeta\kappa t$ accomplishes the proof of Lemma 5.

## References

1. A. Abdi, W. Lau, M.-S. Alouini, et al., "A new simple model for land mobile satellite channels: First and second order statistics" *IEEE Trans. Wireless Commun.*, vol. 2, no. 3, pp. 519–528, 2003.
2. I.S. Gradshteyn and I.M. Ryzhik, *Table of Integrals, Series and Products, 7 Ed.*, San Diego: Academic Press, 2007.
3. Z. Zhu, S. Zhao, Y. Li, et al., "Performance comparison of analogue inter-satellite microwave photonics link using intensity modulation with direct detection and phase modulation with interferometric detection," *IET Optoelectronics*, vol. 9, no. 2, pp. 88–95, 2015.
4. S. Arnon, "Performance of a laser $\mu$satellite network with an optical preamplifier," *J. Opt. Soc. Am. A*, vol. 22, no . 4, pp. 708–715, 2005.
5. A. Polishuk and S. Arnon, "Optimization of a laser satellite communication system with an optical preamplifier," *J. Opt. Soc. Am. A*, vol. 21, no. 7, pp. 1307–1315, 2004.
6. G. Pan, J. Ye, J. An and M. -S. Alouini, "Latency Versus Reliability in LEO Mega-Constellations: Terrestrial, Aerial, or Space Relay?," *IEEE Trans. Mob. Comput.*, vol. 22, no. 9, pp. 5330–5345, 2023.
7. I.S. Gradshteyn and I.M. Ryzhik, *Table of Integrals, Series and Products,* 7th ed. San Diego, CA, USA: Academic Press, 2007.
8. H. Alzer, "On some inequalities for the incomplete gamma function," *Math. Comput.*, vol. 66, pp. 771–778, 1997.
9. N. H. Crisp *et al.*, "System modelling of very low Earth orbit satellites for Earth observation," *Acta Astronautica*, vol. 187, pp. 475–491, 2021.
10. *Technical Characteristics for Telemetry, Tracking and Command in the Space Operation Service Below 1 GHz for Non-GSO Satellites With Short Duration Missions*, document ITU-R SA.2426-0, Sep. 2018. Accessed: May 30, 2022. [Online]. Available: https://www.itu.int/pub/RREP-SA.2426

# Chapter 6
# Summary

**Shuai Wang, Jinpeng Song, Xiaqing Miao, and Gaofeng Pan**

## 6.1 Summary of This Book

Satellite communication holds a crucial position in the development of 5G and the emerging 6G communication systems. This book utilizes stochastic geometry as a mathematical framework to comprehensively explore the communication effectiveness of the satellite-aerial-terrestrial collaborative communication model from diverse viewpoints. The exchange of information between the ground and the satellite can be accomplished either through open space or by employing unmanned aerial vehicles (UAVs) as intermediary transmitters. Thus, the book considers different models of SATNs. Sections 3.1, 3.2, 4.1, 4.2, 5.1, and 5.2 examine the transmission performance of SATNs models, and Sects. 2.2, 2.1, and 2.3 examine the transmission performance of satellite-terrestrial communication models.

A key hurdle in satellite communications lies in the high dynamics caused by the satellites' constant movement. Section 2.2 of this book delves into the non-static nature of both the Earth and satellites, establishing an analytical framework for satellite communication systems that operate under such dynamic conditions. However, radio frequency (RF) signals, being inherently open, often expose receiving equipment to interference from wireless signals transmitted by other legitimate users. Radiofrequency (RF) interference frequently poses a challenge to satellite communication systems, and hence, Sects. 4.1–3.2 of this book devote attention to RF interference emanating from various transmitters within communication systems. Specifically, Sect. 4.1 focuses on RF interference originating from multiple terrestrial users in the downlink, Sect. 2.2 delves into RF interference caused by multiple satellites in the downlink, and Sect. 5.2 examines RF interference arising from multiple unmanned aerial vehicles (UAVs) in a satellite-aerial-terrestrial cooperative system. By addressing these interference sources, the book aims to enhance

S. Wang · J. Song · X. Miao (✉) · G. Pan
Beijing Institute of Technology, Beijing, China
e-mail: xqmiao@bit.edu.cn



S. Wang et al. (eds.), *Understanding Satellite Communications: The Stochastic Geometry Perspective*, https://doi.org/10.1007/978-981-97-7102-8_6

the understanding and mitigation of RF interference in satellite communication systems. With interference in satellite communication systems effectively under control, mega-constellation satellite systems offering low latency and extensive coverage are poised to become a prevailing trend. Amidst the thriving development of satellite mega-constellation communication systems, Sect. 5.1 of this book delves into multi-satellite cooperative communication systems, drawing insightful conclusions and suggesting design principles for satellite mega-constellation communication systems based on these findings. Furthermore, unmanned aerial vehicles (UAVs) can also coalesce into clusters to bolster satellite communications, further enhancing the overall communication capabilities and efficiency. In Sect. 5.2 of this book, signals transmitted from unmanned aerial vehicles (UAVs) traverse either satellite relays or multiple-hop UAV relays before reaching the ground receiver. It is assumed that UAVs within a designated area adhere to the Matérn hard-core point process type II to simulate random interference. Additionally, Sects. 3.1 and 2.3 of the book incorporate security considerations into the analysis of satellite communication performance. These sections calculate the secrecy failure probability as a metric to assess the system's security performance, thereby initiating a foundational discussion on security challenges within satellite communication systems.

## 6.2 Outlook for Future Development

This section offers a concise overview of the anticipated advancements and research trajectories for satellite communications and satellite-based ad hoc networks (SATNs). Broadly speaking, the enhancement of SATNs necessitates advancements in corresponding hardware infrastructure, software functionalities, and system architecture. Consequently, it is anticipated that SATNs will continue to undergo significant evolution in the following areas.

The future of satellite communications holds promise on several fronts. Firstly, New constellation types, such as those deployed by Starlink and OneWeb, are revolutionizing the landscape with their flexible use for global coverage, high data rates, and low latency. Beyond the traditional geostationary earth orbit (GEO) and medium earth orbit (MEO), the deployment of numerous low earth orbit (LEO) satellites is enabling reduced latency and increased throughput, paving the way for seamless communications services worldwide through the collaboration of GEO, MEO, and LEO constellations. Secondly, the focus on enhancing onboard processing capabilities aims to improve the efficiency of satellite networks by reducing the need for data transmission to ground stations, while also enabling real-time data processing and edge intelligence. Lastly, the integration of satellite/aerial communications with terrestrial 5G networks represents a key direction, aiming to provide seamless connectivity and high-speed Internet access to users regardless of their location, even in the remotest parts of the world. Together, these advancements are poised to revolutionize satellite communications, SATNs, and the broader telecommunications industry.

Moreover, SATNs are susceptible to active assaults or passive eavesdropping due to their intricate nature and the inherent openness of wireless communication channels. These vulnerabilities pose significant risks to the transmission of private data. Consequently, it is crucial to address security concerns in SATNs moving forward. Below is a synopsis of the challenges faced and the corresponding strategies to enhance the security of SATNs.

In the upcoming 6G era, SATNs face a range of intricate security challenges. A key concern lies in the deployment of Artificial Intelligence (AI) across multiple network layers. While AI holds promise for enhancing SATN functionality, its inherent insecurity poses a significant threat to the network's stability. Additionally, AI-based attacks can further compromise SATN security, necessitating research into more robust AI algorithms tailored for SATNs, along with detection and prevention mechanisms against such attacks. Cross-layer attacks, where malicious nodes target multiple layers simultaneously, present another challenge. To address this, a cross-layer design approach must be adopted to ensure inherent network security. This involves the joint design and optimization of protocol layers to effectively mitigate cross-layer threats. Finally, covert communication in SATNs is a crucial aspect, particularly in sensitive scenarios like military communications. Traditional communication technologies have primarily focused on information security, relying on cryptography or physical layer security techniques. However, in covert SATNs, it is imperative to conceal the communication behavior itself. Therefore, extending the evaluation of jamming and secrecy loss to include performance metrics specific to covert SATNs will be invaluable in enhancing their overall security posture.

The manufacturer's authorised representative in the EU is Springer Nature Customer Service Centre GmbH, Europaplatz 3, 69115 Heidelberg, Germany. If you have any concerns regarding our products, please contact ProductSafety@springernature.com

Printed and bound by CPI Group (UK) Ltd, Croydon, CR0 4YY
30/09/2026
02225253-0004